Laboratory Manual

Inquiry into Life

Fifteenth Edition

Sylvia S. Mader

Contributor

Linda Smith-Staton

Pellissippi State Community College

 McGraw Hill Education

INQUIRY INTO LIFE LABORATORY MANUAL, FIFTEENTH EDITION

1 2 3 4 5 6 7 8 9 0 RMN/RMN 1 0 9 8 7 6

ISBN 978-1-259-68861-4
MHID 1-259-68861-5

Senior Vice President, Products & Markets: *Kurt L. Strand*
Vice President, General Manager, Products & Markets: *Marty Lange*
Vice President, Content Design & Delivery: *Kimberly Meriwether David*
Managing Director: *Michael Hackett*
Executive Brand Manager: *Michelle Vogler*
Brand Manager: *Chris Loewenberg*
Director, Product Development: *Rose Koos*
Product Developer: *Anne Winch*
Market Development Manager: *Jenna Paleski*
Marketing Manager: *Christine Ho*
Director of Digital Content: *Michael G. Koot, PhD*
Digital Product Analyst: *Christine Carlson*
Director, Content Design & Delivery: *Linda Avenarius*
Program Manager: *Angela R. FitzPatrick*
Content Project Managers: *April R. Southwood/Christina Nelson*
Buyer: *Sandy Ludovissy*
Design: *David W. Hash*
Content Licensing Specialists: *Lori Hancock/Lorraine Buczek*
Cover Image: *hands holding globe ©Leonello Calvetti/Getty Images/RF; Siberian tiger (Panthera tigris altaica) running ©Tom Brakefield/Getty Images/RF; humpback whale (Megaptera novaeangliae), Hawaii, Maui ©M Swiet Productions/Getty Images/RF; lyretail anthias or goldies (Pseudanthias squamipinnis) over coral reef, Egypt, Red Sea ©Georgette Douwma/The Image Bank/Getty Images; Ti plant (Cordyline fruticosa) Carambola Botanical Gardens, Roatan, Honduras ©Danita Delimont/Gallo Images/Getty Images*
Compositor: *Aptara, Inc.*
Printer: *R. R. Donnelley*

mheducation.com/highered

Contents

Preface

To the Instructor

The laboratory exercises in this manual are coordinated with *Inquiry into Life,* a general biology text that covers the entire field of biology. The text emphasizes how we can apply biological knowledge to our own lives and to the biological world in general.

Although each laboratory is referenced to the appropriate chapter(s) in *Inquiry,* this manual may also be used in coordination with other general biology texts. In addition, this laboratory manual can be adapted to a variety of course orientations and designs. There are a sufficient number of laboratories and exercises within each lab to tailor the laboratory experience as desired. Then, too, many exercises may be performed as demonstrations rather than as student activities, thereby shortening the time required to cover a particular concept.

The Exercises

All exercises have been tested for student interest, preparation time, estimated time of completion, and feasibility. The following features are particularly appreciated by adopters:

Integrated opening: Each laboratory begins with a list of Learning Outcomes organized according to the major sections of the laboratory. The major sections of the laboratory are numbered on the opening page and in the laboratory text material. This organization will help students better understand the goals of each laboratory session.

Self-contained content: Each laboratory contains all the background information necessary to understand the concepts being studied and to answer the questions asked. This feature will reduce student frustration and increase learning.

Scientific process: All laboratories stress the scientific process, and many opportunities are given for students to gain an appreciation of the scientific method. The first laboratory of this edition explicitly explains the steps of the scientific method and gives students an opportunity to use them. I particularly recommend this laboratory because it utilizes the pillbug, a living subject.

Student activities: A color bar is used to designate each student activity. Some student exercises are Observations and some are Experimental Procedures. An icon (shaped like a clock) appears whenever a procedure requires a period of time before results can be viewed. Sequentially numbered steps guide students as they perform each activity.

Live materials: Although students work with living material during some part of almost all laboratories, the exercises are designed to be completed within one laboratory session. This facilitates the use of the manual in multiple-section courses.

Laboratory safety: Throughout the laboratories, safety precautions specific to an activity are highlighted and identified by a caution symbol.

Improvements in This Edition

Each Laboratory Review section is completely new for this edition. Fourteen short answer questions now help students identify key vocabulary and basic concepts critical to the topic of the exercise. Questions are organized to match the exercise's order of presentation, facilitating students' ability to identify an appropriate response. Each Laboratory Review also contains three new critical thinking questions that challenge students to make connections between the concepts presented, or to apply information from the exercise to real world examples.

Each laboratory exercise was carefully revised to ensure that directions to students are clear and follow common laboratory standards and practices. For example, the manual now directs students to use inexpensive plastic transfer pipets to measure volume rather than marking off test tubes and filling to a cm.

Exercise 5—Mitosis and Meiosis

This exercise was revised to focus on laboratory-related activities, enabling students to compare and contrast the events and results of mitosis and meiosis more easily.

Exercise 12—Basic Mammalian Anatomy I

The section covering the abdominal cavity was reorganized and revised to present a concise introduction to the abdominal organs and their functions.

Exercise 17—Nervous System and Senses

Directions for experimental procedures were revised to improve clarity.

Exercise 19—Development

The laboratory now utilizes the sea star to illustrate cellular development, the frog for tissue development, and the chick for organ development. Activities related to extraembryonic membranes and human fetal development follow.

Customized Editions

The 31 laboratories in this manual are now available as individual "lab separates," so instructors can custom-tailor the manual to their particular course needs.

Laboratory Resource Guide

The *Laboratory Resource Guide,* an essential aid for instructors and laboratory assistants, free to adopters of the *Inquiry into Life Laboratory Manual,* is online at **http://connect.mheducation.com**. The answers to the Laboratory Review questions are in the Resource Guide.

To the Student

Special care has been taken in preparing the *Inquiry into Life Laboratory Manual* to enable you to **enjoy** the laboratory experience as you **learn** from it. The instructions and discussions are written clearly so that you can understand the material while working through it. Student aids are designed to help you focus on important aspects of each exercise. Student learning aids are carefully integrated throughout this manual:

> **The Learning Outcomes** set the goals of each laboratory session and help you review the material for a laboratory practical or any other kind of exam. The major sections of each laboratory are numbered, and the Learning Outcomes are grouped according to these topics. This system allows you to study the chapter in terms of the outcomes presented.

The Introduction to the laboratory reviews much of the necessary background information required for comprehending the work you will be doing during the laboratory session. Also, specific information, provided before each of the Observations and Experimental Procedures, will assist you in being engaged while you do these activities.

The Observations and Experimental Procedures require your active participation. Space is provided in Tables for you to record the results and conclusions of observations and experiments. Space is also provided for you to answer questions pertaining to these activities.

    Observation: An activity in which models, slides, and preserved or live organisms are observed to achieve a learning outcome.

    Experimental Procedure: An activity in which a series of steps uses laboratory equipment to gather data and come to a conclusion.

At the end of an Observation or Experimental Procedure you are often asked to formulate explanations or conclusions. You should be sure you are truly writing an explanation or conclusion and not just restating the observations made. To do so, you will need to synthesize information from a variety of sources, including the following:

1. Your experimental results and/or the results of other groups in the class. If your data are different from those of other groups in your class, do not erase your answer; add the other groups' answers in parentheses.
2. Your knowledge of underlying principles. Obtain this information from the laboratory Introduction or the appropriate section of the laboratory and from the corresponding chapter of your text.
3. Your understanding of how the experiment was conducted and/or the materials used. (If by chance the results seem inappropriate, it's possible the ingredients were contaminated or you misunderstood the directions. If this occurs, consult with other students and your instructor to see if you should repeat the experiment.)

Color Bars, Time Icon, and Safety Boxes

Observations are identified by the color bar shown to the left. Whenever you see this color you know that the activity will require you to make careful observations and answer questions about these activities.

Experimental procedures are identified by the color bar shown to the left. Whenever you see this color you know that the activity will require you to use laboratory equipment to perform an experiment and answer questions about this experiment.

 A time icon is used to designate when time is needed for a reaction to occur. You may be asked to start these activities at the beginning of the laboratory, proceed to other activities, and return to these when the designated time is up.

A safety icon throughout the manual alerts you to any specific activity that requires a cautionary approach. Read these boxes and follow the advice given in the box and/or your instructor when performing the activity.

The Laboratory Review is a set of questions covering the day's work. Do all the review questions as an aid to understanding the laboratory. Your instructor may require you to hand in these questions for credit.

The Appendices at the end of the book provide useful information on preparing a laboratory report and the metric system. Practical Examination Answer Sheets are available on the *Inquiry into Life* website when and if they are needed.

Laboratory Preparation

It will be very helpful to you to read the entire laboratory chapter before coming to lab. *Study* the introductory material and the Observations and Experimental Procedures so you know ahead of time what you will be doing that week and how it correlates with the lecture material. If necessary, to obtain a better understanding, read the corresponding chapter in your text. If your text is *Inquiry into Life,* by Sylvia S. Mader and Michael Windelspecht, the "text chapter reference" column in the table of contents at the beginning of the *Laboratory Manual* lists the corresponding chapter in the text.

Student Feedback

If you have any suggestions for how this laboratory manual could be improved, you can send your comments to

McGraw-Hill Education
Product Development—General Biology
501 Bell St.
Dubuque, Iowa 52001

Required=Results

McGraw-Hill Connect®
Learn Without Limits

Connect is a teaching and learning platform that is proven to deliver better results for students and instructors.

Connect empowers students by continually adapting to deliver precisely what they need, when they need it, and how they need it, so your class time is more engaging and effective.

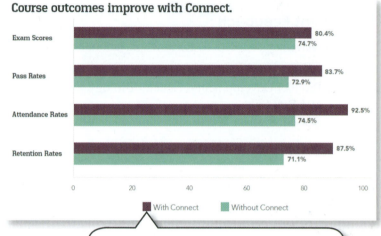

Course outcomes improve with Connect.

	With Connect	Without Connect
Exam Scores	80.4%	74.7%
Pass Rates	83.7%	72.9%
Attendance Rates	92.5%	74.5%
Retention Rates	87.5%	71.1%

Using **Connect** improves passing rates by **10.8%** and retention by **16.4%**.

88% of instructors who use **Connect** require it; instructor satisfaction **increases** by 38% when **Connect** is required.

Analytics

Connect helps students achieve better grades

With Connect	A 36%	B 29.5%	C 22%	D 4.3%	F 6.3%
Without Connect	A 22.2%	B 22.3%	C 25.6%	D 9.8%	F 20%

Based on McGraw-Hill Education Connect Effectiveness Study 2013

Connect Insight®

Connect Insight is Connect's new one-of-a-kind visual analytics dashboard—now available for both instructors and students—that provides at-a-glance information regarding student performance, which is immediately actionable. By presenting assignment, assessment, and topical performance results together with a time metric that is easily visible for aggregate or individual results, Connect Insight gives the user the ability to take a just-in-time approach to teaching and learning, which was never before available. Connect Insight presents data that empowers students and helps instructors improve class performance in a way that is efficient and effective.

Students can view their results for any **Connect** course.

Mobile

Connect's new, intuitive mobile interface gives students and instructors flexible and convenient, anytime–anywhere access to all components of the Connect platform.

Adaptive

THE FIRST AND ONLY **ADAPTIVE READING EXPERIENCE** DESIGNED TO TRANSFORM THE WAY STUDENTS READ

More students earn **A's** and **B's** when they use McGraw-Hill Education **Adaptive** products.

SmartBook®

Proven to help students improve grades and study more efficiently, SmartBook contains the same content within the print book, but actively tailors that content to the needs of the individual. SmartBook's adaptive technology provides precise, personalized instruction on what the student should do next, guiding the student to master and remember key concepts, targeting gaps in knowledge and offering customized feedback, and driving the student toward comprehension and retention of the subject matter. Available on smartphones and tablets, SmartBook puts learning at the student's fingertips—anywhere, anytime.

Over **4 billion questions** have been answered, making McGraw-Hill Education products more intelligent, reliable, and precise.

www.learnsmartadvantage.com

STUDENTS WANT

Mc Graw Hill Education **SMARTBOOK**®

95% of students reported **SmartBook** to be a more effective way of reading material

100% of students want to use the Practice Quiz feature available within **SmartBook** to help them study

100% of students reported having reliable access to off-campus wifi

90% of students say they would purchase **SmartBook** over print alone

95% reported that **SmartBook** would impact their study skills in a positive way

Laboratory Safety

The following is a list of practices required for safety purposes in the biology laboratory and in outdoor activities. Following rules of lab safety and using common sense throughout the course will enhance your learning experience by increasing your confidence in your ability to safely use chemicals and equipment. Pay particular attention to oral and written safety instructions given by the instructor. If you do not understand a procedure, ask the instructor, rather than a fellow student, for clarification. Be aware of your school's policy regarding accident liability and any medical care needed as a result of a laboratory or outdoor accident.

The following rules of laboratory safety should become a habit:

1. Wear safety glasses or goggles during exercises in which glassware and chemical reagents are handled, or when dangerous fumes may be present, creating possible hazards to eyes or contact lenses.
2. Assume that all reagents are poisonous and act accordingly. Read the labels on chemical bottles for safety precautions and know the nature of the chemical you are using. If chemicals come into contact with skin, wash immediately with water.
3. **DO NOT**
    a. ingest any reagents.
    b. eat, drink, or smoke in the laboratory. Toxic material may be present, and some chemicals are flammable.
    c. carry reagent bottles around the room.
    d. pipet anything by mouth.
    e. put chemicals in the sink or trash unless instructed to do so.
    f. pour chemicals back into containers unless instructed to do so.
    g. operate any equipment until you are instructed in its use.
    h. dispose of biological or chemical wastes in regular classroom trash receptacles.
4. **DO**
    a. note the location of emergency equipment such as a first aid kit, eyewash bottle, fire extinguisher, switch for ceiling showers, fire blanket(s), sand bucket, and telephone (911).
    b. be familiar with the experiments you will be doing before coming to the laboratory. This will increase your understanding, enjoyment, and safety during exercises. Confusion is dangerous. Completely follow the procedure set forth by the instructor.
    c. keep your work area neat, clean, and organized. Before beginning, remove everything from your work area except the lab manual, pen, and equipment used for the experiment. Wash hands and desk area, including desk top and edge, before and after each experiment. Use clean glassware at the beginning of each exercise, and wash glassware at the end of each exercise or before leaving the laboratory.
    d. wear clothing that, if damaged, would not be a serious loss, or use aprons or laboratory coats, since chemicals may damage fabrics.
    e. wear shoes as protection against broken glass or spillage that may not have been adequately cleaned up.
    f. handle hot glassware with a test tube clamp or tongs. Use caution when using heat, especially when heating chemicals. Do not leave a flame unattended; do not light a Bunsen burner near a gas tank or cylinder; do not move a lit Bunsen burner; do keep long hair and loose clothing well away from the flame; do make certain gas jets are off when the Bunsen burner is not in use. Use proper ventilation and hoods when instructed.
    g. read chemical bottle labels; be aware of the hazards of all chemicals used. Know the safety precautions for each.
    h. stopper all reagent bottles when not in use. Immediately wash reagents off yourself and your clothing if they spill on you, and immediately inform the instructor. If you accidentally get any reagent in your mouth, rinse the mouth thoroughly, and immediately inform your instructor.
    i. use extra care and wear disposable gloves when working with glass tubing and when using dissection equipment (scalpels, knives, or razor blades), whether cutting or assisting.
    j. administer first aid immediately to clean, sterilize, and cover any scrapes, cuts, and burns where the skin is broken and/or where there may be bleeding. Wear bandages over open skin wounds.
    k. report all accidents to the instructor immediately, and ask your instructor for assistance in cleaning up broken glassware and spills.
    l. report to the instructor any condition that appears unsafe or hazardous.
    m. use caution during any outdoor activities. Watch for snakes, poisonous insects or spiders, stinging insects, poison oak, poison ivy, etc. Be careful near water.
    n. wash your hands thoroughly after handling any preserved biological specimens.

I understand the safety rules as presented above. I agree to follow them and all other instructions given by the instructor.

Name: _____   Date: _____

Laboratory Class and Time: _____

1
Scientific Method

Learning Outcomes

Introduction

This laboratory will provide you with an opportunity to use the scientific method in the same manner as scientists. Today your subject will be the pillbug, *Armadillidium vulgare,* a type of crustacean that lives on land.

   Pillbugs have an exoskeleton consisting of overlapping "armored" plates that make them look like little armadillos. As pillbugs grow, they molt (shed the exoskeleton) four or five times during a lifetime. A pillbug can roll up into such a tight ball that its legs and head are no longer visible, earning it the nickname "roly-poly." They have three body parts: head, thorax, and abdomen. The head bears compound eyes and two pairs of antennae. The thorax bears pairs of walking legs; gills for gas exchange are located at the top of the first five pairs. The gills must be kept slightly moist, which explains why pillbugs are usually found in damp places. The final pair of appendages, the uropods, which are sensory and defensive in function, project from the abdomen of the animal.

   Pillbugs are commonly found in damp leaf litter, under rocks, and in basements or crawl spaces under houses. Following an inactive winter, pillbugs mate in the spring. Several weeks later, the eggs hatch and remain for

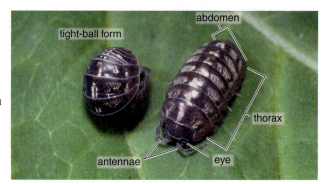

Pillbugs on leaf
© James H. Robinson/Science Source

six weeks in a brood pouch on the underside of the female's body. Once they leave the pouch, they eat primarily dead organic matter, including decaying leaves. Therefore, they are easy to find and to maintain in a moist terrarium with leaf litter, rocks, and wood chips. You are encouraged to collect some for your experiment. Since they live in the same locations as snakes, be careful when collecting them.

1.1   Using the Scientific Method

Some scientists work alone, but often scientists belong to a community of scientists who are working together to study some aspect of the natural world. For example, many scientists from different institutions come together to study Arctic ecology (Fig. 1.1).

Figure 1.1   Scientists work together.
These scientists are studying the ecology of the Arctic.   © Arnulf Husmo/Getty Images

> You will share your study of pillbugs with the other members of the class.

Even though the methodology can vary, scientists often use the **scientific method** (Fig. 1.2) when doing research. The scientific method involves these steps:

Making observations. Observations help scientists begin their study of a particular topic.

> To learn about pillbugs you will visually observe one. You could also do a Google search of the Web or talk to someone who has worked with pillbugs for a long time.

Why does the scientific method begin with observations? _____

_____

Formulating a hypothesis. Based on their observations, scientists come to a tentative decision, called a hypothesis, about their topic. Formulating hypotheses helps scientists decide how an experiment will be conducted.

> Based on your observations you might hypothesize that a pillbug will be attracted to fruit juice.

Now you know what you will actually do. What is the benefit of formulating a hypothesis? _____

_____

Testing the hypothesis. Scientists make further observations or perform experiments in order to test the hypothesis.

> You could decide to expose a pillbug to fruit juice and observe its reaction.

A well-designed experiment must have a **negative control**—that is, a sample or event—that is not exposed to the testing procedure. If the negative control and the test sample produce the same results, either the procedure is flawed or the hypothesis is false.

> Water can substitute for fruit juice and be the control in your experiment.

Scientists call the results of their experiments the **data.** It is very important for scientists to keep accurate records of all their data.

You will record your data in a table that can be easily examined by another person.

When another person repeats the same experiment, and the data is the same, both experiments have merit. Why must a scientist keep a complete record of an experiment? _____

Coming to a conclusion. Scientists come to a conclusion as to whether their data support or do not support the hypothesis.

If a pillbug is attracted to fruit juice, your hypothesis is supported. If the pillbug is not attracted to fruit juice, your hypothesis is not supported.

A scientist never says that a hypothesis has been proven true because, after all, some future knowledge might have a bearing on the experiment. What is the purpose of the conclusion? _____

_____

Developing a scientific theory. A *theory* in science is an encompassing conclusion based on many individual conclusions in the same field. For example, the gene theory states that organisms inherit coded information that controls their anatomy, physiology, and behavior. It takes many years for scientists to develop a theory and, therefore, we will not be developing any theories today. How is a scientific theory different from a conclusion? _____

_____

Figure 1.2 Flow diagram for the scientific method.
Scientists can predict the results by formulating hypotheses that are based on prior observations. Experiments are conducted to test the hypotheses which are supported or not supported by the analysis of data collected. Modification of the hypotheses can lead to more experiments. The conclusions of many related experiments can lead to the development of a theory.

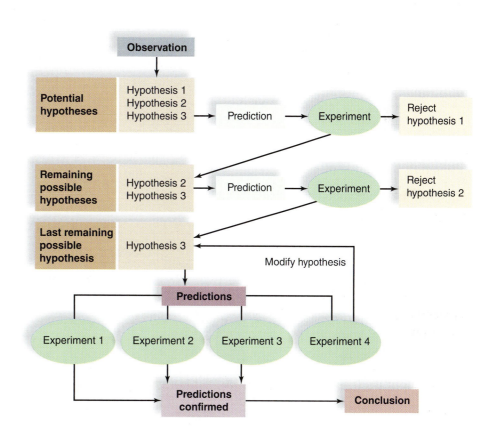

1.2 Observing a Pillbug

Wash your hands before and after handling pillbugs. Please handle them carefully so they are not crushed. When touched, they roll up into a ball or "pill" shape as a defense mechanism. They will soon recover if left alone.

Observation: Pillbug's External Anatomy

Obtain a pillbug that has been numbered with nail polish, white correction fluid, or tape tags. Put the pillbug in a small glass or plastic dish to keep it contained.

1. Examine the exterior of the pillbug with the unaided eye and with a magnifying lens or dissecting microscope.

   - How can you recognize the head end of a pillbug? _____
   - How many segments and pairs of walking legs are in the thorax? _____
   - The abdomen ends in uropods, appendages with a sensory and defense function. (Females have leaflike growths at the base of some legs where developing eggs and embryos are held in pouches.)

2. In the following space, draw an outline of your pillbug (at least 7 cm long). Label the head, thorax, abdomen, antennae, eyes, uropods, and one of the seven pairs of legs.

3. Draw a pillbug rolled into a ball.

1. Watch a pillbug's underside as the pillbug moves up a transparent surface, such as the side of a graduated cylinder or beaker.

    a. Describe the action of the feet and any other motion you see. _____

    b. Allow a pillbug to crawl on your hand. Describe how it feels. _____

    c. Does a pillbug have the ability to move directly forward? _____

    d. Do you see evidence of mouthparts on the underside of the pillbug? _____

2. As you watch the pillbug, identify

    a. the anatomical parts that allow a pillbug to identify and take in food. _____

    b. behaviors that will help the pillbug acquire food. For example, is the ability of the pillbug to move directly forward a help in acquiring food? Explain. _____

    What other behaviors allow a pillbug to acquire food? _____

    c. a behavior that helps a pillbug avoid dangerous situations. _____

    If the pillbug rolls up into a ball, wait a few minutes and it may uncurl itself.

3. Measure the speed of three pillbugs.

    a. Place each pillbug on a metric ruler, and use a stopwatch to measure the number of seconds (sec) it takes for the pillbug to move several centimeters (cm). Quickly record here the number of cm moved and the time in sec.

    pillbug 1 _____

    pillbug 2 _____

    pillbug 3 _____

    b. Knowing that 10 millimeters (mm) are in a cm, convert the number of cm traveled to mm and record this in the first column of Table 1.1. Record the total time taken in the second column.

    c. Use the space above to calculate the speed of each pillbug in mm/sec and record the speed for each pillbug in the last column of Table 1.1.

    d. Add the three speeds together and divide by 3 to determine the average speed for the three pillbugs. Record the average speed of pillbug motion in Table 1.1.

    When you conduct the experiment in section 1.4 you will have to be patient with your pillbug as it moves toward or away from a substance.

Table 1.1 Pillbug Speed			
Pillbug	**Millimeters (mm) Traveled**	**Time (sec)**	**Speed (mm/sec)**
1			
2			
3			
		Average speed:	

1.3 Formulating Hypotheses

You will be testing whether pillbugs are attracted to (move toward and eat), repelled by (move away from), or unresponsive to (don't move away from and do not move toward and eat) the particular substances, which are potential foods. If a pillbug simply rolls into a ball, nothing can be concluded, and you may wish to choose another pillbug or wait a minute or two to check for further response.

1. Choose
   a. two dry substances, such as flour, cornstarch, coffee creamer, or baking soda. Fine sand will serve as a control for dry substances. Record your "dry" choices as 1, 2, and 3 in the first column of Table 1.2.
   b. two liquids, such as milk, orange juice, ketchup, applesauce, or carbonated beverage. Water will serve as a control for liquid substances. Record your "wet" choices as 4, 5, and 6 in the first column of Table 1.2.
2. In the second column of Table 1.2, hypothesize how you expect the pillbug to respond to each substance. Use a plus (+) sign if you hypothesize that the pillbug will move toward and eat the substance; a minus (−) sign if you hypothesize that the pillbug will be repelled by the substance; and a zero (0) if you expect the pillbug to show neither behavior.
3. In the third column of Table 1.2 offer a reason for your hypothesis based on your knowledge of pillbugs from the introduction and your examination of the animal.

Table 1.2 Hypotheses About Pillbug's Response to Potential Foods		
Substance	**Hypothesis About Pillbug's Response**	**Reason for Hypothesis**
1		
2		
3 (control)		
4		
5		
6 (control)		

1.4 Performing the Experiment and Coming to a Conclusion

A good experimental design would be to keep your pillbug in a petri dish to test its reaction to the chosen substances. During your experiment, no substance must be put directly on the pillbug, nor can the pillbug be placed directly onto the substances.

Experimental Procedure: Pillbug's Response to Potential Foods

1. Before testing the pillbug's reaction, fill in the first column of Table 1.3. It will look exactly like the first column of Table 1.2.
2. Since pillbugs tend to walk around the edge of a petri dish, you could put the wet or dry substance there; or for the wet substance you could put liquid-soaked cotton in the pillbug's path.
3. Rinse your pillbug between procedures by spritzing it with distilled water from a spray bottle. Then put it on a paper towel to dry it off.
4. Watch the pillbug's response to each substance, and record it in Table 1.3, using +, −, or 0 as before.

Table 1.3 Pillbug's Response to Potential Foods

Substance	Pillbug's Response	Hypothesis Supported?
1		
2		
3 (control)		
4		
5		
6 (control)		

Conclusion

5. Do your results support your hypotheses? Answer yes or no in the last column of Table 1.3.

6. Are there any hypotheses that were not supported by the experimental results (data)? Does this difference give you more insight into pillbug behavior? Explain. _____

_____

_____

7. **Class Results.** Compare your results with those of other students who tested the same substance. Calculate the proportional response to each potential food (%+, %−, %0) and record your calculations in Table 1.4. As a group, your class can decide what proportion is needed to designate this response as typical. For example, if the pillbugs as a whole were attracted to a substance 70% or more of the time, you can call that response the "typical response."

Table 1.4 Pillbug's Response to Potential Foods: Class Results

Substance		Pillbug's Response			Hypothesis Supported?
1		%+	%−	%0	
2		%+	%−	%0	
3	(control)	%+	%−	%0	
4		%+	%−	%0	
5		%+	%−	%0	
6	(control)	%+	%−	%0	

8. On the basis of the class data do you need to revise your conclusion for any particular pillbug response? ___ Scientists prefer to come to conclusions on the basis of many trials. Why is this the best methodology? ____

_____

9. Did the pillbugs respond as expected to the controls, that is, did not eat them? _____ If they did not respond as expected, what can you conclude about your experimental results? _____

_____

_____ 1. What kind of animal is a pillbug?

_____ 2. What kind of skeleton does a pillbug have?

_____ 3. What structures do pillbugs use for gas exchange, and what condition must these structures be in to function properly?

_____ 4. What do scientists do first when they begin to study a specific topic?

_____ 5. What is a tentative decision about the outcome of an experiment?

_____ 6. What do we call the sample that lacks the factor being tested and goes through all the experimental steps?

_____ 7. What do scientists call the information they collect while doing experiments or making observations?

_____ 8. Which is made by a scientist following experiments and observations, a theory or a conclusion?

_____ 9. What do scientists develop after many years of experimentation and a lot of similar individual conclusions?

_____ 10. If your hypothesis was that your pillbug would be attracted to applesauce and your pillbug moved toward the applesauce, what would you say about your hypothesis?

For 11 and 12, indicate whether the statements are hypotheses, conclusions, or scientific theories.

_____ 11. All organisms are made of cells.

_____ 12. The data show that trans fat intake raises cholesterol and contributes to heart disease.

_____ 13. If a pillbug travels 3 mm in 30 seconds, what is its rate of speed?

_____ 14. If a pillbug moves toward a substance, is it attracted to or repelled by that substance?

Thought Questions

15. Why is a theory more comprehensive than a conclusion?

16. Why is it important to have a control substance for an experiment?

17. Why is it important to test a pillbug's response using one substance at a time?

2

Metric Measurement and Microscopy

Learning Outcomes

2.1 The Metric System
- Use metric units of measurement for length, weight, volume, and temperature.   10–13

2.2 Microscopy
- Describe similarities and differences between the stereomicroscope (dissecting microscope), the compound light microscope, and the electron microscope.   14–15

2.3 Stereomicroscope (Dissecting Microscope)
- Identify the parts and tell how to focus the stereomicroscope.   15–17

2.4 Use of the Compound Light Microscope
- Identify and give the function of the basic parts of the compound light microscope.   17–19
- List, in proper order, the steps for bringing an object into focus with the compound light microscope.   19–20
- Describe how the image is inverted by the compound light microscope.   20
- Calculate the total magnification and the diameter of field for both low- and high-power lens systems.   20–22
- Explain how a slide of colored threads provides information on the depth of field.   22

2.5 Microscopic Observations
- Name and describe the three kinds of cells studied in this exercise.   23–25
- State two differences between human epithelial cells and onion epidermal cells.   23–24
- Examine a wet mount of *Euglena* and pond water. Contrast the organisms observed in pond water.   25

Introduction

This laboratory introduces you to the metric system, which biologists use to indicate the sizes of cells and cell structures. This laboratory also examines the features, functions, and use of the compound light microscope and the stereomicroscope (dissecting microscope). Transmission and scanning electron microscopes are explained, and micrographs produced using these microscopes appear throughout this lab manual. The stereomicroscope and the scanning electron microscope view the surface and/or the three-dimensional structure of an object. The compound light microscope and the transmission electron microscope can view only extremely thin sections of a specimen. If a subject was sectioned lengthwise for viewing, the interior of the projections at the top of the cell, called cilia, would appear in the micrograph. A lengthwise cut through any type of specimen is called a **longitudinal section (ls).** On the other hand, if the subject in Figure 2.1 was sectioned crosswise below the area of the cilia, you would see other portions of the interior of the subject. A crosswise cut through any type of specimen is called a **cross section (cs).**

a. The cell   b. Longitudinal section   c. Cross section

Figure 2.1   Longitudinal and cross sections.
a. Transparent view of a cell. **b.** A longitudinal section would show the cilia at the top of the cell. **c.** A cross section shows only the interior where the cut is made.

2.1 The Metric System

The **metric system** is the standard system of measurement in the sciences, including biology, chemistry, and physics. It has tremendous advantages because all conversions, whether for volume, mass (weight), or length, can be in units of ten. Refer to Appendix B, page B–1, for an in-depth look at the units of the metric system.

Length

Metric units of length measurement include the **meter (m), centimeter (cm), millimeter (mm), micrometer (μm),** and **nanometer (nm)** (Table 2.1). The prefixes milli- (10^{-3}), micro- (10^{-6}), and nano (10^{-9}) are used with length, weight, and volume.

Table 2.1 Metric Units of Length Measurement

Unit	Meters	Centimeters	Millimeters	Relative Size
Meter (m)	1 m	100 cm	1,000 mm	Largest
Centimeter (cm)	0.01 (10^{-2}) m	1 cm	10 mm	
Millimeter (mm)	0.001 (10^{-3}) m	0.1 cm	1.0 mm	
Micrometer (μm)	0.000001 (10^{-6}) m	0.0001 (10^{-4}) cm	0.001 (10^{-3}) mm	
Nanometer (nm)	0.000000001 (10^{-9}) m	0.0000001 (10^{-7}) cm	0.000001 (10^{-6}) mm	Smallest

Experimental Procedure: Length

1. Obtain a small ruler marked in centimeters and millimeters. How many centimeters are

   represented? _____ One centimeter equals how many millimeters? _____ To express the size of small objects, such as cell contents, biologists use even smaller units of the metric system than those on the ruler. These units are the micrometer (μm) and the nanometer (nm).

   According to Table 2.1, 1 μm = _____ mm, and 1 nm = _____ mm.

   Therefore, 1 mm = _____ μm = _____ nm.

2. Measure the diameter (a line passing through the center with endpoints touching the circle) of the circle shown below to the nearest millimeter. This circle's diameter is

   _____ mm = _____ μm = _____ nm.

For example, to convert mm to μm:

$$\underline{\hspace{1cm}} \text{ mm} \times \frac{1,000 \text{ μm}}{\text{mm}} = \underline{\hspace{1cm}} \text{ μm}$$

3. Obtain a meter stick. On one side, find the numbers 1 through 39, which denote inches. One meter equals 39.37 inches; therefore, 1 meter is roughly equivalent to 1 yard. Turn the meter stick over, and observe the metric subdivisions. How many centimeters are in a meter? _____ How many millimeters are in a meter? _____ The prefix *milli* means _____.

4. Use the meter stick and the method shown in Figure 2.2 to measure the length of two long bones from a disarticulated human skeleton. Lay the meter stick flat on the lab table. Place a long bone next to the meter stick between two pieces of cardboard (each about 10 cm × 30 cm), held upright at right angles to the stick. The narrow end of each piece of cardboard should touch the meter stick. The length between the cards is the length of the bone in centimeters. For example, if the bone measures from the 22 cm mark to the 50 cm mark, the length of the bone is _____ cm. If the bone measures from the 22 cm mark to midway between the 50 cm and 51 cm marks, its length is _____ mm, or _____ cm.

5. Record the length of two bones. First bone: _____ cm = _____ mm.
   Second bone: _____ cm = _____ mm.

Figure 2.2 Measurement of a long bone.
How to measure a long bone using a meter stick.

Weight

Two metric units of weight are the **gram (g)** and the **milligram (mg).** A paper clip weighs about 1 g, which equals 1,000 mg. 2 g = _____ mg; 0.2 g = _____ mg; and 2 mg = _____ g.

Experimental Procedure: Weight

1. Use a balance scale to measure the weight of a wooden block small enough to hold in the palm of your hand.

2. Measure the weight of the block to the tenth of a gram. The weight of the wooden block is
   _____ g = _____ mg.

3. Measure the weight of an item small enough to fit inside the opening of a 50 ml graduated cylinder. The item, a(n) _____, is _____ g = _____ mg.

Volume

Two metric units of volume are the **liter (l)** and the **milliliter (ml).** One liter = 1,000 ml.

Experimental Procedure: Volume

1. Volume measurements can be related to those of length. For example, use a millimeter ruler to measure the wooden block used in the previous Experimental Procedure to get its length, width, and depth.

   length = _____ cm; width = _____ cm; depth = _____ cm

   The volume, or space, occupied by the wooden block can be expressed in cubic centimeters

   (cc or cm³) by multiplying: length × width × depth = _____ cm³. For purposes of this Experimental

   Procedure, 1 cubic centimeter equals 1 milliliter; therefore, the wooden block has a volume of _____ ml.

2. In the biology laboratory, liquid volume is usually measured directly in liters or milliliters with appropriate measuring devices. For example, use a 50 ml graduated cylinder to add 20 ml of water to a test tube. First, fill the graduated cylinder to the 20 ml mark. To do this properly, you have to make sure that the lowest margin of the water level, or the **meniscus** (Fig. 2.3), is at the 20 ml mark. Place your eye directly parallel to the level of the meniscus, and add water until the meniscus is at the 20 ml mark. (Having a dropper bottle filled with water on hand can help you do this.) A large, blank, white index card held behind the cylinder can also help you see the scale more clearly. Now pour the 20 ml of water into the test tube.

3. Hypothesize how you could find the total volume of the test tube. _____

   _____

   _____

   What is the test tube's total volume? _____

Figure 2.3 Meniscus.
The proper way to view the meniscus.

meniscus
reading 20 ml

improper position

proper position

improper position

4. Fill a 50 ml graduated cylinder with water to about the 20 ml mark. Hypothesize how you could use this setup to calculate the volume of an object. _____

_____

Now perform the operation you suggested. The object, _____, has a volume of _____ ml.

5. Hypothesize how you could determine how many drops from the pipette of the dropper bottle equal 1 ml. _____

Now perform the operation you suggested. How many drops from the pipette of the dropper bottle equal 1 ml? _____

6. Some pipettes are graduated and can be filled to a certain level as a way to measure volume directly. Your instructor will demonstrate this. Are pipettes customarily used to measure large or small volumes? _____

Temperature

There are two temperature scales: the **Fahrenheit (F)** and **Celsius (centigrade, C)** scales (Fig. 2.4). Scientists use the Celsius scale.

Experimental Procedure: Temperature

1. Study the two scales in Figure 2.4, and complete the following information:

   a. Water freezes at either _____ °F or _____ °C.

   b. Water boils at either _____ °F or _____ °C.

2. To convert from the Fahrenheit to the Celsius scale, use the following equation:

$$°C = (°F - 32°)/1.8$$
$$or$$
$$°F = (1.8°C) + 32$$

   Human body temperature of 98°F is what temperature on the Celsius scale? _____

3. Record any two of the following temperatures in your lab environment. In each case, allow the end bulb of the Celsius thermometer to remain in or on the sample for one minute.

   Room temperature = _____ °C

   Surface of your skin = _____ °C

   Cold tap water in a 50 ml beaker = _____ °C

   Hot tap water in a 50 ml beaker = _____ °C

   Ice water = _____ °C

Figure 2.4 Temperature scales.
The Fahrenheit (°F) scale is on the left, and the Celsius (°C) scale is on the right.

2.2 Microscopy

Because biological objects can be very small, we often use a microscope to view them. Many kinds of instruments, ranging from the hand lens to the electron microscope, are effective magnifying devices. A short description of two kinds of light microscopes and two kinds of electron microscopes follows.

Light Microscopes

Light microscopes use light rays passing through lenses to magnify the object. The **stereomicroscope (dissecting microscope)** is designed to study entire objects in three dimensions at low magnification. The **compound light microscope** is used for examining small or thinly sliced sections of objects under higher magnification than that of the stereomicroscope. The term **compound** refers to the use of two sets of lenses: the ocular lenses located near the eyes and the objective lenses located near the object. Illumination is from below, and visible light passes through clear portions but does not pass through opaque portions. To improve contrast, the microscopist uses stains or dyes that bind to cellular structures and absorb light. Photomicrographs, also called light micrographs, are images produced by a compound light microscope (Fig. 2.5*a*).

Figure 2.5 Comparative micrographs of a lymphocyte.
A lymphocyte is a type of white blood cell. **a.** A photomicrograph (light micrograph) of a lymphocyte shows less detail than a **(b)** transmission electron micrograph (TEM). **c.** A scanning electron micrograph (SEM) of a lymphocyte shows the cell surface in three dimensions.   (a) © Michael Ross/Science Source; (b) © CNRI/SPL/Science Source; (c) © Steve Gschmeissner/Science Source

2,150×
a. Photomicrograph or light micrograph (LM)

3,000×
b. Transmission electron micrograph (TEM)

5,000×
c. Scanning electron micrograph (SEM)

Electron Microscopes

Electron microscopes use beams of electrons to magnify the object. The beams are focused on a photographic plate by means of electromagnets. The **transmission electron microscope** is analogous to the compound light microscope. The object is ultra-thinly sliced and treated with heavy metal salts to improve contrast. Figure 2.5*b* is a micrograph produced by this type of microscope. The **scanning electron microscope** is analogous to the dissecting light microscope. It gives an image of the surface and dimensions of an object, as is apparent from the scanning electron micrograph in Figure 2.5*c*.

The micrographs in Figure 2.5 demonstrate that an object is magnified more with an electron microscope than with a compound light microscope. The difference between these two types of microscopes, however, is not simply a matter of magnification; it is also the electron microscope's ability to show detail. The electron microscope has greater resolving power. **Resolution** is the minimum distance between two objects at which they can still be seen, or resolved, as two separate objects. The use of high-energy electrons rather than light gives electron microscopes a much greater resolving power since two objects that are much closer together can still be distinguished as separate points. Table 2.2 lists several other differences between the compound light microscope and the transmission electron microscope.

Table 2.2 Comparison of the Compound Light Microscope and the Transmission Electron Microscope	
Compound Light Microscope	**Transmission Electron Microscope**
1. Glass lenses	1. Electromagnetic lenses
2. Illumination by visible light	2. Illumination due to beam of electrons
3. Resolution $\cong$ 200 nm	3. Resolution $\cong$ 0.1 nm
4. Magnifies to 2,000×	4. Magnifies to 1,000,000×
5. Costs up to tens of thousands of dollars	5. Costs up to hundreds of thousands of dollars

Conclusions: Microscopy

- Which two types of microscopes view the surface of an object? _____

- Which two types of microscopes view objects that have been sliced and treated to improve

  contrast? _____

- Of the microscopes just mentioned, which one resolves the greater amount of detail?

  _____

2.3 Stereomicroscope (Dissecting Microscope)

The **stereomicroscope (dissecting microscope,** Fig. 2.6) allows you to view objects in three dimensions at low magnifications. It is used to study entire small organisms, any object requiring lower magnification, and opaque objects that can be viewed only by reflected light. It is called a stereomicroscope because it produces a three-dimensional image.

Identifying the Parts

After your instructor has explained how to carry a microscope, obtain a stereomicroscope and a separate illuminator, if necessary, from the storage area. Place it securely on the table. Plug in the power cord, and turn on the illuminator. There is a wide variety of stereomicroscope styles, and your instructor will discuss the specific style(s) available to you. Regardless of style, the following features should be present:

1. **Binocular head:** Holds two eyepiece lenses that move to accommodate for the various distances between different individuals' eyes.
2. **Eyepiece lenses:** The two lenses located on the binocular head. What is the magnification of your eyepieces? _____ Some models have one **independent focusing eyepiece** with a knurled knob to allow independent adjustment of each eye. The nonadjustable eyepiece is called the **fixed eyepiece.**
3. **Focusing knob:** A large, black or gray knob located on the arm; used for changing the focus of both eyepieces together.

Figure 2.6 Binocular dissecting microscope (stereomicroscope).
Label this stereomicroscope with the help of the text material.   © Leica Microsystems GmbH

    b. Low-power objective: This lens is longer than the scanning objective lens and is used to view objects in greater detail. What is the magnifying power of the low-power objective lens on your microscope? _____

    c. High-power objective: If your microscope has three objective lenses, this lens will be the longest. It is used to view an object in even greater detail. What is the magnifying power of the high-power objective lens on your microscope? _____

    d. Oil immersion objective (on microscopes with four objective lenses): Holds a 95× (to 100×) lens and is used in conjunction with immersion oil to view objects with the greatest magnification. Does your microscope have an oil immersion objective? _____ If this lens is available, your instructor will discuss its use when the lens is needed.

6. **Stage:** Platform that holds and supports microscope slides. A mechanical stage is a movable stage that aids in the accurate positioning of the slide. Does your microscope have a mechanical stage? _____

    a. Stage clips: Clips that hold a slide in place on the stage.

    b. Mechanical stage control knobs: Two knobs that control forward/reverse movement and right/left movement, respectively.

7. **Coarse-adjustment knob:** Knob used to bring object into approximate focus; used only with low-power objective.

8. **Fine-adjustment knob:** Knob used to bring object into final focus.

9. **Condenser:** Lens system below the stage to focus the beam of light on the object being viewed.

    a. Diaphragm or **diaphragm control** lever that controls the amount of light passing through the condenser.

10. **Light source:** An attached lamp that directs a beam of light up through the object.

11. **Base:** The flat surface of the microscope on the table.

Rules for Microscope Use

Observe the following rules for using a microscope:

1. The lowest power objective (scanning or low) should be in position at both the beginning and the end of microscope use.
2. Use only lens paper for cleaning lenses.
3. Do not tilt the microscope because the eyepieces could fall or wet mounts could be ruined.
4. Keep the stage clean and dry to prevent rust and corrosion.
5. Do not remove parts of the microscope.
6. Keep the microscope dust-free by covering it after use.
7. Report any malfunctions.

Focusing the Compound Light Microscope—Lowest Power

1. Turn the nosepiece so that the *lowest* power objective on your microscope is in straight alignment over the stage.
2. Always begin focusing with the *lowest* power objective on your microscope (4× [scanning] or 10× [low power]).
3. With the coarse-adjustment knob, lower the stage (or raise the objectives) until it stops.

4. Place a slide of the letter *e* on the stage, and stabilize it with the clips. (If your microscope has a mechanical stage, pinch the spring of the slide arms on the stage, and insert the slide.) Center the *e* as best you can on the stage or use the two control knobs located below the stage (if your microscope has a mechanical stage) to center the *e*.

5. Again, be sure that the lowest-power objective is in place. Then, as you look from the side, decrease the distance between the stage and the tip of the objective lens until the lens comes to an automatic stop or is no closer than 3 mm above the slide.

6. While looking into the eyepiece, rotate the diaphragm (or diaphragm control lever) to give the maximum amount of light.

7. Using the coarse-adjustment knob, slowly increase the distance between the stage and the objective lens until the object—in this case, the letter *e*—comes into view, or focus.

8. Once the object is seen, you may need to adjust the amount of light. To increase or decrease the contrast, rotate the diaphragm slightly.

9. Use the fine-adjustment knob to sharpen the focus if necessary.

10. Practice having both eyes open when looking through the eyepiece, as this greatly reduces eyestrain.

Inversion

Inversion refers to the fact that a microscopic image is upside down and reversed.

Observation: Inversion

1. Draw the letter *e* as it appears on the slide (with the unaided eye, not looking through the eyepiece). _____

2. Draw the letter *e* as it appears when you look through the eyepiece. _____

3. What differences do you notice? _____

4. Move the slide to the right. Which way does the image appear to move? _____

5. Move the slide toward you. Which way does the image appear to move? _____

Focusing the Compound Light Microscope—Higher Powers

Compound light microscopes are **parfocal;** that is, once the object is in focus with the lowest power, it should also be almost in focus with the higher power.

1. Bring the object into focus under the lowest power by following the instructions in the previous section.
2. Make sure that the letter *e* is centered in the field of the lowest objective.
3. Move to the next higher objective (low power [10×] or high power [40×]) by turning the nosepiece until you hear it click into place. Do not change the focus; parfocal microscope objectives will not "hit" normal slides when changing the focus if the lowest objective is initially in focus. (If you are on low power [10×], proceed to high power [40×] before going on to step 4.)
4. If any adjustment is needed, use only the *fine*-adjustment knob. (*Note:* Always use only the fine-adjustment knob with high power, and do not use the coarse-adjustment knob.)
5. On a drawing of the letter *e* to the right, *draw a circle around the portion of the letter that you are now seeing with high-power magnification.* The letter *e* will not disappear because your microscope is parcentric (the focus remains near the center). _____
6. When you have finished your observations of this slide (or any slide), rotate the nosepiece until the lowest-power objective clicks into place, and then remove the slide.

Total Magnification

Total magnification is calculated by multiplying the magnification of the ocular lens (eyepiece) by the magnification of the objective lens. The magnification of a lens is imprinted on the lens casing.

Calculate total magnification figures for your microscope, and record your findings in Table 2.3.

Table 2.3 Total Magnification			
Objective	**Ocular Lens**	**Objective Lens**	**Total Magnification**
Scanning power (if present)			
Low power			
High power			
Oil immersion (if present)			

Field of View

A microscope's **field of view** is the circle visible through the lenses. The **diameter of field** is the length of the field from one edge to the other.

Low-Power (10×) Diameter of Field

1. Place a clear plastic ruler across the stage so that the edge of the ruler is visible as a horizontal line along the diameter of the low-power (not scanning) field. Be sure that you are looking at the millimeter side of the ruler.

2. Estimate the number of millimeters, to tenths, that you see along the field: _____ mm. (*Hint:* Start by placing any millimeter marker at the edge of the field.) Convert the observed number

    of millimeters to micrometers: _____ μm. This is the **low-power diameter of field (LPD)** for your microscope in micrometers.

High-Power (40×) Diameter of Field

1. To compute the **high-power diameter of field (HPD),** substitute these data into the formula given:

    a. LPD = low-power diameter of field (in micrometers) = _____

    b. LPM = low-power total magnification (from Table 2.3) = _____

    c. HPM = high-power total magnification (from Table 2.3) = _____

Example: If the diameter of field is about 2 mm, then the LPD is 2,000 μm. Using the LPM and HPM values from Table 2.3, the HPD would be 500 μm.

$$HPD = LPD \times \frac{LPM}{HPM}$$

$$HPD = (\quad\quad) \times \frac{(\quad\quad\quad)}{(\quad\quad\quad)} = \underline{\quad\quad\quad}$$

Conclusions: Total Magnification and Field of View

- Does low power or high power have a larger field of view (one that allows you to see more of the object)? _____
- Which has a smaller field but magnifies to a greater extent? _____
- To locate small objects on a slide, first find them under _____ ; then place them in the center of the field before rotating to _____ .

Depth of Field

When viewing an object on a slide under high power, the **depth of field** (Fig. 2.8) is the area—from top to bottom—that comes into focus while slowly focusing up and down with the microscope's fine-adjustment knob.

Observation: Depth of Field

1. Obtain a prepared slide with three or four colored threads mounted together, or prepare a wet-mount slide with three or four crossing threads or hairs of different colors. (Directions for preparing a wet mount are given in section 2.5.)
2. With low power, find a point where the threads or hairs cross. Slowly focus up and down. Notice that when one thread or hair is in focus, the others seem blurred. Remember, as the stage moves upward (or the objectives move downward), objects on top come into focus first. Determine the order of the threads or hairs, and complete Table 2.4.
3. Switch to high power, and notice that the depth of field is more shallow with high power than with low power. Focusing up and down with the fine-adjustment knob when viewing a slide with high power will give you an idea of the specimen's three-dimensional form. For example, viewing a number of sections from bottom to top allows reconstruction of the three-dimensional structure, as demonstrated in Figure 2.8.

Figure 2.8 Depth of field.
A demonstration of how focusing at depths 1, 2, and 3 would produce three different images (views) that could be used to reconstruct the original three-dimensional structure of the object.

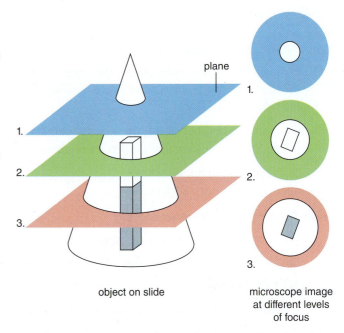

object on slide

microscope image at different levels of focus

Table 2.4 Order of Threads (or Hairs)	
Depth	**Thread (or Hair) Color**
Top	
Middle	
Bottom	

2.5 Microscopic Observations

When a specimen is prepared for observation, the object should always be viewed as a **wet mount.** A wet mount is prepared by placing a drop of liquid on a slide or, if the material is dry, by placing it directly on the slide and adding a drop of water or stain. The mount is then covered with a coverslip, as illustrated in Figure 2.9. Dry the bottom of your slide before placing it on the stage.

Figure 2.9 Preparation of a wet mount.

a. Add drop of liquid or dry object and liquid.

b. Lower coverslip slowly

Onion Epidermal Cells

Epidermal cells cover the surfaces of plant organs, such as leaves. The bulb of an onion is made up of fleshy leaves.

Observation: Onion Epidermal Cells

1. With a scalpel, strip a small, thin, transparent layer of cells from the inside of a fresh onion leaf.

   ⚠ **Scalpel** Exercise care when using a scalpel.

2. Place it gently on a clean, dry slide, and add a drop of iodine solution (or methylene blue). Cover with a coverslip.
3. Observe under the microscope.
4. Locate the cell wall and the nucleus. *Label Figure 2.10.*
5. Count the number of onion cells that line up end to end in a single line across the diameter of the high-power (40×) field.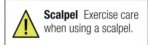

   Based on what you learned in section 2.4 about measuring diameter of field, what is the

   high-power diameter of field (HPD) in micrometers? _____ μm

Figure 2.10 Onion epidermal cells.
Label the cell wall and the nucleus.
© Ted Kinsman/Science Source

1. _____

2. _____

100×

Calculate the length of each onion cell:

(HPD ÷ number of cells): ——————— μm

6. Wash and reuse this slide for the next exercise (Human Epithelial Cells).

Human Epithelial Cells

Epithelial cells (Fig. 2.11) cover the body's surface and line its cavities.

Observation: Human Epithelial Cells

1. Obtain a prepared slide, or make your own as follows:
   a. Obtain a prepackaged flat toothpick (or sanitize one with alcohol or alcohol swabs).
   b. Gently scrape the inside of your cheek with the toothpick, and place the scrapings on a clean, dry slide. Discard used toothpicks in the biohazard waste container provided.
   c. Add a drop of very weak *methylene blue* or *iodine solution,* and cover with a coverslip.
2. Observe under the microscope.
3. Locate the nucleus (the central, round body), the cytoplasm, and the plasma membrane (outer cell boundary). *Label Figure 2.11.*
4. Because your epithelial slides are biohazardous, they must be disposed of as indicated by your instructor.
5. Note some obvious differences between the onion cells and the human cheek cells, and list them in Table 2.5.

⚠ **Methylene blue** Avoid ingestion, inhalation, and contact with skin, eyes, and mucous membranes. If any should spill on your skin, wash the area with mild soap and water. Methylene blue will also stain clothing.

Figure 2.11   Cheek epithelial cells.
Label the nucleus, the cytoplasm, and the plasma membrane.
© Dr. Gopal Murti/Science Source

1. ———————————————

2. ———————————————

3. ———————————————

1,000×

Table 2.5 Differences Between Onion Epidermal and Human Epithelial Cells		
Differences	**Onion Epidermal Cells**	**Human Epithelial Cells (Cheek)**
Shape		
Orientation		
Boundary		

Euglena

Examination of *Euglena* (a unicellular organism with a flagellum to facilitate movement) will test your ability to observe objects with the microscope, to utilize depth of field, and to control illumination to heighten contrast.

Observation: *Euglena*

1. Make a wet mount of *Euglena* by using a drop of a *Euglena* culture and adding a drop of Protoslo® (methyl cellulose solution) onto a slide. The Protoslo® slows the organism's swimming.
2. Mix thoroughly with a toothpick, and add a coverslip.
3. Scan the slide for *Euglena:* Start at the upper left-hand corner, and move the slide forward and back as you work across the slide from left to right. The *Euglena* may be at the edge of the slide because they show an aversion to Protoslo®. Use Figure 2.12 to help identify the structural details of *Euglena.*

Figure 2.12   *Euglena*.
Euglena is a unicellular, flagellated organism.

© M.I. Walker/Science Source

Photomicrograph   200×

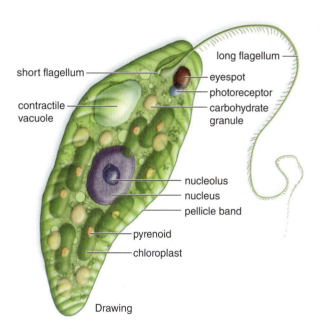

Drawing

4. Experiment by using scanning, low-power, and high-power objective lenses; by focusing up and down with the fine-adjustment knob; and by adjusting the light so that it is not too bright.
5. Compare your *Euglena* specimens with Figure 2.12. List the labeled features that you can actually

   see: _____

   _____

Pond Water

Examination of pond water will also test your ability to observe objects with the microscope, to utilize depth of field, and to control illumination to heighten contrast. See Figure 24.17, page 345, for microorganisms found in pond water.

_____ **1.** 19 mm equals how many cm?

_____ **2.** 880 mm equals how many m?

_____ **3.** 2,700 mg equals how many grams?

_____ **4.** 3.4 l equals how many ml?

_____ **5.** To properly measure 20 ml of water, what must be at the 20 ml mark of the graduated cylinder?

_____ **6.** 90°F equals how many degrees Celsius?

_____ **7.** What kind of microscope would be used to study a whole or opaque object?

_____ **8.** What is the name for the lenses located near the eye?

_____ **9.** If the total magnification of a slide is 400× and the ocular lenses are 10×, what is the magnifying power of the objective being used?

_____ **10.** If the amount of light passing through the condenser needs to be decreased, what microscope part should be adjusted?

_____ **11.** What word describes a microscope that remains in focus when the objective lenses are changed?

_____ **12.** If a slide is being viewed with the high-power objective, which adjustment knob should be used to sharpen the focus?

_____ **13.** Which objective should be in place when the microscope is put away?

_____ **14.** If threads are layered from top to bottom, brown, green, red, which layer will come into focus first if you are using the microscope properly?

Thought Questions

15. Which type of microscope should be used to view a virus that is 50 nm in size? Justify your choice.

16. Justify your choice of an objective to use when starting your observation of the *Euglena* wet mount; if the *Euglena* are swimming to the left, which way should you move your slide to keep the organism in view?

17. If you use the coarse-adjustment knob to focus on an object with the high-power objective, what problems will you encounter? Explain.

3

Chemical Composition of Cells

Introduction

All organisms consist of basic units of matter called **atoms.** Molecules form when atoms bond with one another. Inorganic molecules are often associated with nonliving things. Organic biomolecules are associated with organisms. In this laboratory, you will be studying the biomolecules of cells: **proteins, carbohydrates** (monosaccharides, disaccharides, polysaccharides), and **lipids** (i.e., fat).

Large biomolecules, sometimes called macromolecules, form during dehydration reactions when smaller molecules bond as water is given off. During *hydrolysis reactions,* bonds are broken as water is added.

> ⏱ **Planning Ahead** To save time, your instructor may have you start the boiling water bath needed for the experiments on page 32 and page 38 at the beginning of the laboratory session.

Dehydration reaction   Hydrolysis

A fat contains one glycerol and three fatty acids. Proteins and some carbohydrates (called polysaccharides) are **polymers** because they are made up of smaller molecules called **monomers.** Proteins contain a large number of amino acids (the monomer) joined together by a peptide bond. A polysaccharide, such as starch, contains a large number of glucose molecules joined together. Various chemicals will be used in this laboratory to test for the presence of cellular biomolecules. If a color change is observed, the test is said to be *positive* because it indicates that the molecule is present. If the color change is not observed, the test is said to be *negative* because it indicates that the molecule is not present.

What Is a Control?

The experiments in today's laboratory have both a positive control and a negative control, *which should be saved for comparison purposes until the experiment is complete*. The **positive control** goes through all the steps of the experiment and does contain the substance being tested. Therefore, positive results are expected. The **negative control** goes through all the steps of the experiment except it does not contain the substance being tested. Therefore, negative results are expected.

For example, if a test tube contains glucose (the substance being tested) and Benedict's reagent (blue) is added, a red color develops upon heating. This test tube is the positive control. The red color indicates a positive test for glucose. If a test tube does not contain glucose and Benedict's reagent is added, Benedict's is expected to remain blue. This test tube is the negative control; it tests negative for glucose.

What benefit is a positive control? Positive controls give you a standard by which to tell if the substance being tested is present (or acting properly) in an unknown sample. Negative controls ensure that the experiment is giving reliable results. If a negative control should happen to give a positive result, then the entire experiment may be faulty and unreliable.

3.1   Proteins

Proteins have numerous functions in cells. Antibodies are proteins that combine with pathogens so that the pathogens are destroyed by the body. Transport proteins combine with and move substances from place to place. Hemoglobin transports oxygen throughout the body. Albumin is another transport protein in our blood. Regulatory proteins control cellular metabolism in some way. For example, the hormone insulin regulates the amount of glucose in blood so that cells have a ready supply. Structural proteins include keratin, found in hair, and myosin, found in muscle. **Enzymes** are proteins that speed chemical reactions. A reaction that could take days or weeks to complete can happen within an instant if the correct enzyme is present. Amylase is an enzyme that speeds the breakdown of starch in the mouth and small intestine.

Proteins are made up of **amino acids** (the subunits) joined together. About 20 different common amino acids are found in cells. All amino acids have an acidic group (—COOH) and an amino group (H_2N—). They differ by the **R group** (remainder group) attached to a carbon atom, as shown in Figure 3.1. The *R* groups have varying sizes, shapes, and chemical activities.

A chain of two or more amino acids is called a **peptide,** and the bond between the amino acids is called a **peptide bond.** A **polypeptide** is a very long chain of amino acids. A protein can contain one or more polypeptide chains. A single chain forms insulin, while four chains form hemoglobin. A protein has a particular shape, which is important to its function. The shape comes about because the *R* groups of the polypeptide chain(s) can interact with one another in various ways.

Figure 3.1   Formation of a dipeptide.
During a dehydration reaction, a dipeptide forms when an amino acid joins with an amino acid as a water molecule is removed. The bond between amino acids is called a peptide bond. During a hydrolysis reaction, water is added and the peptide bond is broken.

Test for Proteins

Biuret reagent (blue color) contains a strong solution of sodium or potassium hydroxide (NaOH or KOH) and a small amount of dilute copper sulfate ($CuSO_4$) solution. The reagent changes color in the presence of proteins or peptides because the peptide bonds of the protein or peptide chemically combine with the copper ions in biuret reagent (Table 3.1).

Biuret test for protein and peptides

© David S. Moyer

Table 3.1 Biuret Test for Protein and Peptides		
	Protein	**Peptides**
Biuret reagent (blue)	Purple	Pinkish-purple

Experimental Procedure: Test for Proteins

1. Label four clean test tubes (1 to 4).
2. Using the designated graduated transfer pipets, add 1 ml of the experimental solutions listed in Table 3.2 to the test tubes according to their numbers.
3. Then add five drops of biuret reagent to the tubes, swirling to mix.
4. The reaction is almost immediate. Record your observations in Table 3.2.

> ⚠ **Biuret reagent** Biuret reagent is highly corrosive. Exercise care in using this chemical. If any should spill on your skin, wash the area with mild soap and water. Follow your instructor's directions for its disposal.

Table 3.2 Biuret Test for Protein			
Tube	**Contents**	**Final Color**	**Conclusions (+ or −)**
1	Distilled water		
2	Albumin		
3	Pepsin		
4	Starch		

Conclusions: Proteins

- From your test results, conclude if a protein is present (+) or absent (−). Enter your conclusions in Table 3.2.

- Pepsin is an enzyme. Enzymes are composed of what type of organic molecule? _____

- According to your results, is starch a protein? _____

- Which of the four tubes is the negative control sample? _____ Why? _____
  _____

- Why do experimental procedures include control samples? _____
  _____

- If your results are not as expected, inform your instructor, who will advise you how to proceed.

3.2 Carbohydrates

Carbohydrates include sugars and molecules that are chains of sugars. **Glucose,** which has only one sugar unit, is a monosaccharide; **maltose,** which has two sugar units, is a disaccharide (Fig. 3.2). Glycogen, starch, and cellulose are polysaccharides, made up of chains of glucose units (Fig. 3.3).

Glucose is used by all organisms as an energy source. Energy is released when glucose is broken down to carbon dioxide and water. This energy is used by the organism to do work. Animals store glucose as glycogen and plants store glucose as starch. Plant cell walls are composed of cellulose.

Figure 3.2 Formation of a disaccharide.
During a dehydration reaction, a disaccharide, such as maltose, forms when a glucose joins with a glucose as a water molecule is removed. During a hydrolysis reaction, the components of water are added, and the bond is broken.

Figure 3.3 Starch.
Starch is a polysaccharide composed of many glucose units. **a.** Photomicrograph of starch granules in cells of a potato. **b.** Structure of starch. Starch consists of amylose that is nonbranched and amylopectin that is branched.

a. Photomicrograph   57×   **b.** Structure of starch

Test for Starch

In the presence of starch, iodine solution (yellowish-brown) reacts chemically with starch to form a blue-black color (Table 3.3).

Table 3.3 Iodine Test for Starch	
	Starch
Iodine solution	Blue-black

Experimental Procedure: Test for Starch

1. Label five clean test tubes (1 to 5).
2. Using the designated graduated transfer pipets, add 1 ml of the experimental solutions listed in Table 3.4 to the test tubes according to their numbers.
3. Then add five drops of iodine solution to the tubes at the same time.
4. Note the final color changes and record your observations in Table 3.4.

Iodine test for starch

© Martin Shields/Science Source

Table 3.4 Iodine (IKI) Test for Starch			
Tube	**Contents**	**Color**	**Conclusions**
1	Water		
2	Starch suspension		
3	Onion juice		
4	Potato juice		
5	Glucose solution		

Conclusions: Starch

- From your test results, draw conclusions about what organic compound is present in each tube. Write these conclusions in Table 3.4.

- Does the potato or the onion store glucose as starch? _____ How do you know? _____

  _____

- If your results are not as expected, offer an explanation. Then inform your instructor, who will advise you how to proceed.

Potato

1. With a scalpel, slice a very thin piece of potato. Place it on a microscope slide, add a drop of water and a coverslip, and observe under low power with your compound light microscope. Compare your slide with the photomicrograph of starch granules (see Fig. 3.3*a*). Find the cell wall (large, geometric compartments) and the starch grains (numerous clear, oval-shaped objects).
2. Without removing the coverslip, place two drops of iodine solution onto the microscope slide so that the iodine touches the coverslip. Draw the iodine under the coverslip by placing a small piece of paper towel in contact with the water on the **opposite** side of the coverslip.
3. Microscopically examine the potato again on the side closest to where the iodine solution was applied.

   What is the color of the small, oval bodies? _____

   What is the chemical composition of these oval bodies? _____

Onion

1. Peel a single layer of onion from the bulb. On the inside surface, you will find a thin, transparent layer of onion skin. Peel off a small section of this layer for use on your slide.
2. Add a large drop of iodine solution.
3. Does onion contain starch? _____
4. Are these results consistent with those you recorded for onion juice in Table 3.4? _____

Test for Sugars

Benedict's reagent Benedict's reagent is highly corrosive. Exercise care in using this chemical. If any should spill on your skin, wash the area with mild soap and water. Follow your instructor's directions for disposal of this chemical.

Monosaccharides and some disaccharides will react with **Benedict's reagent** after being heated in a boiling water bath. In this reaction, copper ion (Cu^{2+}) in the Benedict's reagent reacts with part of the sugar molecule, causing a distinctive color change. The color change can range from green to red, and increasing concentrations of sugar will give a continuum of colored products (Table 3.5).

Table 3.5 Benedict's Test for Sugars (Some Typical Reactions)

Chemical	Chemical Category	Benedict's Reagent (After Heating)	
Water	Inorganic	Blue (no change)	
Glucose	Monosaccharide (carbohydrate)	Varies with concentration:	very low—green low—yellow moderate—yellow-orange high—orange very high—orange-red
Maltose	Disaccharide (carbohydrate)	Varies with concentration—see "Glucose"	
Starch	Polysaccharide (carbohydrate)	Blue (no change)	

Benedict's test for sugar

© David S. Moyer

Experimental Procedure: Test for Sugars

1. Prepare a boiling water bath and label five clean test tubes (1 to 5).
2. Using the designated graduated transfer pipets, add 1 ml of the experimental solutions listed in Table 3.6 to the test tubes according to their numbers.
3. Then add five drops of Benedict's reagent to all the tubes at this time.
4. Place the tubes into the boiling water bath at the same time.
5. When, after a few minutes, you see a change of colors, remove all of the tubes from the water bath and record your observations in Table 3.6.
6. Save your tubes for comparison purposes when you do section 3.4.

Table 3.6   Benedict's Test for Sugars

Tube	Contents	Color (After Heating)	Conclusions
1	Water		
2	Glucose solution		
3	Starch suspension		
4	Onion juice		
5	Potato juice		

Conclusions: Sugars

- From your test results, conclude what kind of chemical is present. Enter your conclusions in Table 3.6.

- Which tube served as a negative and which as a positive control? _____

- Compare Table 3.4 with Table 3.6. Sugars are an immediate energy source in cells. In plant cells, glucose (a primary energy molecule) is often stored in the form of starch. Is glucose stored as starch in the potato? _____ Is glucose stored as starch in the onion? _____ Does this explain your results in Table 3.6? _____ Why? _____

3.3 Lipids

Lipids are compounds that are insoluble in water and soluble in solvents, such as alcohol and ether. Lipids include fats, oils, phospholipids, steroids, and cholesterol. Typically, **fat,** such as in the adipose tissue of animals, and **oils,** such as the vegetable oils from plants, are composed of three molecules of fatty acids bonded to one molecule of glycerol (Fig. 3.4). **Phospholipids** have the same structure as fats, except that in place of the third fatty acid there is a phosphate group (a grouping that contains phosphate). **Steroids** are derived from **cholesterol.** These molecules have skeletons of four fused rings of carbon atoms, but they differ by functional groups (attached side chains). Fat, as we know, is long-term stored energy in the human body. Phospholipids are found in the plasma membrane of a cell. Cholesterol, a molecule transported in the blood, has been implicated in causing cardiovascular disease. Regardless, steroids are very important compounds in the body. For example, the sex hormones, like testosterone and estrogen, are steroids.

Figure 3.4   Formation of a fat.
During a dehydration reaction, a fat molecule forms when glycerol joins with three fatty acids as three water molecules are removed. During a hydrolysis reaction, water is added, and the bonds are broken between glycerol and the three fatty acids.

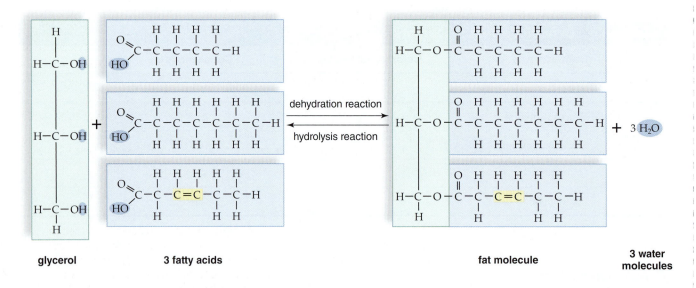

glycerol   3 fatty acids   fat molecule   3 water molecules

Test for Fat

Fats and oils do not evaporate from brown paper or loose-leaf paper; instead, they leave an oily spot.

Experimental Procedure: Paper Test for Fat

1. Place a small drop of water on a square of brown paper or loose-leaf paper. Describe the immediate effect. _____

2. Place a small drop of vegetable oil on a square of the paper. Describe the immediate effect. _____
_____

3. Wait at least 15 minutes for the paper to dry. Evaluate which substance penetrates the paper and which is subject to evaporation. Record your observations and conclusions in Table 3.7. Save the paper for comparison use with section 3.4.

Table 3.7	Paper Test for Fat	
Sample	**Observations**	**Conclusions**
Water spot		
Oil spot		

Emulsification of Oil

Some molecules are **polar,** meaning that they have charged groups or atoms, and some are **nonpolar,** meaning that they have no charged groups or atoms. A water molecule is polar, and therefore, water is a good solvent for other polar molecules. When the charged ends of water molecules interact with the charged groups of polar molecules, these polar molecules disperse in water.

Water is not a good solvent for nonpolar molecules, such as fats. A fat has no polar groups to interact with water molecules. An **emulsifier,** however, can cause a fat to disperse in water. An emulsifier contains molecules with both polar and nonpolar ends. When the nonpolar ends interact with the fat and the polar ends interact with the water molecules, the fat disperses in water, and an **emulsion** results (Fig. 3.5).

Bile salts (emulsifiers found in bile produced by the liver) are used in the digestive tract. Today milk, such as 1% milk, has been homogenized so that fat droplets do not congregate and rise to the top of the container. Homogenization requires the addition of natural emulsifiers such as phospholipids—the phosphate part of the molecule is polar and the lipid portion is nonpolar.

Figure 3.5 Emulsification.
An emulsifier contains molecules with both a polar and a nonpolar end. The nonpolar ends are attracted to the nonpolar fat, and the polar ends are attracted to the water. This causes droplets of fat molecules to disperse.

Experimental Procedure: Emulsification of Lipids

Label three clean test tubes (1 to 3), and use the appropriate graduated pipet to add solutions to the test tubes as follows:

Tube 1
1. Add 3 ml of water and 1 ml of vegetable oil. Shake.
2. Observe for the initial dispersal of oil, followed by rapid separation into two layers. Is vegetable oil soluble in water? _____
3. Let the tube settle for 5 minutes. Label a microscope slide as 1.
4. Use a dropper to remove a sample of the solution that is just below the layer of oil. Place the drop on the slide, add a coverslip, and examine with the low power of your compound light microscope.
5. Record your observations in Table 3.8.

Tube 2
1. Add 2 ml of water, 3 ml of vegetable oil, and 1 ml of the available emulsifier (Tween or bile salts). Shake.
2. Describe how the distribution of oil in tube 2 compares with the distribution in tube 1. _____

   _____

   _____
3. Let the tube settle for 5 minutes. Label a microscope slide as 2.
4. Use a different dropper to remove a sample of the solution that is just below the layer of oil. Place the drop on the slide, add a coverslip, and examine with the low power of your compound light microscope.
5. Record your observations in Table 3.8.

Tube 3
1. Add 1 ml of milk and 2 ml of water. Shake well.
2. Use a different dropper to remove a sample of the solution that is just below the layer of oil. Place the drop on the slide, add a coverslip, and examine with the low power of your compound light microscope.
3. Record your observations in Table 3.8.

Table 3.8	Emulsification		
Tube	**Contents**	**Observations**	**Conclusions**
1	Oil Water		
2	Oil Water Emulsifier		
3	Milk Water		

Conclusions: Emulsification

- From your observations, conclude why the contents of each tube appear as they do under the microscope. Record your conclusions in Table 3.8.
- Explain the correlation between your macroscopic observations (how the tubes look to your unaided eye) and your microscopic observations. _____

   _____

3.4 Testing Foods and Unknowns

It is common for us to associate the term *organic* with the foods we eat, including carbohydrate foods (Fig. 3.6), protein foods (Fig. 3.7), and lipid foods (Fig. 3.8). Though we may recognize foods as being organic, often we are not aware of what specific types of compounds are found in what we eat. In the following Experimental Procedure, you will use the same tests you used previously to determine the composition of everyday foods and unknowns.

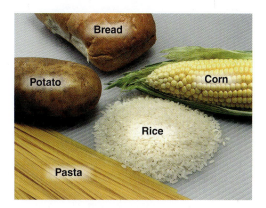

Figure 3.6 Carbohydrate foods.

© John Thoeming/McGraw-Hill Education

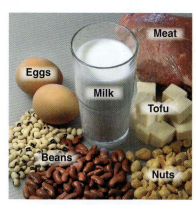

Figure 3.7 Protein foods.

© John Thoeming/McGraw-Hill Education

Figure 3.8 Lipid foods.

© John Thoeming/McGraw-Hill Education

Experimental Procedure: Testing Foods and Unknowns

Your instructor will provide you with several everyday foods including unknowns, and your task is to

1. State how you will test substances for protein (page 29), carbohydrates (pages 31 and 33), and fat (page 35).

   _____

   _____

   _____

   _____

2. Have your instructor okay your procedures, and then conduct the necessary tests.
3. Record your results as positive (+) or negative (−) in Table 3.9.

Table 3.9 Testing Foods and Unknowns

Sample Name	Protein (Biuret)	Starch (Iodine)	Sugar (Benedict's)	Fat (Brown or loose-leaf paper)
Unknown A				
Unknown B				

Conclusions: Testing Foods and Unknowns

- What foods tested positive for only one of the organic compounds? _____

- What does more than one positive test tell you about these foods? _____

_____ 1. What kind of reaction adds water to break large biomolecules into subunits?

_____ 2. What kind of protein is lactase, the biomolecule that speeds up the breakdown of lactose?

_____ 3. What kind of bond forms during a dehydration synthesis reaction involving two amino acids?

_____ 4. If maltose undergoes hydrolysis, what subunits result?

_____ 5. As what molecule do plants store glucose?

_____ 6. If starch is present, what color will the iodine turn?

_____ 7. If a solution contains a large amount of glucose, what color will the Benedict's reagent become?

_____ 8. What kinds of biomolecules are insoluble in water but soluble in alcohol or ether?

_____ 9. If two fatty acids, glycerol, and a phosphate group undergo a dehydration reaction, what biomolecule forms?

_____ 10. What is the name used for plant fats?

_____ 11. What must be present to successfully break down fats during digestion?

_____ 12. If you test an unknown substance with the biuret reagent and it turns purple, what is present?

_____ 13. If you test a sample of potato with the biuret reagent, what do you expect the results to be?

_____ 14. If you want to test a sample to see if glucose is present, what reagent should you use?

Thought Questions

15. If amylase is added to a sample of potato and you test it with Benedict's reagent 10 minutes later, what do you expect the results of the test to be? Justify your reply.

16. After enjoying a meal of fish and chips, you notice your shirt has an oily stain on it. Why won't the oily stain come out when you dab at it with just water?

17. If the sample of water you test with biuret reagent turns purple, what should you conclude about the results of your experiment? Explain.

4

Cell Structure and Function

Introduction

The molecules we studied in Laboratory 3 are not alive—the basic units of life are cells. The **cell theory** states that all organisms are composed of cells and that cells come only from other cells. While we are accustomed to considering the heart, the liver, or the intestines as enabling the human body to function, it is actually cells that do the work of these organs.

Figure 2.11 shows human cheek epithelial cells as viewed by an ordinary compound light microscope available in general biology laboratories. It shows that the content of a cell, called the **cytoplasm,** is enclosed by a **plasma membrane.** The plasma membrane regulates the movement of molecules into and out of the cytoplasm. In this lab, we will study how the passage of water into a cell depends on the difference in concentration of solutes (particles) between the cytoplasm and the surrounding medium or solution. The well-being of cells also depends upon the pH of the solution surrounding them. We will see how a buffer can maintain the pH within a narrow range and how buffers within cells can protect them against damaging pH changes.

> ⏱ **Planning Ahead** To save time, your instructor may have you start a boiling water bath (page 49) and the potato strip experiment (page 53) at the beginning of the laboratory.

Because a photomicrograph shows only a minimal amount of detail, it is necessary to turn to the electron microscope to study the contents of a cell in greater depth. The models of plant and animal cells available in the laboratory today are based on electron micrographs.

4.1 Prokaryotic Versus Eukaryotic Cells

All living cells are classified as either prokaryotic or eukaryotic. One of the basic differences between the two types is that **prokaryotic cells** do not contain nuclei (*pro* means "before"; *karyote* means "nucleus"), while eukaryotic cells do contain nuclei (*eu* means "true"; *karyote* means "nucleus"). Only bacteria (including cyanobacteria) and archaea are prokaryotes; all other organisms are eukaryotes.

Prokaryotes also don't have the organelles found in **eukaryotic cells** (Fig. 4.1). **Organelles** are small, membranous bodies, each with a specific structure and function. Prokaryotes do have **cytoplasm,** the material enclosed by a plasma membrane and cell wall. The cytoplasm contains ribosomes, small granules that coordinate the synthesis of proteins; thylakoids (only in cyanobacteria) that participate in photosynthesis; and innumerable enzymes. Prokaryotes also have a nucleoid, a region in the bacterial cell interior in which the DNA is physically organized but not enclosed by a membrane.

Figure 4.1   Prokaryotic cell.
Prokaryotic cells lack membrane-bound organelles, as well as a nucleus. Their DNA is in a nucleoid region.

(b) © Sercomi/Science Source

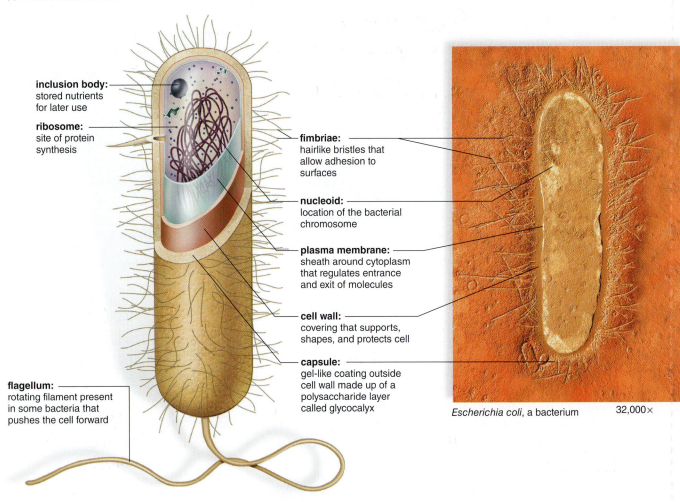

inclusion body:
stored nutrients for later use

ribosome:
site of protein synthesis

fimbriae:
hairlike bristles that allow adhesion to surfaces

nucleoid:
location of the bacterial chromosome

plasma membrane:
sheath around cytoplasm that regulates entrance and exit of molecules

cell wall:
covering that supports, shapes, and protects cell

capsule:
gel-like coating outside cell wall made up of a polysaccharide layer called glycocalyx

flagellum:
rotating filament present in some bacteria that pushes the cell forward

Escherichia coli, a bacterium   32,000×

Observation: Prokaryotic/Eukaryotic Cells

Two microscope slides on display will show you the main difference between prokaryotic and eukaryotic cells.

1. Examine a prepared slide of a bacterium. Is there a nucleus visible? _____

2. Examine a prepared slide of cuboidal cells from a human kidney (see page 141). Is there a nucleus visible? _____

5

Mitosis and Meiosis

Learning Outcomes

5.1 The Cell Cycle
- Name and describe the stages of the cell cycle with emphasis on how the cell prepares for mitosis.   57–58
- Identify the phases of mitosis in animal and plant cells on models and microscopic slides. Explain how DNA replication allows the chromosome number to stay constant during mitosis.   58–62
- Describe animal and plant cell cytokinesis.   63–64

5.2 Meiosis
- Name and describe the phases of meiosis I and meiosis II with attention to the movement of chromosomes. Explain how meiosis results in variation among the gametes and how the chromosome number is reduced.   65–69

5.3 Mitosis Versus Meiosis
- Compare the outcomes of mitosis to those of meiosis.   70
- Contrast the behavior of chromosomes during mitosis with the behavior of chromosomes during meiosis I.   71

Introduction

Dividing cells experience nuclear division, cytoplasmic division, and a period of time between divisions called interphase. During **interphase,** the nucleus appears normal, and the cell is performing its usual cellular functions. Also, the cell is increasing all of its components, including such organelles as the mitochondria, ribosomes, and centrioles, if present. DNA replication (making an exact copy of the DNA) occurs toward the end of interphase. Thereafter, the chromosomes, which contain DNA, are duplicated and contain two chromatids held together at a **centromere** (see Table 5.1). These chromatids are called **sister chromatids.**

During nuclear division, called **mitosis,** the new nuclei receive the same number of chromosomes as the parental nucleus. When the cytoplasm divides, a process called **cytokinesis,** two daughter cells are produced. In multicellular organisms, mitosis permits growth and repair of tissues. In eukaryotic, unicellular organisms, mitosis is a form of asexual reproduction. Sexually reproducing organisms utilize another form of nuclear division, called **meiosis.** In animals, meiosis produces gametes (sex cells). The gametes are sperm in male animals and eggs in female animals. As a result of meiosis, the daughter cells have half the number of chromosomes as the parental cell. Meiosis contributes to recombination of genetic material and to variation among sexually reproducing organisms.

5.1   The Cell Cycle

As stated in the Introduction, the period of time between cell divisions is known as interphase. Because early investigators noted little visible activity between cell divisions, they dismissed this period of time as a resting state. But when later investigators discovered that DNA replication and chromosome duplication occur during interphase, the **cell cycle** concept was proposed. The cell cycle is divided into the four stages noted in Figure 5.1. Match the events described in Figure 5.1 with the stage in which they occur.

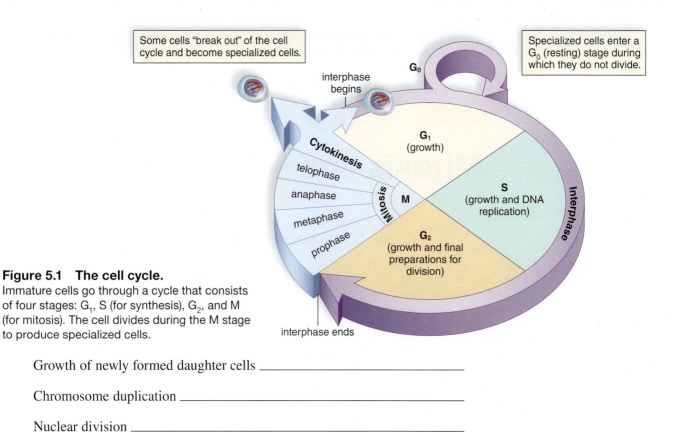

Figure 5.1 The cell cycle.
Immature cells go through a cycle that consists of four stages: G_1, S (for synthesis), G_2, and M (for mitosis). The cell divides during the M stage to produce specialized cells.

Growth of newly formed daughter cells _____

Chromosome duplication _____

Nuclear division _____

Explain why the entire process is called the "cell cycle." _____

Table 5.1	Structures Associated with Mitosis
Structure	**Description**
Nucleus	A large organelle containing the chromosomes and acting as a control center for the cells
Chromosome	Rod-shaped body in the nucleus that is seen during mitosis and meiosis and that contains DNA, and therefore the hereditary units, or genes
Nucleolus	An organelle found inside the nucleus that produces the subunits of ribosomes
Spindle	Microtubule structure that brings about chromosome movement during cell division
Chromatids	The two identical parts of a chromosome following DNA replication
Centromere	A constriction where duplicates (sister chromatids) of a chromosome are held together
Centrosome	The central microtubule-organizing center of cells; consists of granular material; in animal cells, contains two centrioles
Centrioles*	Short, cylindrical organelles in animal cells that contain microtubules and are associated with the formation of the spindle during cell division
Aster*	Short, radiating fibers produced by the centrioles

*Animal cells only

Mitosis

Mitosis is nuclear division that results in two new nuclei, each having the same number of chromosomes as the original nucleus. The **parent cell** is the cell that divides, and the resulting cells are called **daughter cells.** If a parent cell has 22 chromosomes, how many chromosomes does each daughter cell have following mitosis?

_____

When cell division is about to begin, chromatin starts to condense and compact to form visible, rodlike sister chromatids held together at the centromere (Fig. 5.2*a*). Consult Table 5.1 and *label the sister chromatids and centromere in the drawing of a duplicated chromosome in Figure 5.2b.* This illustration represents a chromosome as it would appear just before nuclear division occurs.

1. _____

2. _____

one chromatid

a. SEM   **b.** Drawing

Figure 5.2 Duplicated chromosomes.
DNA replication results in a duplicated chromosome that consists of two sister chromatids held together at a centromere.
(a) © Andrew Syred/Science Source

Spindle

Table 5.1 lists the structures that play a role during mitosis. The **spindle** is a structure that appears and brings about an orderly distribution of chromosomes to the daughter cell nuclei. A spindle has fibers that stretch between two poles (ends). Spindle fibers are bundles of microtubules, protein cylinders found in the cytoplasm that can assemble and disassemble. The **centrosome,** the main microtubule-organizing center of the cell, divides before mitosis so that during mitosis, each pole of the spindle has a centrosome. Animal cells contain two **centrioles** in each centrosome and asters, arrays of short microtubules radiating from the poles (Fig. 5.3). The fact that plant cells lack centrioles suggests that centrioles are not required for spindle formation.

Figure 5.3 The mitotic spindle.
Spindle fibers (bundles of microtubules) help distribute the daughter cells during mitosis.

centrioles in centrosome   duplicated chromosomes   spindle fibers   aster

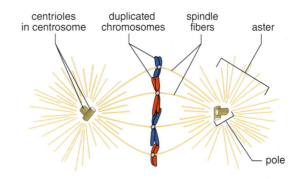

pole

Observation: Animal Mitosis

Animal Mitosis Models

1. Using Figure 5.4 as a guide, identify the phases of animal cell mitosis in models of animal cell mitosis.
2. Each species has its own chromosome number. Counting the number of centromeres tells you the number of chromosomes in the models. What is the number of chromosomes in each of the cells in your model series? _____
3. Do these models show the spindle, illustrated in Figure 5.3? _____
4. What is the shape of the animal cells? _____

Whitefish Blastula Slide

1. Examine a prepared slide of whitefish blastula cells undergoing mitosis. The blastula is an early embryonic stage in the development of animals.
2. Try to find a cell in each phase of mitosis using Figure 5.4 as a guide. Have a partner or your instructor check your identification.

Mitosis Phases

Figure 5.4   Phases of mitosis in animal and plant cells.
The colors signify that half the chromosomes were inherited from one parent and half were inherited from the other parent.

(Animal Cell) Early Prophase: © Ed Reschke; Prophase: © Ed Reschke; Prometaphase: © Michael Abbey/Science Source; Metaphase: © Ed Reschke; Anaphase: © Ed Reschke; Telophase: © Ed Reschke; (Plant Cell) Early Prophase: © Ed Reschke; Prophase: © Kent Wood/Science Source; Prometaphase: © Ed Reschke; Metaphase: © Kent Wood/Science Source; Anaphase: © Kent Wood/Science Source; Telophase: © Kent Wood/Science Source

MITOSIS

centrosome has centrioles

Animal Cell at Interphase

aster   250×

duplicated chromosome   250×

spindle pole   450×

nuclear envelope fragments

centromere

kinetochore

chromatin condenses

nucleolus disappears

spindle fibers forming

kinetochore spindle fiber

polar spindle fiber

kinetochore spindle fiber

Early Prophase
Centrosomes have duplicated. Chromatin is condensing into chromosomes, and the nuclear envelope is fragmenting.

Prophase
Nucleolus has disappeared, and duplicated chromosomes are visible. Centrosomes begin moving apart, and spindle is in process of forming.

Prometaphase
The kinetochore of each chromatid is attached to a kinetochore spindle fiber. Polar spindle fibers stretch from each spindle pole and overlap.

centrosome lacks centrioles

Plant Cell at Interphase

400×

cell wall   chromosomes   900×

500×
spindle pole lacks centrioles and aster

The phases of mitosis are shown in Figure 5.4. Mitosis is the type of nuclear division that occurs when an animal or plant grows larger and when injury heals. Two daughter cells result because there is only one round of division, and it keeps the chromosome number constant (same as the parent cell). The prophase cell in Figure 5.4 has the same number of chromosomes as the telophase nuclei in Figure 5.4. Explain the different appearance of the chromosomes.

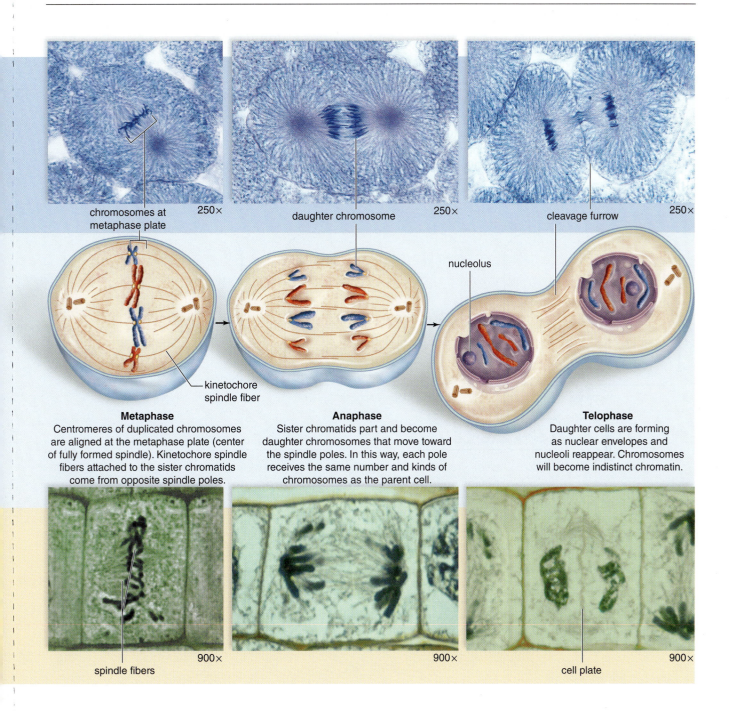

chromosomes at metaphase plate   250×

daughter chromosome   250×

cleavage furrow   250×

nucleolus

kinetochore spindle fiber

Metaphase
Centromeres of duplicated chromosomes are aligned at the metaphase plate (center of fully formed spindle). Kinetochore spindle fibers attached to the sister chromatids come from opposite spindle poles.

Anaphase
Sister chromatids part and become daughter chromosomes that move toward the spindle poles. In this way, each pole receives the same number and kinds of chromosomes as the parent cell.

Telophase
Daughter cells are forming as nuclear envelopes and nucleoli reappear. Chromosomes will become indistinct chromatin.

spindle fibers   900×

900×

cell plate   900×

Plant Mitosis Models

1. Identify the phases of plant cell mitosis using models of plant cell mitosis and Figure 5.4 as a guide.
2. Notice that plant cells do not have centrioles and asters. Plant cells do have centrosomes, and this accounts for the formation of a spindle.
3. What is the number of chromosomes in each of the cells in your model series? _____

Onion Root Tip Slide

1. In plants, the root tip contains tissue that is continually dividing and producing new cells. Examine a prepared slide of onion root tip cells (*Allium*) undergoing mitotic cell division. Try to find the phases that correspond to those shown in Figure 5.4.
2. Using high power, focus up and down on a cell in telophase. You may be able to just make out the cell plate, the region where a plasma membrane is forming between the two prospective daughter cells. Later, cell walls appear in this area.
3. In the boxes provided, draw and label the stages of mitosis as observed in the onion root tip slide.

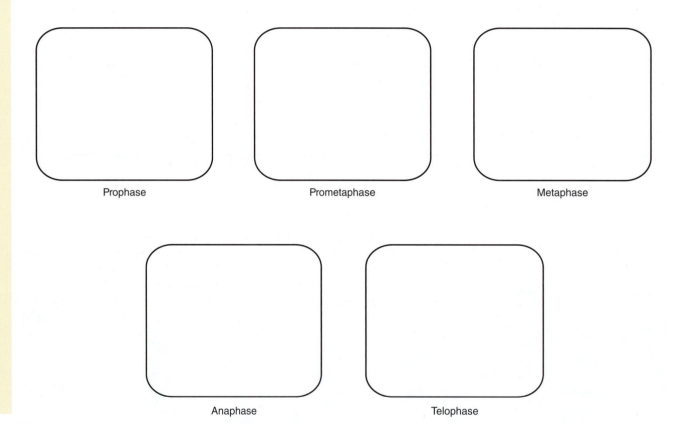

Prophase   Prometaphase   Metaphase

Anaphase   Telophase

Cytokinesis

Cytokinesis, division of the cytoplasm, usually accompanies mitosis. During cytokinesis, each daughter cell receives a share of the organelles that duplicated during interphase. Cytokinesis begins in anaphase, continues in telophase, and reaches completion by the start of the next interphase.

Cytokinesis in Animal Cells

In animal cells, a **cleavage furrow,** an indentation of the membrane between the daughter nuclei, begins as anaphase draws to a close (Fig. 5.5). The cleavage furrow deepens as a band of actin filaments called the contractile ring slowly constricts the cell, forming two daughter cells.

Were any of the cells of the whitefish blastula slide undergoing cytokinesis? _____

What did you observe that indicated cytokinesis was taking place? _____

_____

_____

4,000×
cleavage furrow

contractile ring

4,000×

Figure 5.5   Cytokinesis in animal cells.
A single cell becomes two cells by a furrowing process. A contractile ring composed of actin filaments gradually gets smaller, and the cleavage furrow pinches the cell into two cells.

(top) © National Institutes of Health (NIH)/USHHS; (bottom) © Steve Gschmeissner/SPL/Getty RF

Cytokinesis in Plant Cells

After mitosis, the cytoplasm divides by cytokinesis. In plant cells, membrane vesicles derived from the Golgi apparatus migrate to the center of the cell and form a **cell plate** (Fig. 5.6), which is the location of a new plasma membrane for each daughter cell. Later, individual cell walls appear in this area. Were any of the cells of the onion root tip slide undergoing cytokinesis as shown in Figure 5.4, Telophase, page 61? _____

_____

What did you observe that indicated cytokinesis was taking place? _____

_____

Figure 5.6   Micrograph showing cytokinesis in plant cells.
During plant cell cytokinesis, vesicles fuse to form a cell plate that separates the daughter nuclei. Later, the cell plate gives rise to a new cell wall.

© Biophoto Associates/Science Source

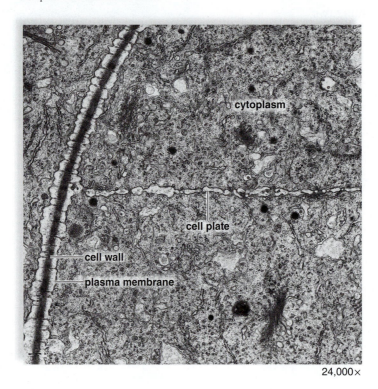

24,000×

Summary of Mitotic Cell Division

1. Do the nuclei in the daughter cells have the same or a different number of chromosomes as the parent cell's nucleus? _____

2. If a parent cell has 16 chromosomes, how many chromosomes do the daughter cells have following mitosis? _____

5.2 Meiosis

Meiosis is a form of nuclear division in which the chromosome number is reduced by half. The nucleus of the parent cell has the **diploid (2n) number** of chromosomes; the daughter cells that result from meiosis have the **haploid (n) number** of chromosomes. In sexually reproducing organisms, meiosis must occur so that the gametes (egg and sperm), which fuse at fertilization, form a zygote with the same number of chromosomes as the parent organisms.

At the start of meiosis, a parent cell is diploid (2n) and contains pairs of chromosomes called **homologues.** Observe the homologues in Figure 5.7. The homologues of each pair look alike and carry the genes for the same traits. Each homologue is a chromosome that has already been duplicated, forming sister chromatids.

Experimental Procedure: Homologues and Producing Variation During Meiosis

First, you will build four chromosomes: two pairs of homologues, as in Figure 5.7. In other words, the parent cell is 2n = 4.

Building Chromosomes

1. Obtain the following materials: 48 pop beads of one color (e.g., red) and 48 pop beads of another color (e.g., blue) for a total of 96 beads; eight magnetic centromeres.
2. Build a homologue pair of duplicated chromosomes using Figure 5.7a as a guide. Each chromatid will have 16 beads. Be sure to bring the centromeres of the same color together so that they form one duplicated chromosome.
3. Build another homologue pair of duplicated chromosomes using Figure 5.7b as a guide. Each chromatid will have eight beads.
4. Note that your chromosomes are the same as those in Figure 5.7.

Figure 5.7 Two pairs of homologues.
The red chromosomes were inherited from one parent, and the blue chromosomes were inherited from the other parent. Color does not signify homologues; size and shape signify homologues.

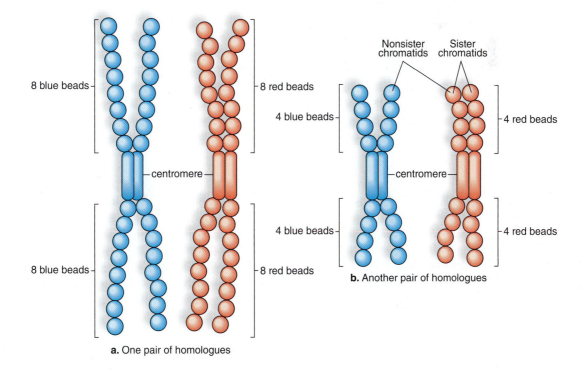

8 blue beads   8 red beads   4 blue beads   4 red beads

Nonsister chromatids   Sister chromatids

—centromere—   —centromere—

8 blue beads   8 red beads   4 blue beads   4 red beads

a. One pair of homologues   **b.** Another pair of homologues

Meiosis I

Meiosis requires two nuclear divisions. The first round is called meiosis I and the second round is called meiosis II. Both meiosis I and meiosis II have stages named the same as in mitosis, but the events are different. Follow these directions to simulate meiosis I.

Prophase I

5. Put all four of your chromosomes in the center of your work area: this area represents the nucleus. Synapsis, a very important event, occurs during meiosis I. To simulate synapsis, place the long blue chromosome next to the long red chromosome and the short blue chromosome next to the short red chromosome to show that the homologues pair up as in Figure 5.8 (Prophase I). Now **crossing-over** occurs. During crossing-over an exchange of genetic material occurs between nonsister chromatids. Perform crossing-over by switching some blue beads for red beads between the inside chromatids of the homologues. The genetic material contains genes, and therefore crossing-over recombines genes.

Genetic Variation Due to Meiosis   As a result of crossing-over the genetic material on a chromosome in a gamete can be different from that in the parent cell.

Metaphase I

6. Keep the homologues together and align them at the metaphase plate. Add centrioles to represent the poles of a spindle apparatus. The presence of centrioles means these are animal cells.

Anaphase I and Telophase I

7. Separate the homologues so that each pole of the spindle receives one member from each pair of homologues. This separation causes each pole to receive the haploid number of chromosomes.

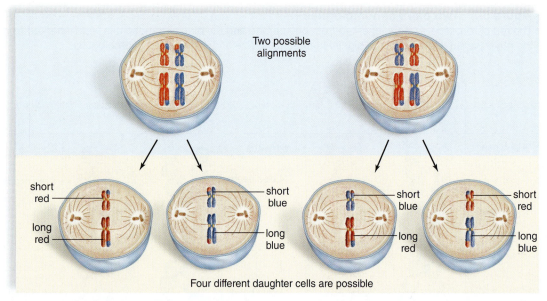

Only two daughter cells result from meiosis I, but two possible alignments of chromosomes can occur during metaphase I. Therefore, four combinations of haploid chromosomes are possible in the daughter cells following meiosis I in a 2n = 4 parent cell. The number of possible combinations increases as the chromosome number increases.

Genetic Variation Due to Meiosis   As a result of homologue separation, all possible combinations of the haploid number of chromosomes can occur among the gametes.

Meiosis II

The second round of nuclear division during meiosis is called **meiosis II** (Fig. 5.8). Follow these directions to simulate meiosis II.

Prophase II

8. Choose one daughter nucleus (see step 7, page 66) to be the parent nucleus undergoing meiosis II.

Metaphase II

9. Move the duplicated chromosomes to the metaphase II metaphase plate, as shown in the art to the right.

Anaphase II

10. Pull the two magnets of each duplicated chromosome apart. What does this action represent? _____

_____

Telophase II

11. Put the chromosomes—each having one chromatid—at the poles near the centrioles. At the end of telophase, the daughter nuclei re-form.

- You chose only one daughter nucleus from meiosis I to be the nucleus that divides. In reality both daughter nuclei go on to divide again. Therefore, how many nuclei are usually present when meiosis II is complete? _____

- In this exercise, how many chromosomes were in the parent cell nucleus undergoing meiosis II? _____

- How many chromosomes are in the daughter nuclei? _____ Explain. _____

_____

Summary of Meiosis

1. Meiosis reduces the chromosome number. If the parent cell is 2n = 4, the daughter cells are

   n = _____. Without meiosis the chromosome number would double with each generation.

   Instead, when a haploid sperm fertilizes a haploid egg, the offspring is 2n.

2. Sexual reproduction results in offspring that can look very different as represented in this illustration.
   Why do the puppies born to these parents show variation?

   © American Images, Inc./Getty Images

   a. During prophase I, the homologues come together and exchange genetic material. Now the inherited chromosomes will be different from those in the parent cell. This process is

   called _____.

   b. During anaphase I, the homologues separate differently and therefore the daughter cells following telophase I can have different _____ of chromosomes.

   c. During fertilization variant sperm fertilize variant eggs, further helping to ensure that the new individual

   inherits different _____ of homologues than a parent had.

Meiosis Phases

Figure 5.8   Meiosis I and II in plant cell micrographs and animal cell drawings.

Crossing-over occurs during meiosis I.

Interphase
chromosome
duplication

Plant Cell

centrosome has
centrioles

**Animal Cell
at Interphase**

MEIOSIS I

2n = 4

kinetochore

Prophase I
Chromosomes have duplicated.
Homologues pair during synapsis
and crossing-over occurs.

Metaphase I
Homologues align
independently
at the metaphase plate.

Anaphase I
Homologues separate
and move toward the poles.

MEIOSIS II

n = 2

n = 2

Prophase II
Cells have one chromosome
from each pair of homologues.

Metaphase II
Chromosomes align
at the metaphase plate.

Anaphase II
Sister chromatids separate and
become daughter chromosomes.

The phases of meiosis are shown in Figure 5.8. If there are prepared slides available of a lily anther where cells are undergoing meiosis I and II, try to find a cell in each phase of meiosis I and II by comparing cells on the slide using Figure 5.8 as a guide.

MEIOSIS I cont'd

Telophase I
Daughter cells have one chromosome
from each pair of homologues.

Interkinesis
Chromosomes still
consist of two chromatids.

n = 2

n = 2

MEIOSIS II cont'd

n = 2

Telophase II
Spindle disappears, nuclei form,
and cytokinesis takes place.

Daughter cells
Meiosis results in four
haploid daughter cells.

n = 2

5.3 Mitosis Versus Meiosis

In comparing mitosis to meiosis it is important to note that meiosis requires two nuclear divisions but mitosis requires only one nuclear division. Therefore mitosis produces two daughter cells and meiosis produces four daughter cells. Following mitosis, the daughter cells are still diploid, but following meiosis, the daughter cells are haploid. Figure 5.9 explains why. Fill in Table 5.2 to indicate general differences between mitosis

Table 5.2 Differences Between Mitosis and Meiosis		
	Mitosis	**Meiosis**
1. Number of divisions		
2. Chromosome number in daughter cells		
3. Number of daughter cells		

Figure 5.9   Meiosis I compared to mitosis.
Compare metaphase I of meiosis I to metaphase of mitosis. Only in metaphase I are the homologues paired at the metaphase plate. Homologues separate during anaphase I, and therefore the daughter cells are haploid. The blue chromosomes were inherited from one parent, and the red chromosomes were inherited from the other parent. The exchange of color between nonsister chromatids represents the crossing-over that occurred during meiosis I.

Prophase I
Synapsis and crossing-over occur.

2n = 4

Metaphase I
Homologues align ndependently at the metaphase plate.

Anaphase I
Homologues separate and move toward the poles.

MEIOSIS I

Prophase

2n = 4

Metaphase
Chromosomes align at the metaphase plate.

Anaphase
Sister chromatids separate and become daughter chromosomes.

MITOSIS

and meiosis and complete Table 5.3 to indicate specific differences between mitosis and meiosis I. Mitosis need be compared only with meiosis I because the same events occur during both mitosis and meiosis II except that the cells are diploid during mitosis and haploid during meiosis II.

Table 5.3　Mitosis Compared with Meiosis I

Mitosis	Meiosis I
Prophase: No pairing of chromosomes	Prophase I: _____
Metaphase: Duplicated chromosomes at metaphase plate	Metaphase I: _____
Anaphase: Sister chromatids separate	Anaphase I: _____
Telophase: Chromosomes have one chromatid	Telophase I: _____

Telophase I
Daughter cells are forming and will go on to divide again.

n = 2

Sister chromatids separate and become daughter chromosomes.

Daughter cells

n = 2

n = 2

Four haploid daughter cells. Their nuclei are genetically different from the parent cell.

MEIOSIS I cont'd　　**MEIOSIS II (not all phases are shown)**

2n = 4

Daughter cells

Telophase
Daughter cells are forming.

Two diploid daughter cells. Their nuclei are genetically identical to the parent cell.

MITOSIS cont'd

Laboratory Review 5

_____ 1. During what stage of the cell cycle does DNA replication occur?

_____ 2. What happens to sister chromatids during anaphase of mitosis?

_____ 3. When are organelles like mitochondria and endoplasmic reticulum divided up between daughter cells?

_____ 4. If asters are observed in cells undergoing mitosis, are the cells animal or plant cells?

_____ 5. What is the structure called that helps chromosomes move and forms during prophase?

_____ 6. During what phase of mitosis does the nuclear envelope re-form and the nucleolus reappear?

_____ 7. What structure gives rise to new plant cell walls?

_____ 8. What kind of nuclear division must occur prior to sexual reproduction to ensure that offspring are 2n?

_____ 9. What do we call the chromosomes that look alike and carry genes for the same traits?

_____ 10. What is the exchange of genetic material between nonsister chromatids called?

_____ 11. During what phase does crossing-over occur?

_____ 12. During what phase do homologues separate?

_____ 13. If a parent cell has 48 chromosomes, how many does each daughter cell have at the end of meiosis II?

_____ 14. What kind of nuclear division is described by 2n $\longrightarrow$ n?

Thought Questions

15. Skeletal muscle cells are multinucleated; each cell has multiple nuclei. Explain how a cell could end up with multiple nuclei.

16. During the life cycle of plants, haploid spores are produced. The spores germinate (divide) to form multicellular haploid gametophytes. What kind of nuclear division (mitosis or meiosis) do the spores undergo to form the gametophytes? Justify your answer.

17. Meiosis I is being simulated with a pair of homologues that are red-long and yellow-long. Why will one resulting daughter cell contain only red-long and the other daughter cell contain only yellow-long? What occurred if each daughter cell contains a red and yellow-long?

6

How Enzymes Function

Learning Outcomes

Introduction
- Describe how an enzyme functions.   73–74

6.1 Catalase Activity
- Identify the substrate, enzyme, and product in the reaction studied today.   74–75

6.2 Effect of Temperature on Enzyme Activity
- Predict the effect of temperature on an enzymatic reaction.   75–76

6.3 Effect of Concentration on Enzyme Activity
- Predict the effect of enzyme and substrate concentration on an enzymatic reaction.   77

6.4 Effect of pH on Enzyme Activity
- Predict the effect of pH on an enzymatic reaction.   78–79

6.5 Factors That Affect Enzyme Activity
- Tell how temperature, concentration of substrate or enzyme, and pH can promote or inhibit enzyme activity.   79

Introduction

The cell carries out many chemical reactions. A possible chemical reaction can be indicated like this:

$$A + B \longrightarrow C + D$$
$$\text{reactants} \qquad \text{products}$$

In all chemical reactions, the **reactants** are molecules that undergo a change, which results in the **products.** The arrow stands for the change that produced the product(s). For example, A + B change and produce C + D. The numbers of reactants and products can vary; in the one you are studying today, a single reactant breaks down to two products. All the reactions that occur in a cell have an enzyme. **Enzymes** are organic catalysts that speed metabolic reactions. Because enzymes are specific and speed only one type of reaction, they are given names. In today's laboratory, you will be studying the action of the enzyme **catalase.** The reactants in an enzymatic chemical reaction are called **substrate(s)** (Fig. 6.1).

Figure 6.1   Enzymatic action.
The reaction occurs on the surface of the enzyme at the active site. The enzyme is reusable. **a.** Degradation: substrate is broken down. **b.** Synthesis: substrates are combined.

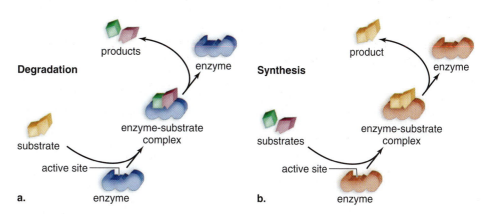

Enzymes are specific because they have a shape that accommodates the shape of their substrates. Enzymatic reactions can be indicated like this:

$$E + S \longrightarrow ES \longrightarrow E + P$$

In this reaction, E = enzyme, ES = enzyme-substrate complex, and P = product.

Two types of enzymatic reactions in cells are shown in Figure 6.1. During a **degradation reaction,** the substrate is broken down to the product(s), and during a **synthesis reaction,** substrates are joined to form a product. A number of other types of reactions also occur in cells. The location where the enzyme and substrate form an enzyme-substrate complex is called the **active site** because the reaction occurs here. At the end of the reaction, the product is released, and the enzyme can then combine with its substrate again. A cell needs only a small amount of an enzyme because enzymes are used over and over. Some enzymes have turnover rates well in excess of a million product molecules per minute.

> **Planning Ahead** To save time, your instructor may have you start a boiling water bath (page 74) at the beginning of the laboratory.

6.1 Catalase Activity

Catalase is involved in a degradation reaction: Catalase speeds the breakdown of hydrogen peroxide (H_2O_2) in nearly all organisms including bacteria, plants, and animals. A cellular organelle called a peroxisome, which contains catalase, is present in every plant and animal organ. This means that we could use any plant or animal organ as our source of catalase today. Commonly, school laboratories use the potato as a source of catalase because potatoes are easily obtained and cut up.

Catalase performs a useful function in organisms because hydrogen peroxide is harmful to cells. Hydrogen peroxide is a powerful oxidizer that can attack and denature cellular molecules like DNA! Knowing its harmful nature, humans use hydrogen peroxide as a commercial antiseptic to kill germs (Fig. 6.2). In reduced concentration, hydrogen peroxide is a whitening agent used to bleach hair and teeth. Skillful technicians use it to provide oxygen to aquatic plants and fish, but it is also used industrially to clean most anything from tubs to sewage. It's even put in glow sticks, where it reacts with a dye that then emits light.

When catalase speeds the breakdown of hydrogen peroxide, water and oxygen are released.

Figure 6.2 Hydrogen peroxide.
Bubbling occurs when you apply hydrogen peroxide to a cut because oxygen is being released when catalase, an enzyme present in the body's cells, degrades hydrogen peroxide.

© Jill Braaten, photographer & Anthony Arena, Chemical Consultant/ McGraw-Hill Education

$$2\,H_2O_2 \xrightarrow{\text{catalase}} 2\,H_2O + O_2$$

hydrogen peroxide   water   oxygen

What is the reactant in this reaction? _____ What is the substrate for catalase? _____

What are the products in this reaction? _____ and _____ Bubbling occurs as the reaction proceeds.

Why? _____

In the experimental procedure that follows, you will use bubble height to indicate the amount of product per unit time and therefore enzyme activity. Examine Table 6.1 and hypothesize which tube (1, 2, or 3) will have the greatest bubble column height. Explain your answer. _____

_____

_____

Experimental Procedure: Catalase Activity

Number three clean test tubes and use the appropriate graduated transfer pipet to add solutions to the test tubes as follows:

Tube 1
1. Add 1 ml of catalase buffered at pH 7.0, the optimum pH for catalase.
2. Add 4 ml of hydrogen peroxide. Swirl well to mix, and wait at least 20 seconds for bubbling to develop.
3. Measure the height of the bubble column (in millimeters), and record your results in Table 6.1.

Tube 2
1. Add 1 ml of water.
2. Add 4 ml of hydrogen peroxide. Swirl well to mix, and wait at least 20 seconds for bubbling to develop.
3. Measure the height of the bubble column (in millimeters), and record your results in Table 6.1.

Tube 3
1. Add 1 ml of catalase buffered at pH 7.0, the optimum pH for catalase.
2. Add 4 ml of sucrose solution. Swirl well to mix, and wait at least 20 seconds for bubbling to develop.
3. Measure the height of the bubble column (in millimeters), and record your results in Table 6.1.

Table 6.1	Catalase Activity		
Tube	**Contents**	**Bubble Column Height (mm)**	**Explanation**
1	Catalase Hydrogen peroxide		
2	Water Hydrogen peroxide		
3	Catalase Sucrose solution		

Conclusions: Catalase Activity

- Which tube showed the amount of bubbling you expected? _____ Record your explanation in Table 6.1.

- Which tube is a negative control? _____ If this tube showed bubbling, what could you conclude about your procedure? _____
Record your explanation in Table 6.1.

- Enzymes are specific; they speed only a reaction that contains their substrate. Which tube exemplifies this characteristic of an enzyme? _____ Record your explanation in Table 6.1.

6.2 Effect of Temperature on Enzyme Activity

The active sites of enzymes increase the likelihood that substrate molecules will find each other and interact. Therefore, enzymes lower the energy of activation (the temperature needed for a reaction to occur). Still, increasing the temperature is expected to increase the likelihood that active sites will be occupied because molecules move about more rapidly as the temperature rises. In this way, a warm temperature increases enzyme activity.

The shape of an enzyme and its active site must be maintained or else they will no longer be functional. A very high temperature, such as the one that causes water to boil, is likely to cause weak bonds of a protein to break. If this occurs, the enzyme **denatures**—it loses its original shape and the active site will no longer function to bring reactants together. Now enzyme activity plummets.

With this information in mind, examine Table 6.2 and hypothesize which tube (1, 2, or 3) will have the most product per unit time as judged by bubble column height. _____ Explain your answer. _____

Experimental Procedure: Effect of Temperature

Number three clean test tubes and use the appropriate graduated pipet to add solutions to the test tubes as follows:

1. To each tube add 1 ml of catalase buffered at pH 7.0, the optimum pH for catalase.
2. Place tube 1 in a refrigerator or cold water bath, tube 2 in an incubator or warm water bath, and tube 3 in a boiling water bath. Complete the second column in Table 6.2. Wait 15 minutes.
3. As soon as you remove the tubes one at a time from the refrigerator, incubator, and boiling water, add 4 ml of hydrogen peroxide.
4. Swirl well to mix, and wait 20 seconds.
5. Measure the height of the bubble column (in millimeters) in each tube, and record your results in Table 6.2. Plot your results in Figure 6.3.

Table 6.2   Effect of Temperature

Tube	Temperature °C	Bubble Column Height (mm)	Explanation
1 Refrigerator			
2 Incubator			
3 Boiling water			

Figure 6.3   Effect of temperature on enzyme activity.

Conclusions: Effect of Temperature

- The bubble column height indicates the degree of enzyme activity. Was your hypothesis supported? _____ Explain in Table 6.2 the degree of enzyme activity per tube.
- What is your conclusion concerning the effect of temperature on enzyme activity? _____

6.3 Effect of Concentration on Enzyme Activity

Consider that if you increase the number of caretakers per number of children, it is more likely that each child will have quality time with a caretaker. So it is if you increase the amount of enzyme per amount of substrate: it is more likely that substrates will find the active site of an enzyme and a reaction will take place. With this in mind, examine Table 6.3 and hypothesize which tube (1, 2, or 3) will have the most product per unit time as judged by bubble column height. _____ Explain your answer. _____

_____

Experimental Procedure: Effect of Enzyme Concentration

Number three clean test tubes and use the appropriate transfer pipet to add solutions to the test tubes as follows:

1. To each tube, add 4 ml of hydrogen peroxide.
2. To tube 1, add 1 ml of water.
3. To tube 2, add 1 ml of buffered catalase.
4. To tube 3, add 3 ml of buffered catalase.
5. Swirl well to mix before measuring the height of the bubble column.
6. Measure the height of the bubble columns, and record your results in Table 6.3.

Table 6.3 Effect of Enzyme Concentration

Tube	Amount of Enzyme	Bubble Column Height (mm)	Explanation
1	none		
2	1 ml		
3	3 ml		

Conclusions: Effect of Concentration

- The bubble column height indicates the degree of enzyme activity. Was your hypothesis supported? _____ Explain in Table 6.3 the degree of enzyme activity per tube.
- If unlimited time were allotted, would the results be the same in all tubes? _____ Explain why or why not. _____
- Would you expect similar results if the substrate concentration were varied in the same manner as the enzyme concentration? _____ Why or why not? _____
- What is your conclusion concerning the effect of concentration on enzyme activity? _____

_____

6.4 Effect of pH on Enzyme Activity

Each enzyme has a pH at which the speed of the reaction is optimum (occurs best). Any higher or lower pH affects hydrogen bonding and the structure of the enzyme, leading to reduced activity.

> ⚠️ **Hydrochloric acid (HCl)** used to produce an acid pH is a strong, caustic acid, and sodium hydroxide (NaOH) used to produce a basic pH is a strong, caustic base. Exercise care in using these chemicals, and follow your instructor's directions for disposal of tubes that contain these chemicals. If any acidic or basic solutions spill on your skin, rinse immediately with clear water.

Catalase is an enzyme found in cells where the pH is near 7 (called neutral pH). Other enzymes prefer different pHs. The pancreas secretes a slightly basic (pH 8–8.3) juice into the digestive tract, and the stomach wall releases a very acidic digestive juice which can be as low as pH 2. With this information about catalase in mind, examine Table 6.4 and hypothesize which tube (1, 2, or 3) will have the most product per unit time as judged by the bubble column height. _____ Explain your answer. _____

_____

Experimental Procedure: Effect of pH

Number three clean test tubes and use the appropriate transfer pipet to add solutions to the test tubes as follows:

1. To each tube, add 1 ml of catalase.
2. To tube 1, add 2 ml of water adjusted to pH 3.
3. To tube 2, add 2 ml of water adjusted to pH 7.
4. To tube 3, add 2 ml of water adjusted to pH 11.
5. Wait 1 minute. Now add 4 ml hydrogen peroxide to each tube and swirl to mix before noting the height of the bubble column.
6. Record your results in Table 6.4 and plot your results in Figure 6.4.

Table 6.4   Effect of pH

Tube	pH	Bubble Column Height (mm)	Explanation
1	3		
2	7		
3	11		

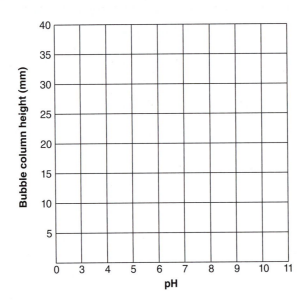

Figure 6.4   Effect of pH on enzyme activity.

Conclusions: Effect of pH

- The amount of bubble column height indicates the degree of enzyme activity. Was your hypothesis

  supported? _____ Explain in Table 6.4 the degree of enzyme activity per tube.

- What is your conclusion concerning the effect of pH on enzyme activity? _____

  _____

6.5   Factors That Affect Enzyme Activity

In Table 6.5, summarize what you have learned about factors that affect the speed of an enzymatic reaction. For example, in general, what type of temperature promotes enzyme activity, and what type inhibits enzyme activity? Answer similarly for enzyme or substrate concentration and pH.

Table 6.5 Factors That Affect Enzyme Activity		
Factors	**Promote Enzyme Activity**	**Inhibit Enzyme Activity**
Enzyme specificity		
Temperature		
Enzyme or substrate concentration		
pH		

Conclusions: Factors That Affect Enzyme Activity

- Why does enzyme specificity promote enzyme activity? _____

  _____

- Why does a warm temperature promote enzyme activity? _____

  _____

- Why does increasing enzyme concentration promote enzyme activity? _____

  _____

- Why does optimum pH promote enzyme activity? _____

  _____

_____ 1. How is "produces" represented in a chemical reaction?

_____ 2. Are the substrates in an enzymatic reaction reactants or products?

_____ 3. What is the name of the location on an enzyme where its substrate or substrates bind?

_____ 4. What cellular organelle contains catalase and is present in all plant and animal organs?

_____ 5. What did measuring the bubble column heights in all the experimental procedures indicate?

_____ 6. What purpose did the tube containing water and hydrogen peroxide serve in the catalase activity experimental procedure?

_____ 7. What word describes the loss of an enzyme's original shape and ability of its active site to function?

_____ 8. What happens to enzyme activity if the temperature increases slightly?

_____ 9. What did the negative control tube lack when the effect of enzyme concentration was studied?

_____ 10. If cells produce more catalase, will there be more or less water and oxygen produced from the breakdown of hydrogen peroxide by catalase?

_____ 11. What word describes the pH at which the speed of a reaction involving an enzyme occurs best?

_____ 12. What do changes in pH (too low or too high) do to an enzyme that reduce its activity?

_____ 13. In the Experimental Procedure on the effect of pH on enzyme activity, which tube contained water with an acidic pH?

_____ 14. Since enzymes are specific for certain substrates and often have similar names, what is the name of the enzyme used in tube 3 of the catalase activity experiment that facilitates the breakdown of sucrose?

Thought Questions

15. People with Crohn's disease may develop lactose intolerance caused by damage to their small intestine. These people may take a commercial product called Lactaid to prevent digestive upset when consuming dairy products. What is in Lactaid that helps these people avoid those symptoms? Explain how this relates to what you learned about the specificity of enzymes for substrates in this lab.

16. High-grade fevers (in humans, 103°F or higher) are of concern, and medical care should be obtained. Explain why, based on your understanding of enzyme function.

17. Pancreatic amylase is a digestive enzyme that digests starch in the small intestine. Pancreatic amylase's optimum pH is slightly basic, which the presence of $NaHCO_3$ provides. Indicate which test tubes (1–3) would show digestion of starch following incubation. Explain why the others would not.

Tube 1   water, starch _____

Tube 2   water, starch, pancreatic amylase, HCl _____

Tube 3   water, starch, pancreatic amylase, $NaHCO_3$ _____

7

Cellular Respiration

Introduction

In this laboratory, you will study **cellular respiration** in germinating soybeans. Cellular respiration is an ATP-generating process that involves the complete breakdown most often of glucose to carbon dioxide and water. In eukaryotes, glucose breakdown begins in the cytoplasm, but is completed in mitochondria. Cellular respiration is aerobic and requires oxygen, and it results in a buildup of ATP (Fig. 7.1). This equation represents cellular respiration:

> **Planning Ahead** You may wish to start the fermentation experiment on page 88 first, to allow time for incubation.

$$\underset{\text{glucose}}{C_6H_{12}O_6} + \underset{\text{oxygen}}{6\,O_2} \longrightarrow \underset{\substack{\text{carbon}\\\text{dioxide}}}{6\,CO_2} + \underset{\text{water}}{6\,H_2O} + ATP$$

Figure 7.1   Cellular respiration.
Both plant cells and animal cells have mitochondria and carry on cellular respiration, a process that utilizes the equation above.

O₂ and glucose enter cells, which release H₂O and CO₂.

Mitochondria use energy from glucose to form ATP from ADP + P.

ADP + P ⟶ ATP

In this laboratory, you will also study **ethanol fermentation,** which is an ATP-generating process. When an organism, such as yeast, breaks down glucose to ethanol and carbon dioxide, only 2 ATP result but the process is anaerobic and does not require oxygen. Fermentation occurs in the cytoplasm and mitochondria are not involved. This reaction represents yeast fermentation:

$$C_6H_{12}O_6 \longrightarrow 2\,CO_2 + 2\,C_2H_5OH + 2\,ATP$$

glucose   carbon   ethanol
                 dioxide

Fermentation by animal cells produces lactate instead of ethanol and carbon dioxide.

7.1 Cellular Respiration

When germination occurs and plants begin to grow, cellular respiration can provide them with the ATP they need to produce all the molecules that allow them to grow. The soybean seed germinating in Figure 7.2 must produce the molecules needed for mitosis and photosynthesis so that a mature soybean plant can develop.

In the Experimental Procedure that follows, we are going to measure the amount of oxygen uptake as evidence that germinating soybeans are carrying on cellular respiration. The need for oxygen by a germinating soybean will be compared to the need by nongerminating soybeans. State a hypothesis for this experiment here:

Hypothesis: _____

Figure 7.2 Germination of a soybean seed.

Experimental Procedure: Cellular Respiration

1. Obtain a volumeter, an apparatus that measures changes in gas volumes. Remove the three vials from the volumeter. Remove the stoppers from the vials. Label the vials 1, 2, and 3.
2. Using the same amounts, place a small wad of absorbent cotton in the bottom of each vial. Without getting the sides of the vials wet, use a dropper to saturate the cotton with 15% potassium hydroxide (KOH). The KOH absorbs CO_2 as it is given off by the soybeans. Place a small wad of dry cotton on top of the KOH-soaked absorbent cotton (Fig. 7.3).

> ⚠ **Potassium hydroxide (KOH)** is a strong, caustic base. Exercise care in using this chemical. If any KOH spills on your skin, rinse immediately with clear water. Follow your instructor's directions for disposal of tubes that contain KOH.

Figure 7.3 Vials.
In this experiment, three vials are filled as noted.

soybeans

dry cotton

KOH-soaked cotton

Vial 1:
no glass beads,
germinating
soybeans

Vial 2:
glass beads,
nongerminating
soybeans

glass beads

Vial 3:
glass beads,
no soybeans

3. Count 25 germinating soybean seeds and add to vial 1. Count 25 dry (nongerminating) soybean seeds and add to vial 2. Add glass beads to vial 2 so that the volume occupied in vial 2 is approximately the same as in vial 1. Add only glass beads to vial 3 to bring to approximately the same volume.

4. Replace the vials in the volumeter and firmly place the stoppers in the vials (Fig. 7.4). Each stopper has a vent (outlet tube) ending in rubber tubing held shut with a clamp. Remove the clamps. Adjust the graduated side arm until only about 5 mm to 1 cm protrudes through the stopper.

5. Use a Pasteur pipet to inject Brodie manometer fluid (or water colored with vegetable dye and a small amount of detergent) into each graduated side arm so that approximately 1 cm of dye is drawn into each side arm.

Figure 7.4 Volumeter containing three respirometers.
In this experiment, the respirometers are vials filled as per Figure 7.3 with graduated side arms attached. Oxygen uptake is measured by movement of a marker drop in each side arm.

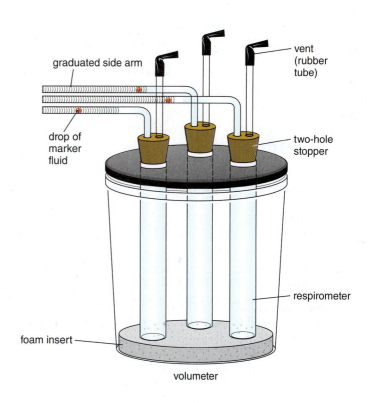

graduated side arm

vent
(rubber tube)

drop of
marker
fluid

two-hole
stopper

respirometer

foam insert

volumeter

6. After waiting one minute for equilibrium, reattach the pinch clamp to the outlet rubber tube and mark the position of the dye with a wax pencil. The dye should be near the outlet and as O_2 is taken up, the marker drop will move toward the vials.

7. Record in Table 7.1 for each vial:
   - the initial position of the marker drop to the nearest 0.01 ml
   - the position of the marker drop after 10 minutes
   - the position of the marker drop after 10 more minutes
   - the net change of position (the distance the marker drop moved from the initial reading)

8. Did the marker drop change in vial 3 (glass beads)? _____ By how much? _____ Enter this number in the "Correction" column of Table 7.1, and use this number to correct the net change you observed in vials 1 and 2. (This is a correction for any change in volume due to atmospheric pressure changes or temperature changes.) This will complete Table 7.1.

Table 7.1	Cellular Respiration						
Vial	Contents	Initial Reading	Reading After 10 Minutes	Reading After 20 Minutes	Net Change	Correction	(Corrected) Net Change
1	Germinating soybeans						
2	Dry (nongerminating) soybeans Glass beads						
3	Glass beads						

Conclusions: Cellular Respiration

- Do your results support or fail to support the hypothesis? _____
  Explain. _____
- Why was it necessary to absorb the carbon dioxide? _____
  _____
- Explain which respirometer in the soybean experiment was a negative control. _____

7.2 Fermentation

When you study the ability of yeast to ferment sugar, you will need to use a respirometer to measure the amount of CO_2 given off. This experimental procedure shows you how to prepare your **respirometer.**

Experimental Procedure: Respirometer Practice

1. Completely fill a small tube (15 × 125 mm) with *water only* (Fig. 7.5).

2. Invert a large tube (20 × 150 mm) over the small tube, and with your finger or a pencil, push the small tube up into the large tube until the upper lip of the small tube is in contact with the bottom of the large tube.

3. Quickly invert both tubes. Do not permit the small tube to slip away from the bottom of the large tube. A little water will leak out of the small tube and be replaced by an air bubble.

4. Practice this inversion until the bubble in the small tube is as small as you can make it.

8

Photosynthesis

Introduction

Photosynthesis involves the use of solar energy to produce a carbohydrate:

$$CO_2 + H_2O \xrightarrow{\text{solar energy}} (CH_2O) + O_2$$

In this equation, (CH_2O) represents any general carbohydrate. Sometimes, this equation is multiplied by 6 so that glucose $(C_6H_{12}O_6)$ appears as an end product of photosynthesis.

Photosynthesis takes place in chloroplasts (Fig. 8.1). Here membranous thylakoids are stacked in grana surrounded by the stroma. During the light reactions, pigments within the thylakoid membranes absorb solar energy, water is split, and oxygen is released. The Calvin cycle reactions occur within the stroma. During these reactions, carbon dioxide (CO_2) is reduced and solar energy is now stored in a carbohydrate (CH_2O).

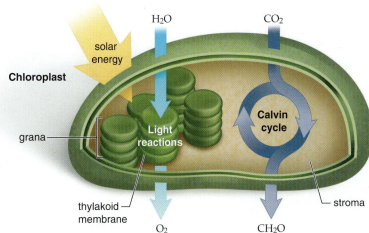

Figure 8.1   Overview of photosynthesis.
Photosynthesis includes the light reactions when energy is collected and O_2 is released and the Calvin cycle reactions when CO_2 is reduced and an energy-rich carbohydrate (CH_2O) is formed.

8.1 Photosynthetic Pigments

Later in this laboratory, we will learn that visible (white) light is composed of different colors of light (see Fig. 8.3). We can hypothesize that leaves contain various pigments that absorb the solar energy of these different colors. Restate this hypothesis here:

Hypothesis: _____

   To test this hypothesis, we will use a technique called chromatography to separate the pigments located in leaves. Chromatography separates molecules from each other on the basis of their solubility in particular solvents. The solvents used in the following Experimental Procedure are petroleum ether and acetone, which have no charged groups and are therefore nonpolar. As a nonpolar solvent moves up the chromatography paper, the pigment moves along with it. The more nonpolar a pigment, the more soluble it is in a nonpolar solvent and the faster and farther it proceeds up the chromatography paper.

Experimental Procedure: Photosynthetic Pigments

1. Assemble a **chromatography apparatus** (Fig. 8.2a):
   - Obtain a large, dry test tube and a cork with a hook.
   - Attach a strip of precut **chromatography paper** (hold from the top) to the hook and test for fit. The paper should hang straight down and barely touch the bottom of the test tube; trim if necessary.
   - Measure 2 cm from the bottom of the paper; place here a small dot with a pencil (not a pen).
   - With the stopper in place, mark the test tube with a wax pencil 1 cm below where the dot is on the paper.
   - Set the chromatography apparatus in a test tube rack.

Figure 8.2 Paper chromatography.
a. For a chromatography apparatus, the paper must be cut to size and arranged to hang down without touching the sides of a dry tube. **b.** The pigment solution is applied to a designated spot. **c.** The chromatogram, which develops after the spotted paper is suspended in the chromatography solution, will show these pigments.

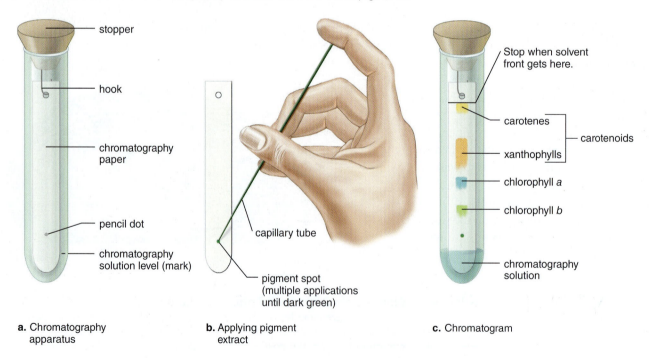

a. Chromatography apparatus

b. Applying pigment extract

c. Chromatogram

Figure 8.4 Volumeter.
A volumeter apparatus is used to study the role of light in photosynthesis.

level after photosynthesis

initial solution level

5. Carefully wrap the tube containing *Elodea* in aluminum foil, and record here the length of time it takes for the edge of the solution in the tubing to move downward 1 mm: _____. Convert your measurement to _____ mm/10 min, and record this value for **cellular respiration** in Table 8.2. (Do not use a minus sign, even though the edge moved downward.) Why does cellular respiration, which occurs in a plant whether it is light or dark, cause the edge to move downward? _____

_____

6. If the *Elodea* had *not* been respiring in step 4, how far would the edge have moved upward?

_____ mm/10 min. This is **gross photosynthesis** (net photosynthesis + cellular respiration). Record this number in Table 8.2.

7. Calculate the **rate of photosynthesis** (mm/hr) by multiplying gross photosynthesis (mm/10 min) by 6 (that is, 10 min × 6 = 60 min = 1 hr): _____ mm/hr. Record this value in Table 8.2.

Table 8.2 Rate of Photosynthesis (White Light)		
	Movement of Edge (mm/10 min)	**Rate of Photosynthesis (mm/hr)**
Net photosynthesis (white light)		_____
Cellular respiration (no light)		_____
Gross photosynthesis (net + cellular respiration)		

Role of Green Light

Green light is only one part of white light (see Fig. 8.3). The photosynthetic pigments absorb certain colors of light better than other colors (Fig. 8.5). According to Figure 8.5, what color light do the chlorophylls absorb best? _____ Least? _____

What color light do the carotenoids (carotenes and xanthophylls) absorb best? _____

Least? _____

Hypothesize which color light is minimally utilized for photosynthesis. _____

The following Experimental Procedure will test your hypothesis.

Figure 8.5 Action spectrum for photosynthesis.
The action spectrum for photosynthesis is the sum of the absorption spectrums for the pigments chlorophyll *a,* chlorophyll *b,* and carotenoids. The peaks in this diagram represent wavelengths of sunlight absorbed by photosynthetic pigments. The chlorophylls absorb predominantly violet-blue and orange-red light and reflect green light. The carotenoids (carotenes and xanthophylls) absorb mostly blue-green light and reflect yellow-red light.

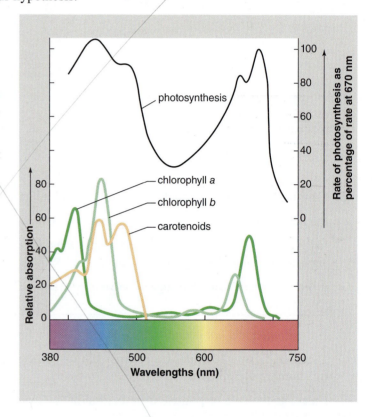

Experimental Procedure: Green Light

1. Add three drops of green dye (or use a green cellophane wrapper) to the beaker of water used in the previous Experimental Procedure until there is a distinctive green color. Remove all previous wax pencil marks from the glass tubing.

2. Record in Table 8.3 your data for gross photosynthesis (mm/10 min) and for rate of photosynthesis for white light (mm/hr) from Table 8.2.

3. Turn on the lamp. Mark the location of the edge of the solution on the glass tubing. As soon as the edge begins to move, time the reaction for 10 minutes. After 10 minutes, mark the edge of the solution, and measure in millimeters the distance the edge moved. Net photosynthesis for green light = _____ mm/10 min. As before, net photosynthesis does not take into account that oxygen was used for cellular respiration.

4. Carefully wrap the tube containing *Elodea* in aluminum foil, and record here the length of time it takes for the edge of the solution in the tubing to recede 1 mm: _____. Convert your measurement to _____ mm/10 min. As before, this reading shows how much oxygen was used for cellular respiration.

5. Calculate gross photosynthesis for green light (mm/10 min) as you did for white light, and record your data in Table 8.3. (As before, add your data for net photosynthesis to that for cellular respiration.)

6. Calculate rate of photosynthesis for green light (mm/hr) as you did for white light, and record your data in Table 8.3. (As before, convert your data for mm/10 min to mm/hr.)

7. Collect the white and green light rates of photosynthesis (mm/hr) from each group in your lab; then average all the rates including your own. In the last column of Table 8.3 record the average for white and green light rates of photosynthesis (mm/hr).

Table 8.3 Rate of Photosynthesis (White and Green Light)		
Gross photosynthesis	Your Data	Class Data
White (from Table 8.2)	mm/10 min	_____
Green	mm/10 min	_____
Rate of photosynthesis		
White (from Table 8.2)	mm/hr	mm/hr
Green	mm/hr	mm/hr

8. Calculate the rate of photosynthesis (green light) as a percentage of the rate of photosynthesis (white light) using this equation.

$$\text{Percentage} = \frac{\text{rate of photosynthesis (green light)}}{\text{rate of photosynthesis (white light)}} \times 100$$

Percentage based on your data recorded in Table 8.3 = _____.

Percentage based on class data recorded in Table 8.3 = _____.

Conclusions: Rate of Photosynthesis

- Do your results support the hypothesis that green light is minimally used by a land plant for photosynthesis? _____ Explain, with reference to Figure 8.5. _____

  _____

  _____

- How does the percentage based on your data differ from that based on class data? _____

  _____

  Explain any difference in the percentage. _____

  _____

8.3 Carbon Dioxide Uptake

During the Calvin cycle reactions of photosynthesis, the plant takes up carbon dioxide (CO_2) and reduces it to a carbohydrate, such as glucose ($C_6H_{12}O_6$). Therefore, the carbon dioxide in the solution surrounding *Elodea* should disappear as photosynthesis takes place.

Experimental Procedure: Carbon Dioxide Uptake

1. Temporarily remove the *Elodea* from the test tube. Empty the sodium bicarbonate ($NaHCO_3$) solution from the test tube, rinse the test tube thoroughly, and fill with a phenol red solution diluted to a faint pink. (Add more water if the solution is too dark.) Phenol red is a pH indicator that turns yellow in an acid and red in a base.

> ⚠️ **Phenol red** Avoid ingestion, inhalation, and contact with skin, eyes, and mucous membranes. Follow your instructor's directions for disposal of this chemical. Use protective eyewear when performing this experiment.

2. Blow *lightly* on the surface of the solution. Stop blowing as soon as the surface color changes to yellow. Then shake the test tube until the rest of the solution turns yellow.

   Blowing onto the solution adds what gas to the test tube? _____ When carbon dioxide combines with water, it forms carbonic acid; therefore, the solution appears yellow.

3. Thoroughly rinse the *Elodea* with distilled water, return it to the test tube, which now contains a yellow solution, and assemble your volumeter as before.

4. The water in the beaker used to absorb heat should be clear.

5. Turn on the lamp, and wait until the edge of the solution just begins to move upward. Note the time.

   Observe until you note a change in color. How long did it take for the color to change? _____

6. Why did the solution eventually turn red? _____

   _____

7. The carbon cycle includes all the many ways that organisms exchange carbon dioxide with the atmosphere. Figure 8.6 notes the relationship between cellular respiration and photosynthesis. Animals produce carbon dioxide used by plants to carry out photosynthesis. Plants produce the food (and oxygen) that they and animals require to carry out cellular respiration. Therefore, the same carbon atoms pass between animals and plants and between plants and animals (Fig. 8.6).

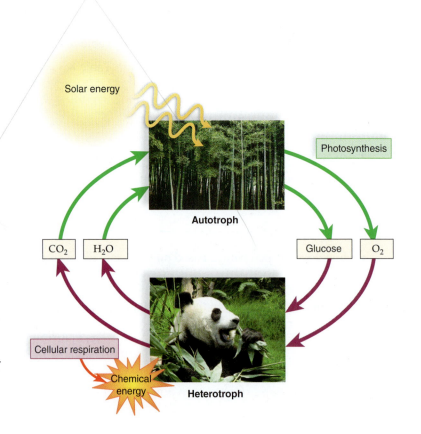

Figure 8.6 Photosynthesis and cellular respiration.
Animals are dependent on plants for a supply of oxygen, and plants are dependent on animals for a supply of carbon dioxide.

(bamboo trees) © Corbis RF; (panda) © Melinda Chan/Getty RF

9

Organization of Flowering Plants

Learning Outcomes

Introduction

Despite their great diversity in size and shape, all flowering plants have three vegetative organs that have nothing to do with reproduction: the root, the stem, and the leaf (Fig. 9.1). Roots anchor a plant and absorb water and minerals from the soil. A stem usually supports the leaves so that they are exposed to sunlight. Leaves carry on photosynthesis and thereby produce the nutrients that sustain a plant and allow it to grow.

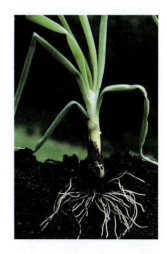

Figure 9.1   Organization of plant body.
Roots, stems, and leaves—the vegetative organs of a plant—are shown in this photo of an onion plant.
© Dwight Kuhn

9.1 External Anatomy of a Flowering Plant

Figure 9.2 shows that a plant has a root system and a shoot system. The **root system** consists of the roots. The **shoot system** consists of the stem and the leaves.

Observation: A Living Plant

Shoot System

What structures are in the shoot system?

_____

_____

_____

The Leaves

Leaves carry on photosynthesis. Which part of a leaf (blade or petiole) is the most expansive part?

_____

_____

The Stem

1. In the **stem,** locate a **node** and an **internode.**
2. Measure the length of the internode in the middle of the stem. Does the internode get larger or smaller toward the apex of the

   stem? _____ Toward the roots? _____
   Based on the fact that a stem elongates as it

   grows, explain your observation. _____

   _____

   _____

   _____

3. Where is the **terminal bud** (also called the shoot

   tip) of a stem?_____

   _____

   Where are axillary buds? _____

   _____

Root System

Observe the root system of a living plant if the root system is exposed. Does this plant have a strong

primary root or many roots of the same size? _____

What structures are in the root system? _____

_____

_____

_____

Where is the root tip? _____

Figure 9.2 Organization of a plant.
Roots, stems, and leaves are vegetative organs. The flower and fruit are reproductive structures studied in Laboratory 10.

9.2 Major Tissues of Roots, Stems, and Leaves

Unlike humans, flowering plants grow in size their entire life because they have an immature tissue called **meristematic tissue** composed of cells that divide. **Apical meristem** is located at the terminal end of the stem, the branches, and at the root tip and the root branches. When apical meristem cells divide, some of the cells differentiate into the mature tissues of a plant:

Dermal tissue: Forms the outer protective covering of a plant organ

Ground tissue: Fills the interior of a plant organ; photosynthesizes and stores the products of photosynthesis

Vascular tissue: Transports water and sugar, the product of photosynthesis, in a plant and provides support

Note in Table 9.1 that roots, stems, and leaves have all three tissues but they are given different specific names.

Table 9.1 Mature Tissues of Vegetative Organs

Tissue Type	Roots	Stems	Leaves
1. Dermal tissue (epidermis)	Protects inner tissues Root hairs absorb water and minerals.	Protects inner tissues	Protects inner tissues Cuticle prevents H_2O loss. Stomata carry on gas exchange.
2. Ground tissue	Cortex: Stores products of photosynthesis Pith: Stores products of photosynthesis	Cortex: Carries on photosynthesis, if green Pith: Stores products of photosynthesis	Mesophyll: Photosynthesis
3. Vascular tissue (xylem and phloem)	Vascular cylinder: Transports water and nutrients	Vascular bundle: Transports water and nutrients	Leaf vein: Transports water and nutrients

Observation: Tissues of Roots, Stems, and Leaves

Meristematic tissue. A slide showing the apical meristem in a shoot tip and another showing the apical meristem in a root tip are on demonstration. Meristematic cells are spherical and stain well when they are dense and have thin cell walls (Fig. 9.3). What is the function of meristematic tissue? _____

_____

Figure 9.3
Apical meristem.
A shoot tip and a root tip contain meristem tissue, which allows them to grow longer the entire life of a plant.

(shoot tip) © Steven P. Lynch; (root tip) © Ray F. Evert, University of Wisconsin

apical meristem

Shoot tip   Root tip

Mature Tissues

1. **Dermal tissue.** In a cross section of a leaf (Fig. 9.4), for example, focus only on the upper or lower epidermis. Epidermal cells tend to be square or rectangular in shape. In a leaf, the epidermis is interrupted by openings called **stomata** (sing., stoma). Later you will have an opportunity to see the epidermis in roots, stems, and leaves. What is a function of epidermis in all three organs (see Table 9.1)? _____

_____

_____

Figure 9.4   Microscopic leaf structure.
Like the stem and root, a leaf contains epidermal tissue, vascular tissue (leaf vein), and ground tissue (mesophyll).
© Ray F. Evert, University of Wisconsin

2. **Ground tissue.** The ground tissue fills the space between epidermis in roots, stems, and leaves. In leaves, for example, ground tissue is called mesophyll (Fig. 9.4). Ground tissue largely contains parenchyma cells and sclerenchyma cells. **Parenchyma cells** can be of different sizes and vary from fairly circular to oval. Those that contain chloroplasts carry on photosynthesis. Those that contain leucoplasts store starches and oils. **Sclerenchyma cells** are usually elongate and have thick walls impregnated with lignin. These dead cells appear hollow, and the presence of lignin means that they stain a red color. Sclerenchyma cells are strong and provide support. Which type of cell (parenchyma or sclerenchyma) carries on photosynthesis or stores the products of photosynthesis in a leaf? _____

Which one lends strength to ground tissue in roots and stems? _____

3. **Vascular tissue.** In a leaf, strands of vascular tissue are called leaf veins (Fig. 9.4). There are two types of vascular tissue, called xylem and phloem. **Xylem** contains hollow dead cells that transport water. The presence of lignin makes the cell walls strong, stains red, and makes xylem easy to spot. **Phloem** contains thin-walled, smaller living cells that transport sugars in a plant. Phloem is harder to locate than xylem, but it is always found in association with xylem. Which type of tissue (xylem or phloem) transports sugars in a plant? _____ Which type of tissue transports water? _____

Monocots Versus Eudicots

Flowering plants are classified into two major groups: **monocots** and **eudicots.** In this laboratory, you will be studying the differences between monocots and eudicots as noted in Figure 9.5. The arrangement of tissues is distinctive enough that you should be able to identify the plant as a monocot or eudicot when examining a slide of a root, stem, or leaf.

Experimental Procedure: Monocot Versus Eudicot

1. Examine the live plant again (see Fig. 9.2). The leaf vein pattern—that is, whether the veins run parallel to one another or whether the veins spread out from a central location (called the net pattern)—indicates that a

	Seed	Root	Stem	Leaf	Flower
Monocots	One cotyledon in seed	Root xylem and phloem in a ring	Vascular bundles scattered in stem	Leaf veins form a parallel pattern	Flower parts in threes and multiples of three
Eudicots	Two cotyledons in seed	Root phloem between arms of xylem	Vascular bundles in a distinct ring	Leaf veins form a net pattern	Flower parts in fours or fives and their multiples

Figure 9.5   Monocots versus eudicots.
The five features illustrated here are used to distinguish monocots from eudicots.

plant is either a monocot or a eudicot. Is the plant in Figure 9.2 a monocot or eudicot? _____ The

leaf pattern in Figure 9.4 appears to be parallel. Is this the leaf of a monocot or eudicot? _____

2. Observe any other available plants, and record in Table 9.2 the name of the plant and note the leaf vein pattern. On the basis of the leaf vein pattern decide if this plant is a monocot or a eudicot.

3. Aside from leaf vein pattern, other external features indicate whether a plant is a monocot or a eudicot. For example, open a peanut if available. The two halves you see are cotyledons. Is the plant that produced the

peanut a monocot or eudicot? _____
If available, examine a flower. If a flower has three petals or six petals, or any multiple of three, is the plant

that produced this flower a monocot or a eudicot? _____

In this laboratory you will have the opportunity to examine the cross sections of roots and stems microscopically; the arrangement of vascular tissue roots and stems also indicates whether a plant is a monocot or a eudicot.

Table 9.2 Monocots Versus Eudicots		
Name of Plant	**Organization of Leaf Veins**	**Monocot or Eudicot?**

9.3 Root System

The **root system** anchors the plant in the soil, absorbs water and minerals from the soil, and stores the products of photosynthesis received from the leaves.

Anatomy of a Root Tip

Primary growth of a plant increases its length. Note the location of the root apical meristem in Figure 9.6. As primary growth occurs, root cells enter zones that correspond to various stages of differentiation and specialization.

Figure 9.6 Eudicot root tip.
In longitudinal section, the root cap is followed by the zone of cell division, zone of elongation, and zone of maturation.

Zone of maturation

root hair

Vascular cylinder

Zone of elongation

procambium

Zone of cell division

root apical meristem protected by root cap

Root cap

106   Laboratory 9   Organization of Flowering Plants

9–6

Anatomy of Eudicot and Monocot Roots

Eudicot and monocot roots differ in the arrangement of their vascular tissue.

Observation: Cross-Section Anatomy of Eudicot and Monocot Roots

Eudicot Root

1. Obtain a prepared cross-section slide of a buttercup (*Ranunculus*) root. Use both low power and high power to identify the **epidermis** (the outermost layer of small cells that gives rise to root hairs). The epidermis protects inner tissues and absorbs water and minerals.

2. Locate the **cortex,** which consists of several layers of thin-walled cells (Fig. 9.7*a, b*). In Figure 9.7*b*, note the many stained starch grains in the cortex cells. The cortex is ground tissue that functions in food storage.

3. Find the **endodermis,** a single layer of cells whose walls are thickened by a layer of waxy material known as the **Casparian strip.** (It is as though these cells are glued together with a waxy glue.) Because of the Casparian strip, the only access to the xylem is through the living endodermal cells. The endodermis regulates what materials that enter a plant through the root? _____

   _____

   Use this illustration to trace the path of water and minerals from the root hairs to xylem. _____

4. Identify the **pericycle,** a layer one or two cells thick just inside the endodermis. Branch roots originate from this tissue.

5. Locate the **xylem** in the vascular cylinder of the root. Xylem has several "arms" that extend like the spokes of a wheel. This tissue conducts water and minerals from the roots to the stem.

6. Find the **phloem,** located between the arms of the xylem. Phloem conducts organic nutrients from the leaves to the roots and other parts of the plant.

Figure 9.7 Eudicot root cross section.
The vascular cylinder of a dicot root contains the vascular tissue. Xylem is typically star-shaped, and phloem lies between the points of the star.
a. Drawing. **b.** Micrograph.

(b) © Carolina Biological Supply Company/Phototake

Laboratory 9   Organization of Flowering Plants   **107**

Monocot Root

1. Obtain a prepared cross-section slide of a corn (*Zea mays*) root (Fig. 9.8*a*, *b*). Use both low power and high power to identify the six tissues mentioned for the eudicot root.

2. In addition, identify the **pith,** a centrally located ground tissue that functions in food storage.

Figure 9.8   Monocot root cross section.
a. Micrograph of a monocot root cross section. **b.** An enlarged portion.

(a) © Dr. Keith Wheeler/Science Source; (b) © George Ellmore, Tufts University

Comparison

Contrast the arrangement of vascular tissue (xylem and phloem) in the vascular cylinder of monocot roots and eudicot roots by writing "monocot" or "eudicot" on the appropriate line.

_____ Xylem has the appearance of a wheel. Phloem is between the spokes of the wheel.

_____ Ring of xylem (inside) and phloem (outside) surrounds pith.

Root Diversity

Roots are quite diverse, and we will take this opportunity to become acquainted with only a few select types.

Observation: Root Diversity

Taproots and fibrous roots. Most plants have either a taproot or a fibrous root. Note in Figure 9.9*a* that carrots have a **taproot.** The main root is many times larger than the branch roots. Grasses such as the type shown in Figure 9.9*b* have a **fibrous root:** All the roots are approximately the same size.

Examine the taproots on display, and name one or two in which the taproot is enlarged for storage. _____

_____

The dandelion on display has a taproot. Describe. _____

_____

Taproots in particular function in food storage.

Adventitious roots. Some plants have **adventitious** roots. Roots that develop from nonroot tissues, such as nodes of stems, are called adventitious roots. Examples include the prop roots of corn (Fig. 9.9*c*) and the aerial roots of ivy that attach this plant to structures such as stone walls.

Which plants on display have adventitious roots? _____

Other types of roots. Mangroves and other swamp-dwelling trees have roots called pneumatophores that rise above the waterline. Pneumatophores have numerous lenticels, which are openings that allow gas exchange to occur.

What root modifications not noted here are on display in your laboratory? _____

_____

Figure 9.9 Root diversity.
a. Carrots have a taproot. **b.** Grass has a fibrous root system. **c.** A corn plant has prop roots, and (**d**) black mangroves have pneumatophores to increase their intake of oxygen.

(a) © Johanna Parkin/Getty Images; (b) © Evelyn Jo Johnson/McGraw-Hill Education; (c) © NokHoOkNoi/iStock/360/GettyRF; (d) © Tim Laman/National Geographic/Getty RF

a. Taproot system

b. Fibrous root system

c. Prop roots, a type of adventitious root

d. Pneumatophores of black mangrove trees

Laboratory 9   Organization of Flowering Plants   **109**

9.4 Stems

Stems are usually found aboveground where they provide support for leaves and flowers. Vascular tissue extends from the roots through the stem and its branches to the leaves. Therefore, what function do botanists assign to stems in addition to support for branches and leaves? _____

_____ Explain why a branch cannot live if severed from the rest of the plant.

_____

Stems that do not contain wood are called **herbaceous,** or nonwoody, stems. Usually, monocots remain herbaceous throughout their lives. Some eudicots, such as those that live a single season, are also herbaceous. Other eudicots, namely trees, become woody as they mature.

Anatomy of Herbaceous Stems

Herbaceous stems undergo primary growth. **Primary growth** results in an increase in length due to the activity of the apical meristem located in the terminal bud (see Fig. 9.3) of the shoot system.

Observation: Anatomy of Eudicot and Monocot Herbaceous Stems

Eudicot Herbaceous Stem

1. Examine a prepared slide of a eudicot herbaceous stem (Fig. 9.10), and identify the **epidermis** (the outer protective layer). *Label the epidermis in Figure 9.10a.*
2. Locate the **cortex,** which may photosynthesize or store nutrients.
3. Find a **vascular bundle,** which transports water and organic nutrients. The vascular bundles in a eudicot herbaceous stem occur in a ring pattern. *Label the vascular bundle in Figure 9.10a.* Which vascular tissue (xylem or phloem) is closer to the surface?_____
4. Label the central **pith,** which stores organic nutrients. Both cortex and pith are composed of which tissue type listed in Table 9.1? _____

Figure 9.10   Eudicot herbaceous stem.
The vascular bundles are in a definite ring in this photomicrograph of a eudicot herbaceous stem. Complete the labeling as directed by the Observation.

(a) © Ed Reschke; (b) © Ray F. Evert, University of Wisconsin

a.

b.

Monocot Herbaceous Stem

1. Examine a prepared slide of a monocot herbaceous stem (Fig. 9.11). Locate the epidermal, ground, and vascular tissues in the stem.

2. The vascular bundles in a monocot herbaceous stem are said to be scattered. Explain. _____

_____

Figure 9.11 Monocot stem.
The vascular bundles, one of which is enlarged, are scattered in this photomicrograph of a monocot herbaceous stem.
(left) © Carolina Biological Supply Company/Phototake; (right) © Kingsley Stern

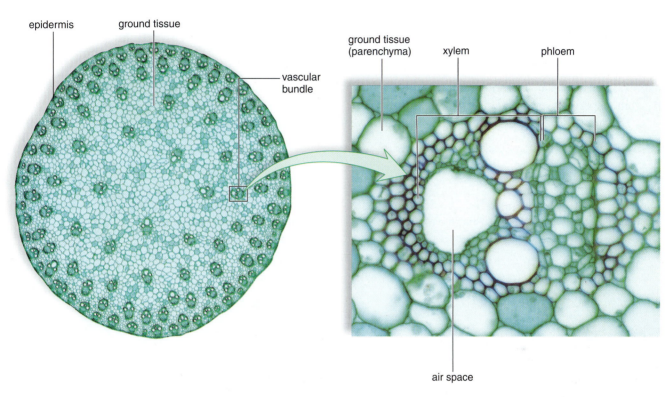

Comparison

1. Compare the arrangement of ground tissue in eudicot and monocot stems. _____

_____

2. Compare the arrangement of vascular bundles in the stems of eudicots and monocots. _____

_____

Stem Diversity

Stems can be quite diverse as well. We take this opportunity to become acquainted with those that allow a plant to accomplish vegetative reproduction and/or function in food storage. Several of these and other types of stems may be on display in the laboratory.

Stolons. The strawberry plant in Figure 9.12*a* has a horizontal aboveground stem called a runner or **stolon.** The stolon produces adventitious roots and new shoots at nodes. *Label an adventitious root and a new shoot in Figure 9.12a.*

List any other plants on display that spread and produce new shoots by sending out stolons. _____

_____

Rhizomes. An iris has a belowground horizontal stem called a **rhizome,** which functions as a fleshy food storage organ (Fig. 9.12*b*). New plants can grow from a single piece of the rhizome.

List any other plants on display that have the same belowground horizontal stems (rhizomes) as

the iris. _____

Tubers. A white potato has a belowground rhizome that gives off food storage **tubers** (Fig. 9.12*c*). Each eye is a node that can produce new plants.

List any other types of plants on display whose belowground stems have tubers. _____

_____

Corms. A gladiolus has a belowground vertical stem called a **corm,** which functions in food storage and has thin, papery leaves (Fig. 9.12*d*).

List any other types of plants on display that have a vertical stem called a corm. _____

_____

Figure 9.12 Stem diversity.
a. Stolon of a strawberry plant. **b.** Rhizome of an iris. **c.** Tuber of a potato. **d.** Corm of a gladiolus.

(a) © Evelyn Jo Johnson/McGraw-Hill Education; (b) © Science Pictures Limited/Science Source; (c) © Carlyn Iverson/McGraw-Hill Education; (d) © Carlyn Iverson/McGraw-Hill Education

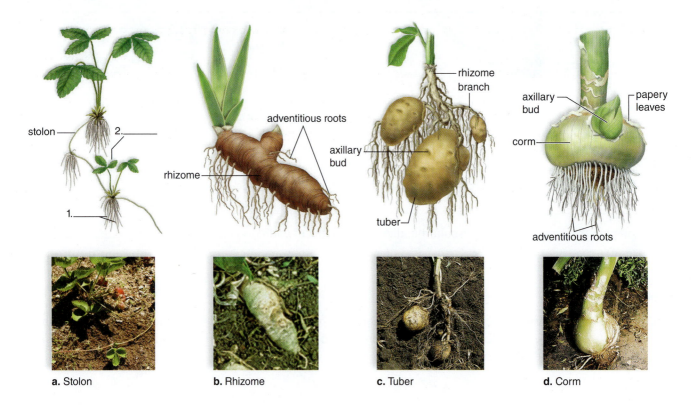

a. Stolon   **b.** Rhizome   **c.** Tuber   **d.** Corm

Anatomy of Woody Stems

Woody stems undergo both primary growth (increase in length) and secondary growth (increase in girth). When *primary growth* occurs, the apical meristem within a terminal bud is active. When *secondary growth* occurs, the vascular cambium is active. **Vascular cambium** is meristem tissue, which produces new xylem and phloem called **secondary xylem** and **phloem** each year. The buildup of secondary xylem year after year is called **wood**. Complete Table 9.3 to distinguish between primary growth and secondary growth of a stem.

Table 9.3 Primary Growth Versus Secondary Growth		
	Primary Growth	**Secondary Growth**
Active meristem		
Result		

Observation: Anatomy of a Winter Twig

1. A winter twig typically shows several past years' primary growth. Examine several examples of winter twigs (Fig. 9.13), and identify the **terminal bud** located at the tip of the twig. This is where new primary growth will originate. During the next growing season, the terminal bud produces new tissues including vascular bundles and leaves.
2. Locate a **terminal bud scar.** These are marks left on a stem from terminal bud scales (modified leaves protecting the bud). The distance between two adjacent terminal bud scars equals one year's primary growth.
3. Find a **leaf scar.** Mark where a leaf was attached to the stem.
4. Note the **bundle scars.** Complete this sentence: Marks left in the leaf scar where the vascular tissue

   _____.

5. Identify a **node.** This is the region where you find leaf scars and bundle scars. The region between nodes is

   called an _____.

6. Locate an **axillary bud.** This is where new branch growth can occur.
7. Note the numerous lenticels, breaks in the outer surface where gas exchange can occur.

Figure 9.13   External structure of a winter twig.
Counting the terminal bud scars tells the age of a particular branch.

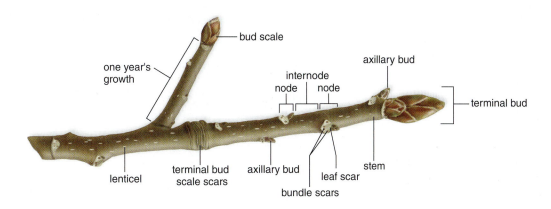

Observation: Anatomy of Woody Stem

1. Examine a prepared slide of a cross section of a woody stem (Fig. 9.14), and identify the **bark** (the dark outer area), which contains **cork,** a protective outer layer; **cortex,** which stores nutrients; and **phloem,** which transports organic nutrients.

Figure 9.14   Woody eudicot stem cross section.
Because xylem builds up year after year, it is possible to count the annual rings to determine the age of a tree. This tree is three years old.

(b) © Carolina Biological Supply Company/Phototake

a. Drawing

b. Photomicrograph

36×

2. Locate the **vascular cambium** at the inner edge of the bark, between the bark and the wood. Vascular cambium is meristem tissue whose activity accounts for secondary growth, which causes increased girth of a tree. Secondary phloem (which disappears) and secondary xylem (which builds up) are produced by vascular cambium each growing season.
3. Find the **wood,** which contains annual rings. An **annual ring** is the amount of xylem added to the plant during one growing season. Rings appear to be present because spring wood has large xylem vessels and looks light in color, while summer wood has much smaller vessels and appears much darker. How old is the stem you are observing? _____ Are all the rings the same width? _____
4. Identify the **pith,** a ground tissue at the center of a woody stem that stores organic nutrients and may disappear.
5. Locate **rays,** groups of small, almost cuboid cells that extend out from the pith laterally.

9.5 Leaves

A **leaf** is the organ that produces food for the plant by carrying on photosynthesis. Leaves are generally broad and quite thin. An expansive surface facilitates the capture of solar energy and gas exchange. Water and nutrients are transported to the cells of a leaf by leaf veins, extensions of the vascular bundles from the stem.

Anatomy of Leaves

Observation: Anatomy of Leaves

1. Examine a model of a leaf. With the help of Figure 9.15, identify the waxy **cuticle,** the outermost layer that protects the leaf and prevents water loss.
2. Locate the **upper epidermis** and **lower epidermis,** single layers of cells at the upper and lower surfaces. Trichomes are hairs that grow from the upper epidermis and help protect the leaf from insects and water loss.
3. Find the leaf veins in your model. The bundle sheath is the outer boundary of a vein; its cells surround and protect the vascular tissue. If this is a model of a monocot, all the leaf veins will be _____

   _____. If this is a model of a eudicot, some leaf veins will be circular and some will

   be oval. Why? _____
4. Identify the **palisade mesophyll,** located near the upper epidermis. These cells contain chloroplasts and carry on most of the plant's photosynthesis. Locate the **spongy mesophyll,** located near the lower epidermis. These cells have air spaces that facilitate the exchange of gases across the plasma membrane. *Label the layers of mesophyll in Figure 9.15.* Collectively, the mesophyll represents which of the three types

   of tissue found in all parts of a plant (see Table 9.1)? _____
5. *Label the two layers of epidermis in Figure 9.15.* Find a **stoma** (pl., stomata), an opening through which gas exchange occurs and water escapes. Stomata are more numerous in the lower epidermis. A stoma has two guard cells that regulate its opening and closing.

Figure 9.15 Leaf anatomy.

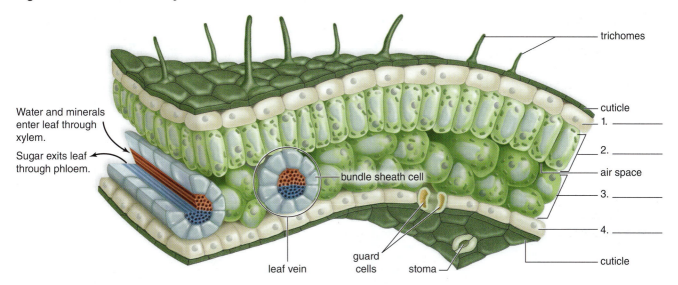

Leaf Diversity

A eudicot leaf consists of a flat blade and a stalk, called the petiole. An axillary bud appears at the point where the petiole attaches a leaf to the stem. In other words, an axillary bud is a tip-off that you are looking at a single leaf.

Observation: Leaf Diversity

Several types of eudicot leaves will be on display in the laboratory. Examine them using these directions.

- A leaf may be **simple,** in which case it consists of a single blade; or a leaf may be **compound,** meaning its single blade is divided into leaflets. *In Figure 9.16a write "simple" or "compound" before the word* leaf *in 1–3.* (Among the leaves on display, find one that is simple and one that is compound.)
- A compound leaf can be **palmately** compound, meaning the leaflets are spread out from one point. Which leaf in Figure 9.16*a* is palmately compound? *Add "palmately" in front of "compound" where appropriate.* (See if you can find a palmately compound leaf among those on display.)
- A compound leaf can be **pinnately** compound, meaning the leaflets are attached at intervals along the petiole. Which leaf in Figure 9.16*a* is pinnately compound? *Add "pinnately" in front of "compound" where appropriate.* (See if you can find a pinnately compound leaf among those on display.)
- As shown in Figure 9.16*b*, leaves can be in various positions on a stem. On which stem in Figure 9.16*b*, 4–6, are the leaves **opposite** one another? *Write the word "opposite" where appropriate.* On which stem do the leaves **alternate** along the stem? *Write the word "alternate" where appropriate.* On which stem do the leaves **whorl** about a node? *Add the word "whorl" where appropriate in Figure 9.16b.* (See if you can find different arrangements of leaves among the stems on display.)

Figure 9.16   Classification of leaves.

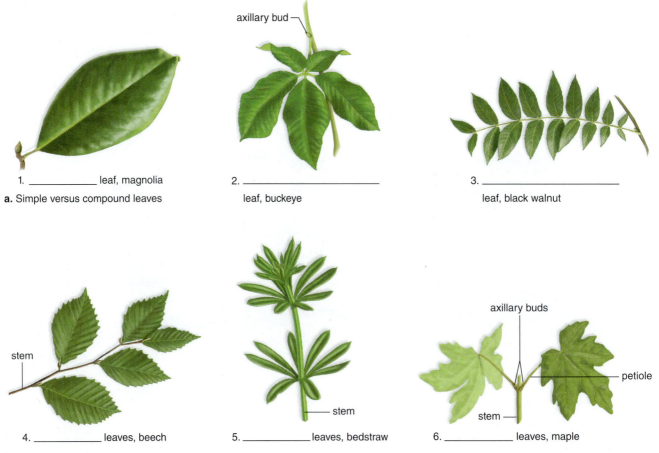

1. _____ leaf, magnolia

a. Simple versus compound leaves

axillary bud

2. _____
leaf, buckeye

3. _____
leaf, black walnut

stem

4. _____ leaves, beech

b. Arrangement of leaves on stem

stem

5. _____ leaves, bedstraw

axillary buds

petiole

stem

6. _____ leaves, maple

9.6 Xylem Transport

Xylem (Fig. 9.17), which transports water from the roots to the leaves, contains two types of conducting cells: tracheids and vessel elements. Both types of conducting cells are hollow and nonliving; the vessel elements are larger, they lack transverse end walls, and they are arranged to form a continous pipeline for water and mineral transport.

Water Column

The water column in xylem is continuous because water molecules are cohesive (they cling together) and because water molecules adhere to the sides of xylem cells. Evaporation of water from the leaves when the stomata are open is called transpiration. This water loss creates a force that pulls the water column upward from roots to leaves.

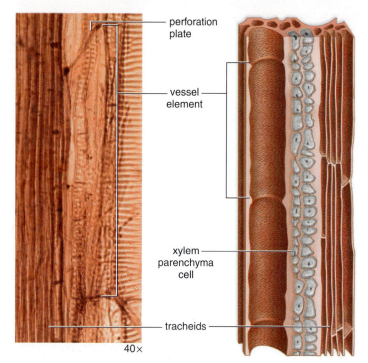

40×

Figure 9.17   Xylem structure.
Xylem contains two types of conducting cells: tracheids and vessel elements. Tracheids have pitted walls, but vessel elements are larger and form a continuous pipeline from the roots to the leaves.

(left) © Garry DeLong/Science Source

Experimental Procedure: The Water Column

1. Place a small amount of red-colored water in two beakers. Label one beaker "wet" and the other beaker "dry."
2. Transfer a stalk of celery (which was cut and then immediately placed in a container of water) into the "wet" beaker so that the large end is in the colored water.
3. Transfer a stalk of celery of approximately the same length and width (that was kept in the air after being cut) into the "dry" beaker so that the large end is in the colored water.
4. With scissors, cut off the top end of each stalk, leaving about 10 cm total length.
5. Time how long it takes for the red-colored water to reach the top of each stalk, and record these data in Table 9.4.
6. In which celery stalk was the water column broken? _____

   Use this information to write a conclusion in Table 9.4.
7. Make a cross-sectional wet mount of the stalk in the "wet" beaker. Observe this slide under the microscope.

   What type of tissue has been stained by the dye? _____

Table 9.4 Celery Stalk Experiment		
Stalk	**Speed of Dye (Minutes)**	**Conclusion**
Cut end placed in water prior to experiment		
Cut end kept in air prior to experiment		

_____ 1. What kind of plant tissue is composed of cells that divide?

_____ 2. What term is used to identify ground tissue in a leaf?

_____ 3. What kind of plant, monocot or eudicot, has the xylem and phloem in its leaves arranged in a parallel pattern?

_____ 4. In what zone of a eudicot root will tracheids and vessel elements be found?

_____ 5. What type of tissue gives rise to root hairs?

_____ 6. How is xylem typically arranged in a eudicot root cross section?

_____ 7. What term refers to stems that are nonwoody?

_____ 8. What kind of plant, monocot or eudicot, has stem vascular bundles that occur in a ring?

_____ 9. What kind of growth do woody plants undergo when the diameter of their trunk increases?

_____ 10. Where does new primary growth occur on a winter twig?

_____ 11. What type of vascular tissue forms the annual rings in a woody stem?

_____ 12. What is the outermost layer of a leaf that protects and prevents water loss?

_____ 13. What cells are on either side of a stoma and regulate whether it is open or closed?

_____ 14. If its leaf veins form a net pattern and there is one blade attached to a petiole, is the leaf simple or compound, and is it from a monocot or a eudicot?

Thought Questions

15. Where is glucose produced in a leaf? What type of tissue carries glucose away from a leaf, and where is the glucose stored in a plant like a beet?

16. Stomata are more numerous in the lower epidermis of leaves. Explain why it would be detrimental to the plant to have more stomata in the upper epidermis of leaves.

17. Compare and contrast monocots and eudicots by identifying two similarities and two differences between them.

10

Reproduction in Flowering Plants

Learning Outcomes

Introduction

The evolution of the flower and the involvement of animals in the plant life cycle helps account for the great success of flowering plants, also called **angiosperms.** The flower is the center of sexual reproduction for angiosperms and is involved in the production of male and female gametophytes, which are gametes and embryos enclosed within seeds. The flower also gives rise to fruits that cover the seeds.

Flowering plants are stationary, but they have evolved a successful mechanism for pollination that helps them complete their life cycle. Motile animals, often insects, visit flowers and collect the pollen. Pollen moves sperm that fertilizes an egg, which triggers the development of a seed. The relationship between the pollinator and the flower (Fig. 10.1) is mutualistic because the pollinator acquires nectar, a nutrient substance produced by the flower. Animals also help flowering plants disperse their seeds. When animals feed on fruits they inadvertently take the seeds to new locations.

Figure 10.1   Pollinator on a landing platform.
This composite flower contains many individual flowers. The bee is attracted to this flower because of its bright color and sweet smell, and the bee uses the flower's reproductive parts as a place to land. Pollen will be collected by the bee when it feeds on nectar.

© Steven P. Lynch

10.1   The Flower

The structure of a flower allows a flowering plant to produce seeds whose dispersal accounts for the success of these plants.

Structure of a Flower

The flower contains the reproductive structures necessary to the life cycle of a flowering plant.

Observation: Structure of a Flower

1. Examine a model and identify these parts of a flower (Fig. 10.2).

Figure 10.2   Flower structure.

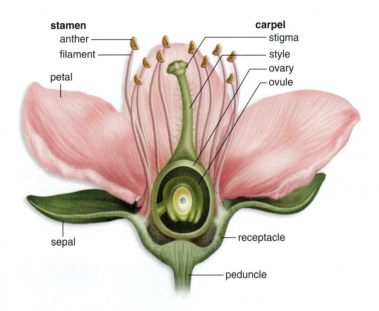

Peduncle:	The flower stalk.
Receptacle:	The portion of a flower that bears the other parts.
Sepals:	Collectively called the **calyx,** the **sepals** protect the flower bud before it opens. The sepals may drop off or may be colored like the petals. Usually, sepals are green and remain attached to the receptacle.
Petals:	Collectively called the **corolla,** the **petals** are quite diverse in size, shape, and color. Their color or arrangement can attract a pollinator.
Stamens:	Each **stamen** consists of two parts—the **anther,** and the **filament,** a slender stalk. Pollen grains develop in the anther.
Carpel:	A vaselike structure with three major regions from top to bottom: the **stigma,** an enlarged, sticky knob; the **style,** a slender stalk; and the **ovary,** an enlarged base that contains one or more ovules. **Ovules** develop into seeds. The ovule wall becomes the seed coat, which protects an embryonic plant and its stored food. An ovary becomes a fruit. Fruit is instrumental in the distribution of seeds. For example, after a bird eats a berry, it will probably fly to a new location, where it defecates the seed.

Flower parts in threes and multiples of three

Table 10.1 Monocots and Eudicots	
Monocots	**Eudicots**
One cotyledon in seed	Two cotyledons in seed
Flower parts in threes or multiples of three	Flower parts in fours or fives or their multiples
Usually herbaceous	Woody or herbaceous
Usually parallel venation in leaves	Usually net venation in leaves
Scattered vascular bundles in stem	Ring of vascular bundles in stem

Flower parts in fours or fives and their multiples

2. As we studied in the previous lab, some flowering plants are monocots (monocotyledons) and some are eudicots (eudicotyledons) as described in Table 10.1. Note the difference between monocot and eudicot flowers. Is the flower model you have been examining a monocot or a eudicot? _____

Observation: Live Flower

1. Obtain a live flower similar to the one in Figure 10.3. It should be possible to see the sepals, petals, stamens, and carpel(s) of this flower as in Figure 10.3. Sketch the overall appearance of the whole flower here and label as much as possible using the terms in Figure 10.2.

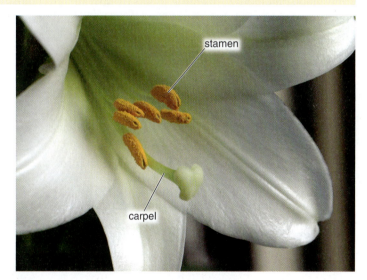

Figure 10.3   Lily (*Lilium longiflorum*).
© Arlene Gee/Getty RF

2. Describe the color and scent of your flower, if any. _____
_____

3. Count the number of petals and sepals (see Table 10.1). Is this flower a monocot or a eudicot? _____
_____ Explain. _____

4. Determine where the petals and sepals are attached and remove them. What color are the sepals? _____
_____ Do they resemble the petals in any way? _____ If yes, how?
_____

5. Count the number of stamens. Is this the same as the petal number or a multiple of the petal

   number? _____ Remove the stamens and examine one to locate the anther and filament. Place a stamen on a slide and use a scalpel to remove and open the anther to disperse pollen on the slide. Remove all but a few pollen grains from the slide. Add a coverslip and examine microscopically, using high power. Can you find

   two different nuclei? _____ At this point, a pollen grain contains two cells. One of these cells will divide to produce the sperm and the other will produce the pollen tube through which the sperm will travel to the embryo sac in an ovule of an ovary.

6. Before removing the carpel, identify the stigma, style, and ovary. Remove a carpel and place it on a slide.

   Use a scalpel to slit it longitudinally. Do you see any ovules? _____

   Describe. _____

   Use the scalpel to make a cross section of the ovary. Does this ovary contain chambers, each

   with ovules? _____

7. If available, examine the fruit produced by this type of flower. The exterior wall came from what part of the

   carpel? _____ The seeds came from what structure? _____

10.2 The Flowering Plant Life Cycle

Land plants, including flowering plants, follow the **alternation-of-generation life cycle** (Fig. 10.4). Just as your arms are shaped differently from your legs, land plants have two generations that look different and have different functions:

1. The **sporophyte** (diploid generation) produces haploid spores in a structure called a sporangium by meiosis. **Spores** undergo mitosis and develop into the haploid generation.
2. The **gametophyte** (haploid generation) produces gametes (eggs and sperm) by mitosis.
   If fertilization occurs, the gametes unite to form a diploid zygote (a young sporophyte).

In humans, meiosis is involved in the production

of _____. What does meiosis

produce in plants? _____

_____

Alternation of Generation in Flowering Plants

What do the sporophyte and gametophyte of a flowering plant look like? The sporophyte is the dominant generation in flowering plants. It is the generation we can easily see because we recognize it as the plant. An oak tree, a rose bush, a daffodil plant—each is the sporophyte generation. A flowering plant produces spores in sporangia,

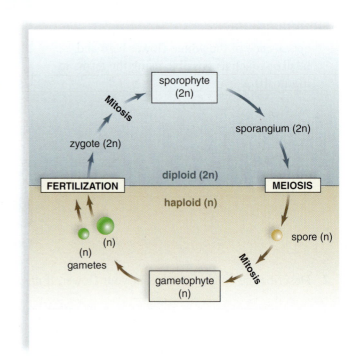

Figure 10.4   Alternation of generations.

Use Figure 10.6 as a guide to describe the life cycle of flowering plants.

1. The parts of the flower involved in reproduction are the _____ and the _____.

2. The anther at the top of the stamen has two _____ sacs, which produce _____ by meiosis.

3. Within an ovule, a megaspore mother cell undergoes meiosis to produce four _____, three of which die.

4. A microspore undergoes mitosis and becomes a _____, the male gametophyte.

5. One megaspore undergoes mitosis and develops into a(n) _____, which contains, in addition to other cells, two _____ nuclei and the _____ cell. (These terms are shaded.)

6. **Pollination** is the transport of pollen by wind or by animals such as insects (see Fig. 10.1) from the pollen sacs to the stigmas of carpels. After a pollen grain lands on the stigma of a carpel, it develops a pollen tube that passes down the style and takes _____ sperm to the embryo sac.

7. During double fertilization, one sperm from the pollen tube fertilizes the egg within the embryo sac, and the other joins with the two _____ nuclei.

8. The fertilized egg becomes an _____, and the joining of polar nuclei and sperm becomes the triploid (3n) **endosperm.** The ovule wall becomes the seed coat. A **seed** contains a sporophyte embryo, stored food, and a seed coat. In angiosperms, seeds are enclosed by _____ (not shown). A fruit assists in the dispersal of seeds (e.g., when animals eat fleshy fruits, they may ingest the seeds and disperse them by defecation sometime later).

9. After dispersal of the seed, it germinates and begins to grow, eventually becoming an adult sporophyte that bears flowers.

10.3 Pollination and Development of the Embryo

Pollination, the movement of pollen from the pollen sacs to the stigmas of carpels, is an important event in flowering plants. In some plants, wind accomplishes pollination and these plants produce copious amounts of pollen (Fig. 10.7). Why would you expect wind to carry out pollination for a

tree and not for a garden plant? _____

_____

_____

_____

Figure 10.7 Wind pollination of a grass, with SEM of pollen grains.

(grass releasing pollen) © Tim Gainey/Alamy; (pollen) © Eye of Science/Science Source

Laboratory 10   Reproduction in Flowering Plants

Plants and Their Pollinators

Through the course of evolution, many plants—even trees—have come to depend on animals (e.g., beetles, bats, insects) called **pollinators** to carry out pollination. The relationship is mutualistic because both the plant and the pollinator benefit from the relationship. **Nectar** is a sugary substance produced by the plant for the pollinator. The pollinator goes from plant to plant to gather food (nectar and even pollen) and in the process carries pollen from flower to flower.

It might seem as if a pollinator such as a bee goes to all types of flowers, but it doesn't. Bees visit only certain flowers—the ones that provide them with nectar, a sugary liquid that serves as their food. Bee-pollinated flowers have a shape and appearance preferred by bees, even a particular type of bee. Particular bees and particular flowers are suited to one another because there has been interaction between them over time that caused coevolution to occur.

How is this advantageous to both the plant and the bee? _____

_____

_____

Observation: Plants and Their Pollinators

1. The description you gave in the live flower observation (section 10.1) can help you decide what type of pollinator would be attracted to your flower. See also Figure 10.8.

   - Bees and moths can smell. But bees like a delicate, sweet smell, while moths prefer a strong smell that can allow them to find flowers in the dark. Moths are nocturnal and feed at night. If the flower has a smell, which of these two pollinators might pollinate your flower? _____

   - Bees and butterflies generally need a landing platform. Bees can land on a small petal, but butterflies typically walk around on a cluster of flowers.

   - Moths and hummingbirds do not need a landing platform because they hover (flap their wings to stay in one place). Moths and hummingbirds have a long tongue to reach nectar at the bottom of the floral tube. Based on this information, which type of pollinator mentioned so far might pollinate your flower? _____

   - Moth-pollinated flowers are typically white; bees can't see red but can see yellow; butterflies like brightly colored flowers; and hummingbirds prefer the color red. Which of these pollinators might prefer your flower? _____

2. Your instructor may have suggested that you bring other flowers to lab. Examine several available flowers and tell how each of the following features help decide the pollinator. These features will help you decide the pollinator for each type of flower. If a flower is the same color as the rest of the plant (green or brown) and hangs down from the end of a branch, it likely produces pollen that is windblown. Now complete Table 10.2.

   How does each of these flower features help decide the pollinator?

   Landing platform _____

   Color _____

   Smell _____

   Shape of flower _____

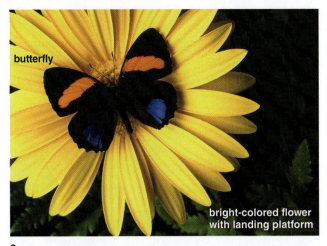

butterfly

bright-colored flower
with landing platform

a.

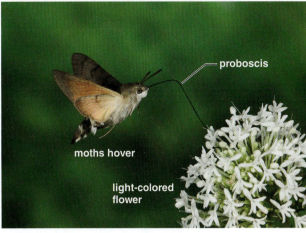

proboscis

moths hover

light-colored
flower

b.

long, thin beak

hummingbirds
hover

floral tube with
curved back margins

c.

Figure 10.8 Pollinators.
a. The butterfly-pollinated flower is often a brightly colored composite containing many individual flowers. The broad exposure provides room for the butterfly to land. **b.** A moth-pollinated flower is usually light in color; a moth depends on scent to find the flower it prefers in the dark. **c.** Hummingbird-pollinated flowers are curved back, allowing the bird to insert its beak to reach the rich supply of nectar.

(a) © Corbis RF; (b) © Andrew Darrington/Alamy; (c) © Anthony Mercieca/Science Source

Table 10.2 Plants and Their Pollinators		
Description of Flower	**Possible Pollinator**	**Explanation**
1		
2		
3		
4		
5		
6		

Development of Eudicot Embryo

Stages in the development of a eudicot embryo are shown in Figure 10.9. During development, the **suspensor** anchors the embryo and transfers nutrients to it from the mature plant. The **cotyledons** store nutrients that the embryo uses as nourishment. An embryo consists of the **epicotyl,** which becomes the leaves; the **hypocotyl,** which becomes the stem; and the **radicle,** which becomes the roots.

Figure 10.9 Development of a eudicot embryo.
Embryogenesis consists of these stages.

(proembryo) © Dr. Chun-Ming Liu; (globular) © Dr. Chun-Ming Liu; (heart) © Dr. Chun-Ming Liu; (torpedo) © Biology Media/Science Source; (mature) © Steven P. Lynch

Arabidopsis thaliana

endosperm nucleus

zygote

1. Zygote stage: The single-celled zygote lies beneath the endosperm nucleus.

endosperm

basal cell

2. Proembryo stage: Multicellular proembryo (green) is located above the suspensor (purple). The endosperm is a mass of tissue surrounding the embryo.

Observation: Development of Eudicot Embryo

Study prepared slides and identify the stages described in Figure 10.9. List the stages you were able to identify. _____

_____

_____

_____

_____

endosperm

3. Globular stage: As cell division continues, the proembryo (green) becomes globe-shaped. The stalklike suspensor (purple) anchors the embryo.

A. thaliana

cotyledons appearing

4. Heart stage: The embryo becomes heart-shaped as the cotyledons begin to appear.

Capsella

shoot apical meristem

bending cotyledons

endosperm

root apical meristem

5. Torpedo stage: The embryo becomes torpedo-shaped as the cotyledons enlarge. The endosperm lessens, and tissues become differentiated.

Capsella

hypocotyl (root axis)

epicotyl (shoot apical meristem)

seed coat

radicle (root apical meristem)

cotyledons

6. Mature embryo stage: The embryo consists of the epicotyl (represented here by the shoot apex), the hypocotyl, and the radicle (which contains the root apex).

10.4 Fruits

Because fruits are defined as the mature ovary, what people normally call a vegetable can be a fruit to a botanist. For example, in addition to a blackberry, an almond, a pea pod, and a tomato are fruits (Fig. 10.10).

As a fruit develops from an ovary, the ovary wall thickens to become the **pericarp.** The pericarp can have as many as three layers: exocarp, mesocarp, and endocarp. The **exocarp** forms the outermost skin of a fruit. The **mesocarp** between the exocarp and endocarp can be fleshy. The receptacle of a flower often contributes

a. Almond, *Prunus*

seed

b. Tomato, *Lycopersicon*

one chamber of ovary

c. Pea, *Pisum*

developing
fruit

seed

d. Blackberry, *Rubus*

many
carpels

Figure 10.10   Fruit diversity.
a. The almond fruit is fleshy, with a single seed enclosed by a hard covering. **b.** The tomato is derived from a compound ovary. **c.** Dry fruit of the pea plant develops from a simple ovary. **d.** The blackberry fruit is an aggregate fruit derived from a flower that had many ovaries. Each ovary produces one of the "berries" in the aggregate fruit.

to the flesh of a fruit also. The **endocarp** serves as the boundary around the seed(s). The endocarp may be fleshy (as in tomatoes), hard (as in peach pits), or papery (as in apples). Botanists classify fruits in the manner described in the following key. This is a simplified key, but it does contain some of the most common types of fruits.

Dichotomous Key to Major Types of Fruit

I. Fleshy fruits

    A. Simple fruits (i.e., from a single ovary)

        1. Flesh mostly of ovary tissue, particularly the mesocarp

            a) Endocarp is hard and stony; ovary superior and single-seeded (cherry, olive, coconut): **drupe**

            b) Endocarp is fleshy or slimy; ovary usually many-seeded (tomato, grape, green pepper): **berry**

        2. Flesh mostly of receptacle tissue (apple, pear, quince): **pome**

    B. Complex fruits (i.e., from more than one ovary)

        1. Fruit from many carpels on a single flower (strawberry, raspberry, blackberry): **aggregate fruit**

        2. Fruit from carpels of many flowers fused together (pineapple, mulberry): **multiple fruit**

II. Dry fruits

    A. Fruits that split open at maturity (usually more than one seed)

        1. Split occurs along two seams in the ovary. Seeds borne on one of the halves of the split ovary (pea and bean pods, peanuts): **legume**

        2. Seeds released through pores or multiple seams (poppies, irises, lilies): **capsule**

    B. Fruits that do not split open at maturity (usually one seed)

        1. Pericarp hard and thick, with a cup at its base (acorn, chestnut, hickory): **nut**

        2. Pericarp thin and winged (maple, ash, elm): **samara**

        3. Pericarp thin and not winged (sunflower, buttercup): **achene** (cereal grains): **caryopsis**

Source: Vodopich and Moore: *Biology Laboratory Manual* 8/e, p. 344.

To use this key understand that at each step you have at least two choices and you pick one of these in order to proceed further. Therefore, first determine if the fruit is fleshy or dry. If fleshy, proceed further with I and if dry, proceed further with II. Next choose either A or B and so forth, until you arrive at the particular type of fruit. The last column in Table 10.3 will be one of the boldface terms from the key.

Observation: Fruits

1. Again examine the fruit of your flower. Use the key to determine the fruit type and list this fruit as an example in Table 10.3.

2. Examine an apple that has been sliced to give a longitudinal and a cross-section view of the interior. The flesh of an apple is from the receptacle and only the core of an apple is from the ovary. Locate the outer limit of the pericarp and the limit of the endocarp. An ovary can be simple (have one chamber) or can be compound (have more than one chamber). What type of ovary does an apple have? _____

   How could an animal, such as a deer, help disperse the seeds of an apple? _____

   _____

   Use the apple as an example of a fruit in Table 10.3.

3. Examine the pod of a string bean or pea plant. How many seeds (beans or peas) are in the

pod? _____ Would it help disperse the seeds of a pea plant if an animal were to eat the

peas? _____ Why or why not? _____

Split the pea or bean and look for the embryo. Is this plant a monocot or eudicot plant? _____

How do you know? _____

Use a pea pod as an example of a fruit in Table 10.3.

4. Examine a sunflower fruit and remove the seed. The outer coat of a sunflower seed is actually _____

_____. How can examining the seed tell you that the sunflower plant is a

eudicot? _____

Except for the apple, all the fruits you have examined are dry fruits. What does this mean? _____

_____

Add sunflower fruit to Table 10.3.

5. Examine other available fruits and complete Table 10.3.

Table 10.3 Identification of Simple Fruits			
Common Name	**Fleshy or Dry?**	**Eaten as a Vegetable, Fruit, Other?**	**Type of Fruit (from Dichotomous Key)**
1			
2			
3			
4			
5			
6			
7			
8			
9			
10			

10.5 Seeds

The seeds of flowering plants develop from ovules. As noted previously in Figures 10.6 and 10.9, a seed contains an embryonic plant, stored food, and a seed coat. Monocot seeds have one **cotyledon** (seed leaf); eudicots have two cotyledons.

Observation: Eudicot and Monocot Seeds

Bean Seed

1. Obtain a presoaked bean seed (eudicot). Carefully dissect it, using Figure 10.11 to help you identify the following:

   a. **Seed coat:** The outer covering. Remove the seed coat with your fingernail.

   b. **Cotyledons:** Food storage organs. The endosperm was absorbed by the cotyledons during development. What is the function of cotyledons? _____

   _____

   c. **Epicotyl:** The small portion of the embryo located above the attachment of the cotyledons. The first true leaves (**plumules**) develop from the epicotyl.

   d. **Hypocotyl:** The small portion of embryo located below the attachment of the cotyledons. The lower end develops into the embryonic root, or **radicle.**

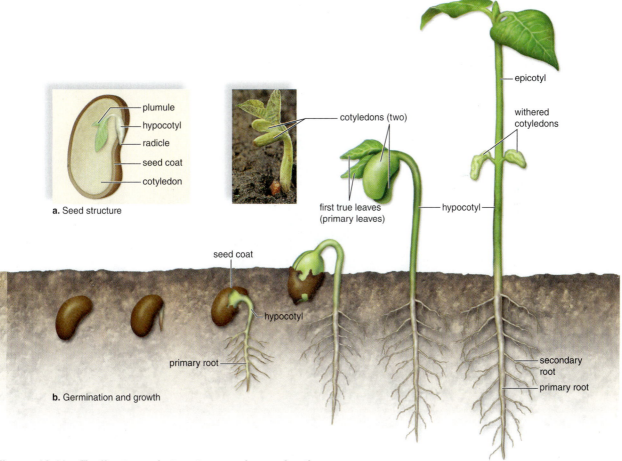

Figure 10.11   Eudicot seed structure and germination.
a. A seed contains an embryo, stored food, and a seed coat as exemplified by a bean seed. A eudicot seed has two cotyledons. Following (**b**) germination, a bean seedling grows to become a mature plant.
(top) © Ed Reschke

2. As observed in Figure 10.11, which organ emerges first from a dicot seed—the plumule or the radicle (i.e., root)? _____

   Of what advantage is this to the plant? _____

   _____

3. The hypocotyl is the first part to emerge from the soil. What is the advantage of the hypocotyl pulling the plumule up out of the ground instead of pushing it up through the ground? _____

   _____

4. Do cotyledons stay beneath the ground in eudicots? _____

Corn Kernel

1. Obtain a presoaked corn kernel (monocot). Lay the seed flat, and with a razor blade or scalpel, carefully slice it in half. A corn kernel is a fruit, and the seed coat is tightly attached to the pericarp (Fig. 10.12).
2. Identify the cotyledon, plumule, and radicle. In addition, identify the
   a. **Endosperm:** Stored food for the embryo; the nutrients pass into the cotyledon as the seedling grows.
   b. **Coleoptile:** A sheath that covers the emerging leaves.
3. As observed in Figure 10.12, does the cotyledon of a corn seed, our example of a monocot, stay beneath the ground? _____

a. Corn kernel

b. Germination and growth

Figure 10.12   Monocot seed structure and germination.
a. A monocot seed has only one cotyledon as exemplified by a corn kernel. A corn kernel is a fruit—the seed is covered by a pericarp. Following (**b**) germination, a corn seedling grows to become a mature plant.

(top) © Ed Reschke/Getty Images

Seed Germination

Mature seeds contain an embryo that does not resume growth until after germination, which requires the proper environmental conditions. Mature seeds are dry, and for germination to begin, the dry tissues must take up water in a process called **imbibition.** After water has been imbibed, enzymes break down the food source into small molecules that can provide energy or be used as building blocks until the seedling is ready to photosynthesize.

Experimental Procedure: Seed Germination

Your instructor has placed sunflower seeds in five containers and watered them. The seedlings are in increasing stages of growth.

1. Order the seedlings in a series according to increasing stages of growth. What criteria did you use to order the seedlings? _____

    _____

2. Is this plant a monocot or a eudicot? _____ What criteria did you use to decide? _____

    _____

3. Can you see the cotyledons? _____ Explain why the cotyledons of a eudicot seedling shrivel as the seedling grows. _____

4. Use the space below to draw two contrasting stages of growth and add these labels to your drawings: hypocotyl, cotyledons, epicotyl, leaves, stem, terminal bud, node. For terminal bud and node, see Figure 9.2.

_____ 1. In what flower structures are pollen grains produced?

_____ 2. What do the stigma, style, and ovary of a flower collectively form?

_____ 3. What do a flower's ovule and ovary become?

_____ 4. What kind of angiosperm makes flowers with parts in threes or multiples of three?

_____ 5. What is the haploid generation of the alternation of generations called?

_____ 6. What process occurs to forms gametes in plants?

_____ 7. What is the name of the female gametophyte of flowering plants?

_____ 8. What process occurs to form microspores and megaspores in flowering plants?

_____ 9. What transports sperm from the flower's stigma to the embryo sac?

_____ 10. What does the fusion of the polar nuclei and sperm form, and how many n is it?

_____ 11. What is the likely pollinator for a night blooming, white flower with a strong scent?

_____ 12. What is an example of a simple fruit that has a fleshy endocarp?

_____ 13. What term describes a fruit that develops from more than one ovary?

_____ 14. What is the name for the part of the embryo above the attachment of the cotyledons?

Thought Questions

15. If a flower has five united petals, five anthers, and is red in color without a scent, is it made by a monocot or a eudicot plant, and what is the mostly likely pollinator for its flowers? Justify your choice.

16. Why is the relationship between a flowering plant and its pollinator considered mutualistic?

17. How do angiosperms increase the likelihood that their embryos are moved away from the parent plant? Why does moving the embryos away from the parent plant increase the success of the embryo?

11

Animal Organization

Learning Outcomes

11.1 Epithelial Tissue
- Identify slides and models of various types of epithelium.   140–142
- Explain where particular types of epithelium are located in the body, and describe their function.   140–142

11.2 Connective Tissue
- Identify slides and models of various types of connective tissue.   143–146
- Explain where particular connective tissues are located in the body, and describe their function.   143–146

11.3 Muscular Tissue
- Identify slides and models of three types of muscular tissue.   147–148
- Explain where each type of muscular tissue is located in the body, and describe their function.   147–148

11.4 Nervous Tissue
- Identify a slide and model of a neuron.   149
- Explain where nervous tissue is located in the body, and describe its function.   149

11.5 Organ Level of Organization
- Identify a slide of the intestinal wall and the layers in the wall. Describe the function of each tissue.   150–151
- Identify a slide of skin and the two regions of skin. Describe the function for each region of skin.   151

Introduction

Humans, as well as all other organisms, are made up of **cells.** Groups of cells that have the same structural characteristics and perform the same functions are called **tissues.** Figure 11.1 shows the four categories of tissues in the human body. An **organ** is composed of different types of tissues, and various organs form **organ systems.** Humans thus have the following levels of biological organization: cells ⟶ tissues ⟶ organs ⟶ organ systems.

The photomicrographs of tissues in this laboratory were obtained by viewing prepared slides with a light microscope. Preparation required the following sequential steps:

1. **Fixation:** The tissue is immersed in a preservative solution to maintain the tissue's existing structure.
2. **Embedding:** Water is removed with alcohol, and the tissue is impregnated with paraffin wax.
3. **Sectioning:** The tissue is cut into extremely thin slices by an instrument called a microtome. When the section runs the length of the tissue, it is called a longitudinal section (l.s.); when the section runs across the tissue, it is called a cross section (c.s.).
4. **Staining:** The tissue is immersed in dyes that stain different structures. The most common dyes are hematoxylin and eosin stains (H & E). They give a differential blue and red color to the basic and acidic structures within the tissue. Other dyes are available for staining specific structures.

Figure 11.1   The major tissues in the human body.

The many kinds of tissues in the human body are grouped into four types: epithelial tissue, muscular tissue, nervous tissue, and connective tissue.

(simple squamous) © Ed Reschke; (pseudostratified) © Ed Reschke; (simple cuboidal) © Ed Reschke; (simple columnar) © Ed Reschke; (cardiac) © Ed Reschke; (smooth) © Dennis Strete/McGraw-Hill Education; (skeletal) © Ed Reschke; (nervous) © Ed Reschke; (blood) © Al Telser/McGraw-Hill Education; (adipose) © Al Telser/McGraw-Hill Education; (bone) © Dennis Strete/McGraw-Hill Education; (cartilage) © Ed Reschke; (dense) © Ed Reschke/Photolibrary/Getty Images

Epithelial tissue

Simple squamous epithelium

cilia

Pseudostratified ciliated columnar epithelium

microvilli

Simple cuboidal epithelium

Simple columnar epithelium

Muscular tissue

muscle fiber

intercalated disk

Cardiac muscle

muscle fiber

Smooth muscle

muscle fiber

Skeletal muscle

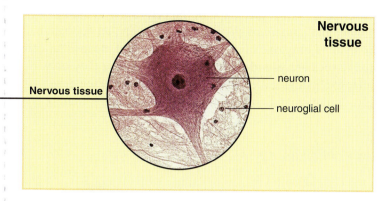

Nervous tissue

Nervous tissue

neuron

neuroglial cell

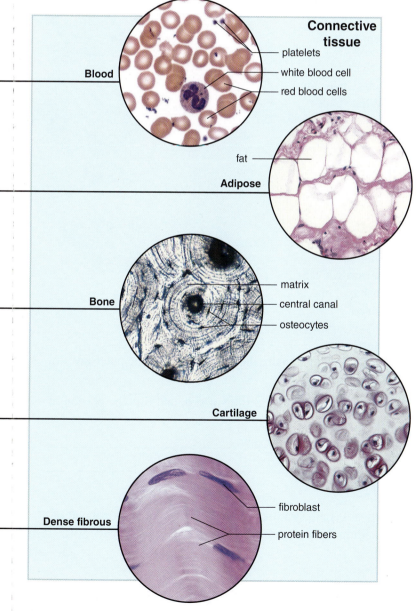

Connective tissue

Blood

platelets

white blood cell

red blood cells

fat

Adipose

matrix

central canal

osteocytes

Bone

Cartilage

fibroblast

protein fibers

Dense fibrous

11.1 Epithelial Tissue

Epithelial tissue (epithelium) forms a continuous layer, or sheet, over the entire body surface and most of the body's inner cavities. Externally, it forms a covering that protects the animal from infection, injury, and drying out. Some epithelial tissues produce and release secretions. Others absorb nutrients.

The name of an epithelial tissue includes two descriptive terms: the shape of the cells and the number of layers. The three possible shapes are *squamous, cuboidal,* and *columnar.* With regard to layers, an epithelial tissue may be simple or stratified. **Simple** means that there is only one layer of cells; **stratified** means that cell layers are placed on top of each other. Some epithelial tissues are **pseudostratified,** meaning that they only appear to be layered. Epithelium may also have cellular extensions called **microvilli** or hairlike extensions called **cilia.** A **basement membrane** consisting of glycoproteins and collagen fibers joins an epithelium to underlying connective tissue.

Observation: Simple and Stratified Squamous Epithelium

Simple Squamous Epithelium

Simple squamous epithelium is a single layer of thin, flat, many-sided cells, each with a central nucleus. It lines internal cavities, the heart, and all the blood vessels. It also lines parts of the urinary, respiratory, and male reproductive tracts.

1. Study a model or diagram of simple squamous epithelium.

   What does squamous mean? _____

2. Examine a prepared slide of squamous epithelium. Under low power, note the close packing of the flat cells. What shapes

   are the cells? _____

   _____

3. Under high power, examine an individual cell, and identify the plasma membrane, cytoplasm, and nucleus.
4. Add a sketch of this tissue to Table 11.1 (page 142).

Simple squamous
• lining of lungs, blood vessels
• protects

250×

basement membrane

© Ed Reschke

Stratified Squamous Epithelium

As would be expected from its name, stratified squamous epithelium consists of many layers of cells. The innermost layer produces cells that are first cuboidal or columnar in shape, but as the cells push toward the surface, they become flattened.

The outer region of the skin, called the epidermis, is stratified squamous epithelium. As the cells move toward the surface, they flatten, begin to accumulate a protein called **keratin,** and eventually die. Keratin makes the outer layer of epidermis tough, protective, and able to repel water.

1. Either now or when you are studying skin in section 11.2, examine a slide of skin and find the portion of the slide that is stratified squamous epithelium.

2. Approximately how many layers of cells make up this portion of skin? _____

3. Which layers of cells best represent squamous epithelium? _____

4. Add a sketch of this tissue to Table 11.1 (page 142).

The linings of the mouth, throat, anal canal, and vagina are stratified epithelium. The outermost layer of cells surrounding the cavity is simple squamous epithelium. In these organs, this layer of cells remains soft, moist, and alive.

Simple cuboidal epithelium is a single layer of cube-shaped cells, each with a central nucleus. It is found in tubules of the kidney and in the ducts of many glands, where it has a protective function. It also occurs in the secretory portions of some glands—that is, where the tissue produces and releases secretions.

1. Study a model or diagram of simple cuboidal epithelium.
2. Examine a prepared slide of simple cuboidal epithelium. Move the slide until you locate cube-shaped cells that line a lumen (cavity). Are these cells ciliated? _____
3. Add a sketch of this tissue to Table 11.1 (page 142).

Simple cuboidal
- lining of kidney tubules, various glands
- absorbs molecules

250×

basement membrane

© Ed Reschke

Observation: Simple Columnar Epithelium

Simple columnar epithelium is a single layer of tall, cylindrical cells, each with a nucleus near the base. This tissue, which lines the digestive tract from the stomach to the anus, protects, secretes, and allows absorption of nutrients.

1. Study a model or diagram of simple columnar epithelium.
2. Examine a prepared slide of simple columnar epithelium. Find tall and narrow cells that line a lumen. Under high power, focus on an individual cell. Identify the plasma membrane, the cytoplasm, and the nucleus. Epithelial tissues are attached to underlying tissues by a basement membrane composed of extracellular material containing protein fibers.
3. The tissue you are observing contains mucus-secreting cells. Search among the columnar cells until you find a **goblet cell,** so named because of its goblet-shaped, clear interior. This region contains mucus, which may be stained a light blue. In the living animal, the mucus is discharged into the gut cavity and protects the lining from digestive enzymes.
4. Add a sketch of this tissue to Table 11.1 (page 142).

Simple columnar
- lining of small intestine, oviducts
- absorbs nutrients

250×

goblet cell secretes mucus

basement membrane

© Ed Reschke

Pseudostratified ciliated columnar epithelium appears to be layered, while actually all cells touch the basement membrane. Many cilia are located on the free end of each cell. In the human trachea, the cilia wave back and forth, moving mucus and debris up toward the throat so that they cannot enter the lungs. Smoking destroys these cilia, but they will grow back if smoking is discontinued.

1. Study a model or diagram of pseudostratified ciliated columnar epithelium.
2. Examine a prepared slide of pseudostratified ciliated columnar epithelium. Concentrate on the part of the slide that resembles the model. Identify the cilia.
3. Add a sketch of this tissue to Table 11.1.

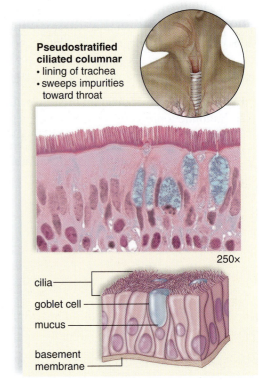

Pseudostratified ciliated columnar
- lining of trachea
- sweeps impurities toward throat

250×

cilia
goblet cell
mucus
basement membrane

© Ed Reschke

Summary of Epithelial Tissue

Add a sketch of each type of epithelial tissue in the third column of Table 11.1. Recognizing that structure suits function, state a universal function for epithelial tissue. _____

Table 11.1   Epithelial Tissue

Sketch	Structure	Sketch	Location
Simple squamous	Tightly packed thin flat cells		Walls of capillaries, lining of blood vessels, air sacs of lungs, lining of internal cavities
Stratified squamous	Innermost layers are cuboidal or columnar; outermost layers are flattened dead cells.		Skin, linings of mouth, throat, anal canal, vagina
Simple cuboidal	Tightly packed cuboidal cells that often have microvilli at one end		Surface of ovaries, linings of ducts and glands, lining of kidney tubules
Simple columnar	Columnlike—tall, cylindrical nucleus at base; may contain goblet cells.		Lining of uterus, tubes of digestive tract
Pseudostratified ciliated columnar	Tightly packed columnar cells appear to be layered but are not. Cells are ciliated and may contain goblet cells.		Linings of respiratory passages

11.2 Connective Tissue

Connective tissue joins different parts of the body together. There are four general classes of connective tissue: connective tissue proper, bone, cartilage, and blood. All types of connective tissue consist of cells surrounded by a matrix that usually contains fibers. Elastic fibers are composed of a protein called elastin. Collagenous fibers contain the protein collagen.

Observation: Connective Tissue

There are several different types of connective tissue. The accompanying illustrations will help you understand the name of each tissue, where it occurs, and its functions. We will study loose fibrous connective tissue, dense fibrous connective tissue, adipose tissue, bone, cartilage, and blood. **Loose fibrous connective tissue,** so named because it has space between components, supports epithelium and also many internal organs, such as muscles, blood vessels, and nerves. Its loose construction allows organs to freely move. **Dense fibrous connective tissue** contains many collagenous fibers packed closely together, as in tendons, which connect muscles to bones, and in ligaments, which connect bones to other bones at joints.

1. Examine a slide of loose fibrous connective tissue, and compare it to the figure below (*left*). What is the function of loose fibrous connective tissue? _____

2. Examine a slide of dense fibrous connective tissue, and compare it to the figure below (*right*). What two kinds of structures in the body contain dense fibrous connective tissue? _____
_____

3. Add sketches of these tissues to Table 11.2 (page 146).

Loose fibrous connective tissue
• has space between components.
• occurs beneath skin and most epithelial layers.
• functions in support and binds organs.

fibroblast

elastic fiber   collagen fiber   250×

© Ed Reschke

Dense fibrous connective tissue
• has collagenous fibers closely packed.
• is in dermis of skin, tendons, ligaments.
• functions in support.

collagen fibers   nuclei of fibroblasts   400×

© Dennis Strete/McGraw-Hill Education

Observation: Adipose Tissue

In **adipose tissue,** the cells have a large, central, fat-filled vacuole that causes the nucleus and cytoplasm to be at the perimeter of the cell. Adipose tissue occurs beneath the skin, where it insulates the body, and around internal organs, such as the kidneys and heart. It cushions and helps protect these organs.

1. Examine a prepared slide of adipose tissue. Why is the nucleus pushed to one side? _____
   _____

2. State a location for adipose tissue in the body. _____
   _____
   _____

   What are two functions of adipose tissue at this location? _____
   _____
   _____

3. Add a sketch of this tissue to Table 11.2 (page 146).

Adipose tissue
• cells are filled with fat.
• occurs beneath skin, around heart and other organs.
• functions in insulation, stores fat.

250×

nucleus

© Al Telser/McGraw-Hill Education

Observation: Compact Bone

Compact bone is found in the bones that make up the skeleton. It consists of **osteons** (Haversian system) with a **central canal,** and concentric rings of spaces called **lacunae,** connected by tiny crevices called **canaliculi.** The central canal contains a nerve and blood vessels, which service bone. The lacunae contain bone cells called **osteocytes,** whose processes extend into the canaliculi. Separating the lacunae is a matrix that is hard because it contains minerals, notably calcium salts. The matrix also contains collagenous fibers.

1. Study a model or diagram of compact bone. Then look at a prepared slide and identify the central canal, lacunae, and canaliculi.

2. What is the function of the central canal and canaliculi? _____
   _____
   _____

3. Add a sketch of this tissue to Table 11.2 (page 146).

Compact bone
• has cells in concentric rings.
• occurs in bones of skeleton.
• functions in support and protection.

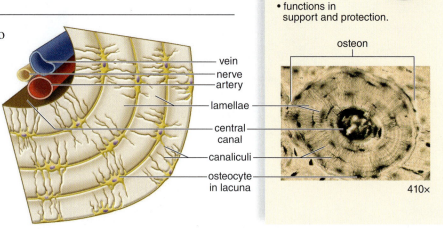

vein
nerve
artery
lamellae
central canal
canaliculi
osteocyte in lacuna

osteon

410×

© Ed Reschke/Getty Images

Observation: Hyaline Cartilage

In **hyaline cartilage,** cells called **chondrocytes** are found in twos or threes in lacunae. The lacunae are separated by a flexible matrix containing weak collagenous fibers.

1. Study the diagram and photomicrograph of hyaline cartilage in the figure at the right. Then study a prepared slide of hyaline cartilage, and identify the matrix, lacunae, and chondrocytes.

2. Compare compact bone and hyaline cartilage. Which of these types of connective tissue is more organized? _____

   Why? _____

3. Which of these two types of connective tissue lends more support to body parts? Why? _____

   _____

4. Add a sketch of this tissue to Table 11.2 (page 146).

Hyaline cartilage
- has cells in lacunae.
- occurs in nose and walls of respiratory passages; at ends of bones, including ribs.
- functions in support and protection.

chondrocyte within lacunae — matrix — 250×

© Ed Reschke

Observation: Blood

Blood is a connective tissue in which the matrix is an intercellular fluid called **plasma. Red blood cells** (erythrocytes) have a biconcave appearance and lack a nucleus. These cells carry oxygen combined with the respiratory pigment hemoglobin. **White blood cells** (leukocytes) have a nucleus and are typically larger than the more numerous red blood cells. These cells fight infection.

1. Study a prepared slide of human blood. With the help of Figure 11.2, identify the red blood cells and the white blood cells, which appear faint because of the stain.
2. Try to identify a neutrophil, which has a multilobed nucleus, and a lymphocyte, which is the smallest of the white blood cells, with a spherical or slightly indented nucleus.
3. Add a sketch of this tissue to Table 11.2 (page 146).

Figure 11.2 Blood cells.
Red blood cells are more numerous than white blood cells. White blood cells can be separated into five distinct types. If you have blood work done that includes a complete blood count (CBC), the doctor is getting a count of each of these types of white blood cells. (a-e: Magnification 1,050×)

(a) © Ed Reschke; (b) © Ed Reschke; (c) © Ed Reschke; (d) © Ed Reschke; (e) © Ed Reschke/Getty Images

— red blood cell
— white blood cell
— plasma

a. Neutrophil

b. Lymphocyte

c. Eosinophil

d. Basophil

e. Monocyte

Summary of Connective Tissue

1. Add a sketch of each type of connective tissue in the third column of Table 11.2. Recognizing that structure suits function, state a universal function for connective tissue. _____
_____

Table 11.2	Connective Tissue		
Type	**Structure**	**Sketch**	**Location**
Loose fibrous connective tissue	Fibers are widely separated.		Between the muscles; beneath the skin; beneath most epithelial layers
Dense fibrous connective tissue	Fibers are closely packed.		Tendons, ligaments
Adipose	Large cells with fat-filled vacuoles; nuclei are pushed to one side.		Beneath the skin; around the kidney and heart; in the breast
Compact bone	Concentric circles of cells separated by a hard matrix		Bones of skeleton
Hyaline cartilage	Cells in lacunae within a flexible matrix		Nose; ends of bones; rings in walls of respiratory passages; between ribs and sternum
Blood	Red and white cells floating in plasma		Blood vessels

2. Working with others in a group, decide how the structure of each connective tissue suits its function.

Loose fibrous connective tissue _____

Dense fibrous connective tissue _____

Adipose tissue _____

Compact bone _____

Hyaline cartilage _____

Blood _____

11.3 Muscular Tissue

Muscular (contractile) tissue is composed of cells called muscle fibers. Muscular tissue has the ability to contract, and contraction usually results in movement. The body contains skeletal, cardiac, and smooth muscle.

Observation: Skeletal Muscle

Skeletal muscle occurs in the muscles attached to the bones of the skeleton. The contraction of skeletal muscle is said to be **voluntary** because it is under conscious control. Skeletal muscle is striated; it contains light and dark bands. The striations are caused by the arrangement of contractile filaments (actin and myosin filaments) in muscle cells often called fibers. Each fiber contains many nuclei, all peripherally located.

1. Study a model or diagram of skeletal muscle, and note that striations are present. You should see several muscle fibers, each marked with striations.
2. Examine a prepared slide of skeletal muscle. The striations may be difficult to make out, but bringing the slide in and out of focus may also help.
3. Each skeletal muscle cell is not only multinucleated, but also cylindrical and elongated. Counting partial cells, how many muscle cells are in the adjoining micrograph?_____

_____

Skeletal muscle
- has striated cells with multiple nuclei.
- occurs in muscles attached to skeleton.
- functions in voluntary movement of body.

striation   nucleus   250x

© Ed Reschke

Observation: Cardiac Muscle

Cardiac muscle is found only in the heart. It is called **involuntary** because its contraction does not require conscious effort. Cardiac muscle is striated in the same way as skeletal muscle. However, the fibers are branched and bound together at **intercalated disks,** where their folded plasma membranes touch. This arrangement aids communication between fibers.

1. Study a model or diagram of cardiac muscle, and note that striations are present.
2. Examine a prepared slide of cardiac muscle. Find an intercalated disk. What is the function of cardiac muscle? _____
_____
_____

3. Aside from being striated, cardiac muscle cells are rectangular and branched. What is the benefit of this arrangement? _____
_____

Cardiac muscle
- has branching, striated cells, each with a single nucleus.
- occurs in the wall of the heart.
- functions in the pumping of blood.
- is involuntary.

intercalated disk   nucleus   250x

© Ed Reschke

Smooth muscle is sometimes called **visceral muscle** because it makes up the walls of the internal organs, such as the intestines and the blood vessels. Smooth muscle is involuntary because its contraction does not require conscious effort.

1. Study a model or diagram of smooth muscle, and note the shape of the cells and the centrally placed nucleus. Smooth muscle has spindle-shaped cells. What does *spindle-shaped* mean? _____

   _____

   _____

2. Examine a prepared slide of smooth muscle. Distinguishing the boundaries between the different cells may require you to take the slide in and out of focus.

Summary of Muscular Tissue

1. Complete Table 11.3 to summarize your study of muscular tissue.
2. How does it benefit an animal that skeletal muscle is voluntary while cardiac and smooth muscle are involuntary?_____

   _____

   _____

Smooth muscle
- has spindle-shaped cells, each with a single nucleus.
- cells have no striations.
- functions in movement of substances in lumens of body.
- is involuntary.
- is found in blood vessel walls and walls of the digestive tract.

smooth muscle cell   nucleus   400×

© Dennis Strete/McGraw-Hill Education

Table 11.3 Muscular Tissue			
Type	Striations (Yes or No)	Branching (Yes or No)	Conscious Control (Yes or No)
Skeletal			
Cardiac			
Smooth			

11.4 Nervous Tissue

Nervous tissue is found in the brain, spinal cord, and nerves. Nervous tissue receives and integrates incoming stimuli before conducting nerve impulses, which control the glands and muscles of the body. Nervous tissue is composed of two types of cells: **neurons** that transmit messages and **neuroglia** that support and nourish the neurons. Motor neurons, which take messages from the spinal cord to the muscles, are often used to exemplify typical neurons. Motor neurons have several **dendrites,** processes that take signals to a **cell body,** where the nucleus is located, and an **axon** that takes nerve impulses away from the cell body.

Observation: Nervous Tissue

1. Study a model or diagram of a neuron, and then examine a prepared slide. Most likely, you will not be able to see neuroglia because they are much smaller than neurons.

2. Identify a cell body, the nucleus, a dendrite, and the axon in Figure 11.3a and *label the micrograph. Also label the neuroglia surrounding the neuron.*

3. Explain the appearance and function of the parts of a motor neuron:

   a. Dendrites _____

   _____

   b. Cell body _____

   _____

   c. Axon _____

   _____

a. Drawing

Figure 11.3   Motor neuron anatomy.

(b) © Ed Reschke

b. Photomicrograph of a neuron   200×

11.5 Organ Level of Organization

Organs are structures composed of two or more types of tissue that work together to perform particular functions. You may tend to think that a particular organ contains only one type of tissue. For example, muscular tissue is usually associated with muscles and nervous tissue with the brain. However, muscles and the brain also contain other types of tissue—for example, loose connective tissue and blood. Here we will study the compositions of two organs—the intestine and the skin.

Intestine

The **intestine**, a part of the digestive system, processes food and absorbs nutrient molecules.

Observation: Intestinal Wall

Study a slide of a cross section of intestinal wall. With the help of Figure 11.4, identify the following layers:

1. **Mucosa** (mucous membrane layer): This layer, which lines the central lumen (cavity), is made up of columnar epithelium overlying the submucosa. This epithelium is glandular—that is, it secretes mucus from goblet cells and digestive enzymes from the rest of the epithelium. The membrane is arranged in deep folds (fingerlike projections) called **villi,** which increase the small intestine's absorptive surface.
2. **Submucosa** (submucosal layer): This loose fibrous connective tissue layer contains nerve fibers, blood vessels, and lymphatic vessels. The products of digestion are absorbed into these blood and lymphatic vessels.
3. **Muscularis** (smooth muscle layer): Circular muscular tissue and then longitudinal muscular tissue are found in this layer. Rhythmic contraction of these muscles causes **peristalsis,** a wavelike motion that moves food along the intestine.
4. **Serosa** (serous membrane layer): In this layer, a thin sheet of loose fibrous connective tissue underlies a thin, outermost sheet of squamous epithelium. This membrane is part of the **peritoneum,** which lines the entire abdominal cavity.

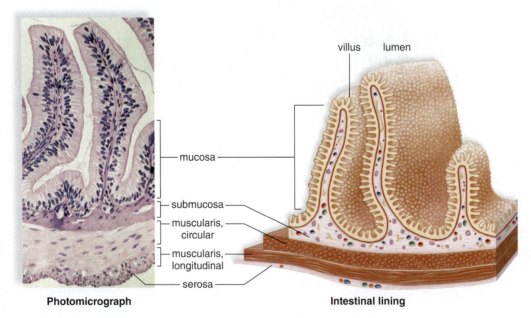

Photomicrograph   Intestinal lining

Figure 11.4   Tissues of the intestinal wall.

© Ed Reschke

In Table 11.4, list the types of tissue found in the layers of the intestinal wall.

Table 11.4 Tissues of the Intestinal Wall				
	Mucosa	Submucosa	Muscularis	Serosa
Tissue(s)				

Skin

The skin covers the entire exterior of the human body. Skin functions include protection, water retention, sensory reception, body temperature regulation, and vitamin D synthesis.

Observation: Skin

Study a model or diagram and also a prepared slide of the skin. Identify these two skin regions and the subcutaneous layer.

1. **Epidermis:** This region is composed of stratified squamous epithelial cells. The outer cells of the epidermis are nonliving and create a waterproof covering that prevents excessive water loss. These cells are always being replaced because an inner layer of the epidermis is composed of living cells that constantly produce new cells.

2. **Dermis:** This region is a connective tissue containing blood vessels, nerves, sense organs, and the expanded portions of oil (sebaceous) and sweat glands and hair follicles.

   List the structures you can identify on your slide: _____

   _____

   _____

3. **Subcutaneous layer:** This is a layer of loose connective tissue and adipose tissue that lies beneath the skin proper and serves to insulate and protect inner body parts.

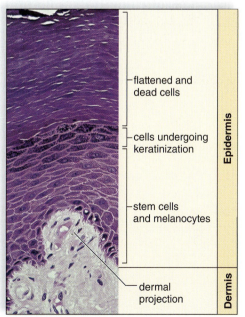

flattened and dead cells

cells undergoing keratinization

stem cells and melanocytes

Epidermis

dermal projection

Dermis

Photomicrograph of skin

© Ed Reschke/Getty Images

_____ **1.** What are cells with similar structural features and a common function called?

_____ **2.** If organ systems are present in an animal, what else must be present?

_____ **3.** What kind of tissue covers the entire body surface and most inner cavities?

_____ **4.** What term is used to describe cell tissues that only appear to be layered?

_____ **5.** What is the name of the protein that makes the outer layer of epidermis tough, protective, and able to repel water?

_____ **6.** What kinds of cells produce mucus that protects the simple columnar epithelium lining in the stomach from digestive enzymes?

_____ **7.** What type of connective tissue has a hard matrix that contains calcium salts?

_____ **8.** What type of connective tissue has a fluid matrix called plasma?

_____ **9.** Which type of muscle is controlled voluntarily?

_____ **10.** What feature aids communication between cardiac muscle fibers?

_____ **11.** Which type of muscle is found in the bladder?

_____ **12.** What is the name of the cells in nervous tissue that transmit messages?

_____ **13.** When two or more tissues work together to perform a specific function, what is the structure formed by the multiple tissues called?

_____ **14.** In what region of the skin are blood vessels, nerves, sense organs, glands, and hair follicles located?

Thought Questions

15. Why is a muscle, like the biceps brachii in your arm, considered an organ and not just a tissue? What kind of muscle tissue is part of the biceps brachii?

16. Why is an increased consumption of dairy products recommended after someone breaks a bone?

17. Based on your understanding of the various functions of epithelium, explain why long-time smokers tend to cough a lot.

12

Basic Mammalian Anatomy I

Introduction

In this laboratory, you will dissect a fetal pig. Alternately your instructor may choose to have you observe a pig that has already been dissected. Both pigs and humans are mammals; therefore, you will be studying mammalian anatomy. The period of pregnancy, or gestation, in pigs is approximately 17 weeks (compared with an average of 40 weeks in humans). The piglets used in class will usually be within 1 to 2 weeks of birth.

The pigs may have a slash in the right neck region, indicating the site of blood drainage. A red latex solution may have been injected into the **arterial system,** and a blue latex solution may have been injected into the **venous system** of the pigs. If so, when a vessel appears red, it is an artery, and when a vessel appears blue, it is a vein.

As a result of this laboratory, you should gain an appreciation of which organs work together. For example, the liver and the pancreas help to digest fat in the small intestine.

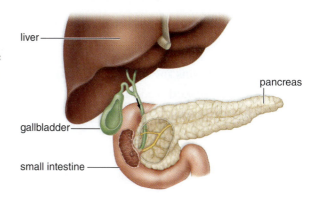

liver

pancreas

gallbladder

small intestine

12.1 External Anatomy

Mammals are characterized by the presence of mammary glands and hair. Mammals also occur in two distinct sexes, males and females, often distinguishable by their external **genitals,** the reproductive organs.

Both pigs and humans are placental mammals, which means that development occurs within the uterus of the mother. An **umbilical cord** stretches externally between the fetal animal and the **placenta,** where carbon dioxide and organic wastes are exchanged for oxygen and organic nutrients.

Pigs and humans are tetrapods—that is, they have four limbs. Pigs walk on all four of their limbs; in fact, they walk on their toes, and their toenails have evolved into hooves. In contrast, humans walk only on the feet of their legs.

Observation: External Anatomy

Body Regions and Limbs

1. Place your animal in a dissecting pan, and observe the following body regions: the rather large head; the short, thick neck; the cylindrical trunk with two pairs of appendages (forelimbs and hindlimbs); and the short tail (Fig. 12.1*a*). The tail is an extension of the vertebral column.

> ⚠ **Latex gloves:** Wear protective latex gloves when handling preserved animal organs. Use protective eyewear and exercise caution when using sharp instruments during this experiment. Wash hands thoroughly upon completion of this experiment.

2. Examine the four limbs, and feel for the joints of the digits, wrist, elbow, shoulder, hip, knee, and ankle.
3. Determine which parts of the forelimb correspond to your arm, elbow, forearm, wrist, and hand.
4. Do the same for the hindlimb, comparing it with your leg.
5. The pig walks on its toenails, which would be like a ballet dancer on "tiptoe." Notice how your heel touches

the ground when you walk. Where is the heel of the pig? _____

Umbilical Cord

1. Locate the umbilical cord arising from the ventral (toward the belly) portion of the abdomen.
2. Note the cut ends of the umbilical blood vessels. If they are not easily seen, cut the umbilical cord near the end and observe this new surface.
3. What is the function of the umbilical cord? _____

_____

Nipples and Hair

1. Locate the small **nipples,** the external openings of the **mammary glands.** The nipples are *not* an indication

of sex, since both males and females possess them. How many nipples does a pig have? _____ When is

it advantageous for a pig to have so many nipples? _____

_____

2. Can you find hair on the pig? _____ Where? _____

_____

> **Directional Terms for Dissecting Fetal Pig**
>
> Anterior: toward the head end   Ventral: toward the belly
>
> Posterior: toward the hind end   Dorsal: toward the back

Figure 12.1 External anatomy of the fetal pig.

a. Body regions and limbs. **b, c.** The sexes can be distinguished by the external genitals.

a. Lateral view, male

b. Ventral view, female

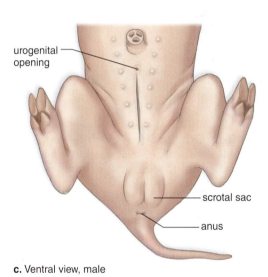

c. Ventral view, male

Anus and External Genitals

1. Locate the **anus** under the tail. Name the organ system that ends in the opening called the anus. _____
2. In females, locate the **urogenital opening,** just anterior to the anus, and a small, fleshy **urogenital papilla** projecting from the urogenital opening (Fig. 12.1*b*).
3. In males, locate the urogenital opening just posterior to the umbilical cord (Fig. 12.1*c*). The duct leading to it runs forward from between the legs in a long, thick tube, the **penis,** which can be felt under the skin. In males, the urinary system and the genital system are always joined.
4. You are responsible for identifying pigs of both sexes. What sex is your pig? _____
   Be sure to look at a pig of the opposite sex that another group of students is dissecting.

12.2 Oral Cavity and Pharynx

The **oral cavity** is the space in the mouth that contains the tongue and the teeth. The **pharynx** is dorsal to the oral cavity and has three openings: The **glottis** is an opening through which air passes on its way to the **trachea** (the windpipe) and lungs. The **esophagus** is a portion of the digestive tract that leads through the neck and thorax to the stomach. The **nasopharynx** leads to the nasal passages.

Observation: Oral Cavity and Pharynx

Oral Cavity

1. Insert a sturdy pair of scissors into one corner of the specimen's mouth, and cut posteriorly (toward the hind end) for approximately 4 cm. Repeat on the opposite side until the mouth is open as in Figure 12.2.
2. Place your thumb on the tongue at the front of the mouth, and gently push downward on the lower jaw. This will tear some of the tissue in the angles of the jaws so that the mouth will remain partly open (Fig. 12.2).
3. Note small, underdeveloped teeth in both the upper and lower jaws. Care should be taken because teeth can be very sharp. Other embryonic, nonerupted teeth may also be found within the gums. The teeth are used to chew food.
4. Examine the tongue, which is partly attached to the lower jaw region but extends posteriorly and is attached to a bony structure at the back of the oral cavity (Fig. 12.2). The tongue manipulates food for swallowing.
5. Locate the hard and soft palates (Fig. 12.2). The **hard palate** is the ridged roof of the mouth that separates the oral cavity from the nasal passages. The **soft palate** is a smooth region posterior to the hard palate. An extension of the soft palate—the **uvula**—hangs down into the throat in humans. (A pig does not have a uvula.)

Figure 12.2 Oral cavity of the fetal pig.
The roof of the oral cavity contains the hard and soft palates, and the tongue lies above the floor of the oral cavity.

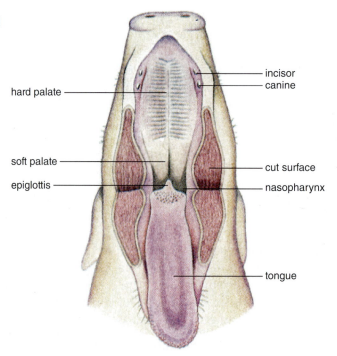

Pharynx

1. Push down on the tongue until you open the jaws far enough to see a slightly pointed flap of tissue pointing dorsally (toward the back) (Fig. 12.2). This flap is the **epiglottis,** which covers the glottis. The **glottis** leads to the trachea (Fig. 12.3*a*).

2. Posterior and dorsal to the glottis, find the opening into the **esophagus,** a tube that takes food to the stomach. Note the proximity of the glottis and the opening to the esophagus. Each time the pig—or a human—swallows, the epiglottis instantly closes to keep food and fluids from going into the lungs via the trachea.

3. Insert a blunt probe into the glottis, and note that it enters the trachea. Remove the probe, insert it into the esophagus, and note the position of the esophagus beneath (dorsal to) the trachea.

4. Make a midline cut in the soft palate from the epiglottis to the hard palate. Then make two lateral cuts at the edge of the hard palate.

5. Posterior to the soft palate, locate the openings to the nasal passages.

6. Explain why it is correct to say that the air and food passages cross in the pharynx. _____

_____

_____

_____

Figure 12.3   Air and food passages in the fetal pig.
The air and food passages cross in the pharynx. **a.** Drawing. **b.** Dissection of specimen.   (b) © Ken Taylor/Wildlife Images

a.

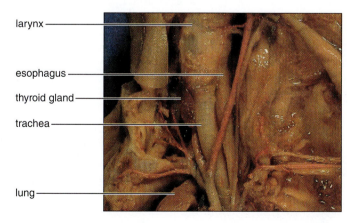

b.

12.3 Thoracic and Abdominal Incisions

First, prepare your pig according to the following directions, and then make thoracic and abdominal incisions so that you will be able to study the internal anatomy of your pig.

Preparation of Pig for Dissection

1. Place the fetal pig on its back in the dissecting pan.
2. Tie a cord around one forelimb, and then bring the cord around underneath the pan to fasten back the other forelimb.
3. Spread the hindlimbs in the same way.
4. With scissors always pointing up (never down), make the following incisions to expose the thoracic and abdominal cavities. The incisions are numbered on Figure 12.4 to correspond with the following steps.

Thoracic Incisions

1. Starting at the diaphragm, a structure that separates the thoracic cavity from the abdominal cavity, cut anteriorly until you reach the hairs in the throat region.
2. Make two lateral cuts, one on each side of the midline incision anterior to the forelimbs, taking extra care not to damage the blood vessels around the heart.
3. Make two lateral cuts, one on each side of the midline just posterior to the forelimbs and anterior to the diaphragm, following the ends of the ribs. Pull back the flaps created by these cuts (do not remove them) to expose the **thoracic cavity.** List the organs you find in the thoracic cavity. _____

_____

Abdominal Incisions

4. With scissors pointing up, cut posteriorly from the diaphragm to the umbilical cord.
5. Make a flap containing the umbilical cord by cutting a semicircle around the cord and by cutting posteriorly to the left and right of the cord.
6. Make two cuts, one on each side of the midline incision posterior to the diaphragm. Examine the diaphragm, attached to the chest wall by radially arranged muscles. The central region of the diaphragm, called the **central tendon,** is a membranous area.
7. Make two more cuts, one on each side of the flap containing the umbilical cord and just anterior to the hindlimbs. Pull back the side flaps created by these cuts to expose the **abdominal cavity.**
8. Lifting the flap with the umbilical cord requires cutting the **umbilical vein.** Before cutting the umbilical vein, tie a thread on each side of where you will cut to mark the vein for future reference.
9. Rinse out your pig as soon as you have opened the abdominal cavity. If you have a problem with excess fluid, obtain a disposable plastic pipet to suction off the liquid.
10. Name the two cavities separated by the diaphragm. _____
11. List the organs located in the abdominal cavity. _____

_____

Figure 12.4 Ventral view of the fetal pig indicating incisions.

These incisions are to be made preparatory to dissecting the internal organs. They are numbered here in the order they should be done.

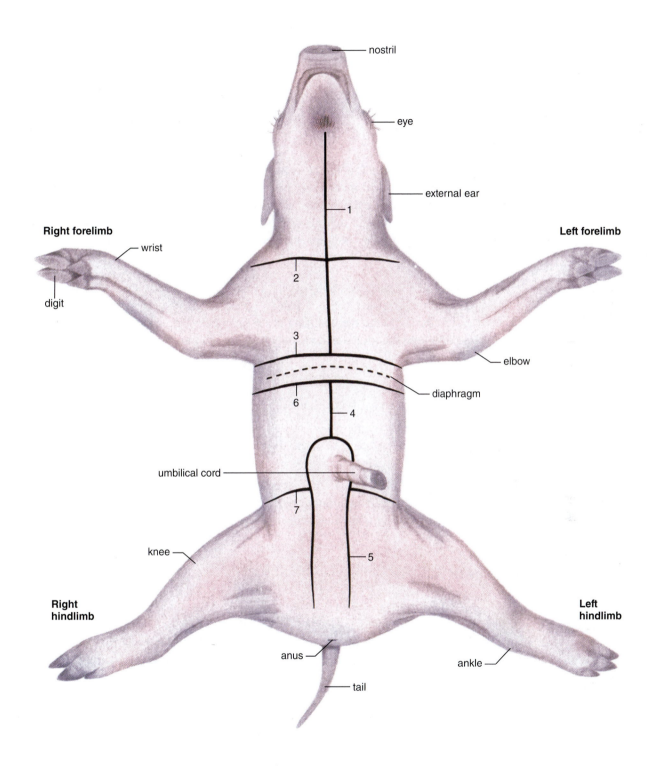

12.4 Neck Region

You will locate several organs in the neck region. Use Figures 12.3*b* and 12.5 as a guide, but *keep all the flaps* in order to close the thoracic and abdominal cavities at the end of the laboratory session.

The **thymus gland** is a part of the lymphatic system. Certain white blood cells called T (for thymus) lymphocytes mature in the thymus gland and help us fight disease. The **larynx,** or voice box, sits atop (anterior to) the **trachea,** or windpipe. The esophagus is a portion of the digestive tract that leads to the stomach. The **thyroid gland** secretes hormones that travel in the blood and act upon other body cells. These hormones (e.g., thyroxine) regulate the rate at which metabolism occurs in cells.

Observation: Neck Region

Thymus Gland

1. Move the skin apart in the neck region just below the hairs mentioned earlier. If necessary, cut the body wall laterally to make flaps. You will most likely be viewing exposed muscles.
2. *Cut through and clear away muscle* to expose the thymus gland, a diffuse gland that lies among the muscles. Later you will notice that the glandular thymus flanks the thyroid and overlies the heart (Fig. 12.5). The thymus is particularly large in fetal pigs, since their immune systems are still developing.

Larynx, Trachea, and Esophagus

1. Probe down into the deeper layers of the neck. Medially (toward the center), beneath several strips of muscle, find the hard-walled larynx and the trachea, which are parts of the respiratory passage to be examined later in Laboratory 15. Dorsal to the trachea, find the esophagus.
2. Open the mouth and insert a probe into the glottis and esophagus from the pharynx to better understand the orientation of these two organs.

Thyroid Gland

Locate the thyroid gland just posterior to the larynx, lying ventral to (on top of) the trachea.

12.5 Thoracic Cavity

As previously mentioned, the body cavity of mammals, including humans, is divided by the diaphragm into the thoracic cavity and the abdominal cavity. The heart and lungs are in the thoracic cavity (Figs. 12.5 and 12.6). The **heart** is a pump for the cardiovascular system, and the **lungs** are organs of the respiratory system where gas exchange occurs.

Observation: Thoracic Cavity

Heart and Lungs

1. If you have not yet done so, fold back the chest wall flaps. To do this, you will need to tear the thin membranes that divide the thoracic cavity into three compartments: the **left pleural cavity** containing the left lung, the **right pleural cavity** containing the right lung, and the **pericardial cavity** containing the heart.
2. Examine the lungs. Locate the four lobes of the right lung and the three lobes of the left lung. The trachea, dorsal to the heart, divides into the **bronchi,** which enter the lungs. Later (Laboratory 15), when the heart is removed, you will be able to see the trachea and bronchi.
3. Sequence the organs of the respiratory tract to trace the path of air from the nasal passages to the lungs.

_____

_____

Figure 12.5 Internal anatomy of the fetal pig.

The major organs are featured in this drawing. In the fetal pig, a red color tells you a vessel is an artery, and a blue color tells you it is a vein. (It does not tell you whether this vessel carries O_2-rich or O_2-poor blood.) Contrary to this drawing, *keep all the flaps on your pig* so you can close the thoracic and abdominal cavities at the end of the laboratory session.

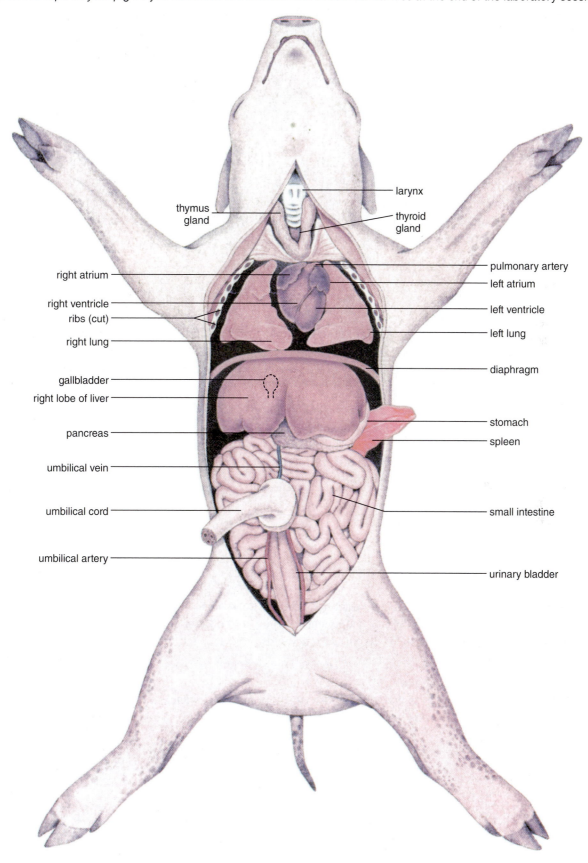

12.6 Abdominal Cavity

The abdominal wall and organs are lined by a membrane called **peritoneum,** consisting of epithelium supported by connective tissue. Double-layered sheets of peritoneum, called **mesenteries,** project from the body wall and support the organs.

Observation: Abdominal Cavity

If your pig is partially filled with dark, brownish material, take your animal to the sink and rinse it out. This material is clotted blood. Consult your instructor before removing any red or blue latex masses, since they may enclose organs you will need to study.

Liver

The **liver,** the largest organ in the abdomen (Fig. 12.6), performs numerous vital functions, including (1) disposing of worn-out red blood cells, (2) producing bile, (3) storing glycogen, (4) maintaining the blood glucose level, and (5) producing blood proteins.

1. Locate the liver, a large, brown organ. Its anterior surface is smoothly convex and fits snugly into the concavity of the diaphragm.

2. Name several functions of the liver. _____

_____

Stomach and Spleen

The organs of the digestive tract include the stomach, small intestine, and large intestine. The **stomach** (see Fig. 12.5) stores food and has numerous gastric glands. These glands secrete a juice that digests protein. The **spleen** (see Fig. 12.5) is a lymphoid organ in the lymphatic system that contains both white and red blood cells. It purifies blood and disposes of worn-out red blood cells.

1. Push aside and identify the stomach, a large sac dorsal to the liver on the left side.
2. Locate the point near the midline of the body where the **esophagus** penetrates the diaphragm and joins the stomach.
3. Find the spleen, a long, flat, reddish organ attached to the stomach by mesentery.
4. The stomach is a part of what system? _____

   What is its function? _____

5. The spleen is a part of what system? _____

   What is its function? _____

Small Intestine

The **small intestine** is the part of the digestive tract that receives secretions from the pancreas and gallbladder. Besides being an area for the digestion of all components of food, carbohydrate, protein, and fat, the small intestine absorbs the products of digestion: glucose, amino acids, glycerol, and fatty acids.

1. Look posteriorly where the stomach makes a curve to the right and narrows to join the anterior end of the small intestine called the **duodenum.**
2. From the duodenum, the small intestine runs posteriorly for a short distance and is then thrown into an irregular mass of bends and coils held together by a common mesentery.
3. The small intestine is a part of what system? _____

   What is its function? _____

Figure 12.6 Internal anatomy of the fetal pig.

Most of the major organs are shown in this photograph. The stomach has been removed. The spleen, gallbladder, and pancreas are not visible. *Do not* remove any organs or flaps from your pig. © Ken Taylor/Wildlife Images

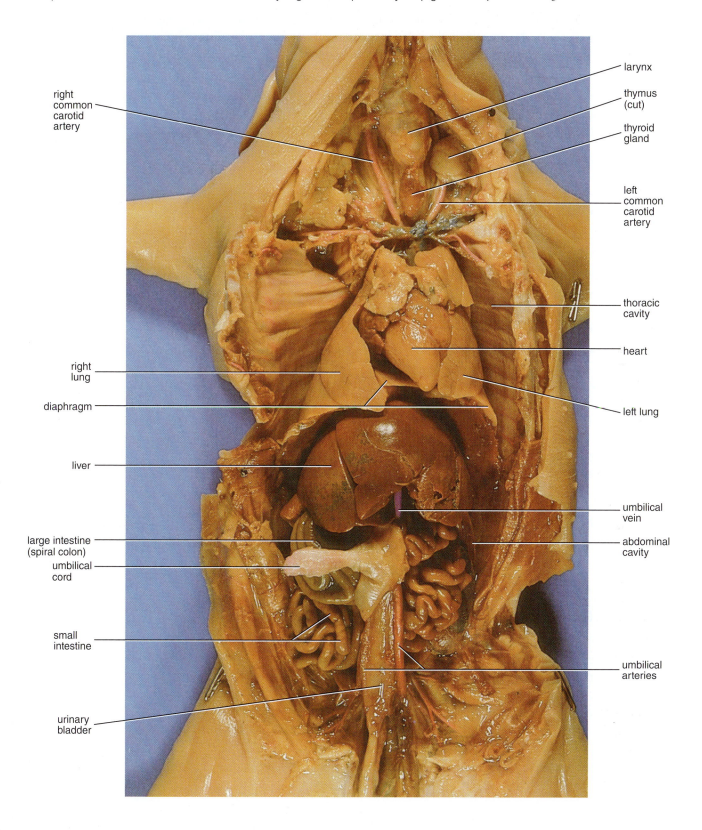

Gallbladder and Pancreas

The **gallbladder** stores and releases bile, which aids in the digestion of fat. The **pancreas** (see Fig. 12.5) is both an exocrine and an endocrine gland. As an exocrine gland, it produces and secretes pancreatic juice, which digests all the components of food in the small intestine. Both bile and pancreatic juice enter the duodenum by way of ducts. As an endocrine gland, the pancreas secretes the hormones insulin and glucagon into the bloodstream. Insulin and glucagon regulate blood glucose levels.

1. Locate the **bile duct,** which runs in the mesentery stretching between the liver and the duodenum. Find the gallbladder, embedded in the liver on the underside of the right lobe. It is a small, greenish sac.
2. Lift the stomach and locate the pancreas, the light-colored, diffuse gland lying in the mesentery between the stomach and the small intestine. The pancreas has a duct that empties into the duodenum of the small intestine.
3. What is the function of the gallbladder? _____

_____

4. What is the function of the pancreas? _____

_____

Large Intestine

The **large intestine** is the part of the digestive tract that absorbs water and prepares feces for defecation at the anus. The first part of the large intestine, called the **cecum,** has a projection called the vermiform (meaning wormlike) appendix.

1. Locate the distal (far) end of the small intestine, which joins the large intestine posteriorly, in the left side of the abdominal cavity (right side in humans). At this junction, note the cecum, a blind pouch.
2. Compare the large intestine of your pig to Figure 12.7. The organ does not have the same appearance in humans.
3. Follow the main portion of the large intestine, known as the **colon,** as it runs from the point of juncture with the small intestine into a tight coil (spiral colon), then out of the coil anteriorly, then posteriorly again along the midline of the dorsal wall of the abdominal cavity. In the pelvic region, the **rectum** is the last portion of the large intestine. The rectum leads to the **anus.**
4. The large intestine is a part of what system? _____
5. What is the function of the large intestine? _____
6. Sequence the organs of the digestive system to trace the path of food from the mouth to the anus. _____

_____

Storage of Pigs

1. Before leaving the laboratory, place your pig in the plastic bag provided.
2. Expel excess air from the bag, and tie it shut.
3. Write your *name* and *section* on the tag provided, and attach it to the bag. Your instructor will indicate where the bags are to be stored until the next laboratory period.
4. Clean the dissecting tray and tools, and return them to their proper location.
5. Wipe off your goggles.
6. Wash your hands.

12.7 Human Anatomy

Humans and pigs are both mammals, and their organs are similar. A human torso model shows the exact location of the organs in humans (Fig. 12.7). You should learn to associate each human organ with its particular system. Six systems are color-coded in Figure 12.7.

Figure 12.7   Human internal organs.

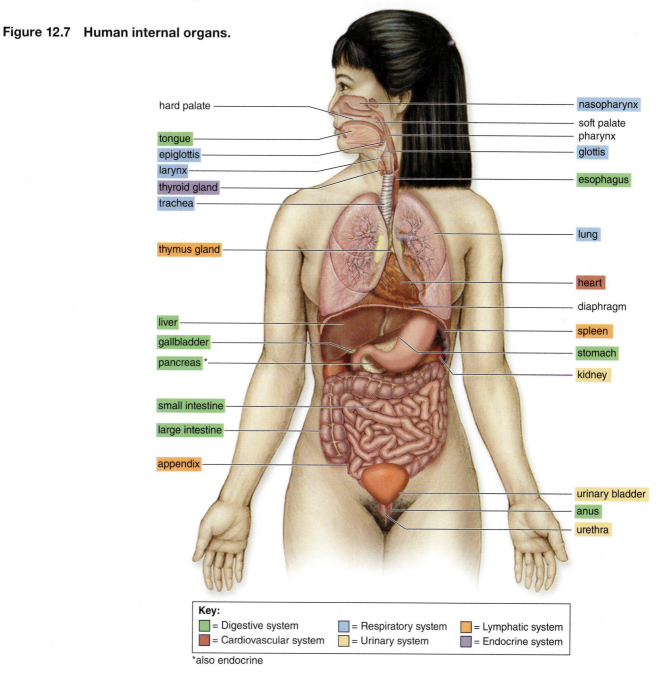

Key:
- = Digestive system
- = Cardiovascular system
- = Respiratory system
- = Urinary system
- = Lymphatic system
- = Endocrine system

*also endocrine

Observation: Human Torso

1. Examine a human torso model, and using Figure 12.7 as a guide, locate the same organs just dissected in the fetal pig.
2. In your studies so far, have you seen any major differences between pig internal anatomy and human internal anatomy? _____

_____

Laboratory Review 12

_____ 1. What directional term refers to the head end of the fetal pig?

_____ 2. If the urogenital opening is just anterior to the anus, is the fetal pig male or female?

_____ 3. What space is dorsal to the oral cavity and has three openings?

_____ 4. What is the name of the flap that closes during swallowing to prevent food and fluids from entering the trachea?

_____ 5. What is the structure that separates the thoracic cavity from the abdominal cavity?

_____ 6. What glandular organ is especially large in fetal pigs since their immune systems are still developing?

_____ 7. Is the trachea ventral or dorsal to the esophagus?

_____ 8. In what thoracic cavities are the lungs located?

_____ 9. What is the largest organ in the abdominal cavity?

_____ 10. What organ do secretions from the gallbladder and pancreas enter?

_____ 11. What is the name of the tissue that holds the small intestine together?

_____ 12. What is the name of the first part of the large intestine?

_____ 13. What organ produces digestive enzymes and hormones that regulate blood glucose?

_____ 14. To what organ system do the kidneys belong?

Thought Questions

15. Why would someone born without a thymus gland be likely to suffer from a greater number of infections?

16. Treatment for a person's laryngeal cancer might require surgical removal of the epiglottis. What would postsurgical therapy for this person involve?

17. Based on the location of the spleen, explain why it is damaged so often in car accidents, especially when the driver's side is struck.

13

Chemical Aspects of Digestion

Learning Outcomes

Introduction
- Sequence the organs of the digestive tract from the mouth to the anus.   168
- State the contribution of each organ, if any, to the process of chemical digestion.   168

13.1 Protein Digestion by Pepsin
- Associate the enzyme pepsin with the ability of the stomach to digest protein.   169
- Explain why stomach contents are acidic and how a warm body temperature aids digestion.   169–170

13.2 Fat Digestion by Pancreatic Lipase
- Associate the enzyme lipase with the ability of the small intestine to digest fat.   171
- Explain why the emulsification process assists the action of lipase. 171
- Explain why a change in pH indicates that fat digestion has occurred.   171–172
- Explain the relationship between time and enzyme activity.   172

13.3 Starch Digestion by Pancreatic Amylase
- Associate the enzyme pancreatic amylase with the ability of the small intestine to digest starch.   173–174

13.4 Requirements for Digestion
- Assuming a specific enzyme, list four factors that can affect the activity of all enzymes.   175
- Explain why the operative procedure that reduces the size of the stomach causes an individual to lose weight.   175

Introduction

In Laboratory 12, you examined the organs of digestion in the fetal pig. Now we wish to further our knowledge of the digestive process by associating certain digestive enyzmes with particular organs, as shown in Figure 13.1. This laboratory will also give us an opportunity to study the action of enzymes, much as William Beaumont did when he removed food samples through a hole in the stomach wall of his patient, Alexis St. Martin. Every few hours, Beaumont would see how well the food had been digested.

In Laboratory 6 we learned that enzymes are very specific and usually participate in only one type of reaction. The active site of an enzyme has a shape that accommodates its substrate, and if an environmental factor such as a boiling temperature or a wrong pH alters this shape, the enzyme loses its ability to function well, if at all. We will have an opportunity to make these observations with controlled experiments. The box on page 168 reviews what is meant by a controlled experiment.

> **⏱ Planning Ahead** Be advised that protein digestion (page 169) requires 1½ hours and fat digestion (page 172) requires 1 hour. Also a boiling water bath is required for starch digestion (page 173).

Figure 13.1 Organs of the digestive tract (right) and accessory organs (left).

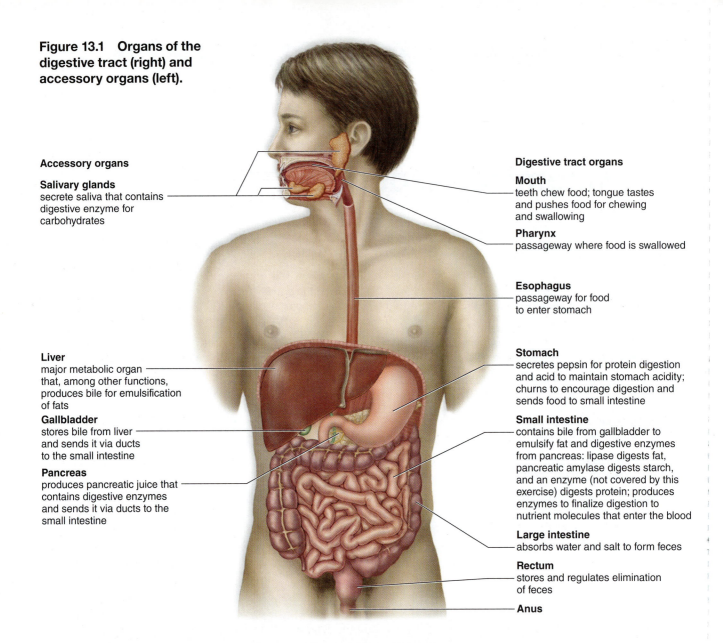

Accessory organs

Salivary glands
secrete saliva that contains digestive enzyme for carbohydrates

Liver
major metabolic organ that, among other functions, produces bile for emulsification of fats

Gallbladder
stores bile from liver and sends it via ducts to the small intestine

Pancreas
produces pancreatic juice that contains digestive enzymes and sends it via ducts to the small intestine

Digestive tract organs

Mouth
teeth chew food; tongue tastes and pushes food for chewing and swallowing

Pharynx
passageway where food is swallowed

Esophagus
passageway for food to enter stomach

Stomach
secretes pepsin for protein digestion and acid to maintain stomach acidity; churns to encourage digestion and sends food to small intestine

Small intestine
contains bile from gallbladder to emulsify fat and digestive enzymes from pancreas: lipase digests fat, pancreatic amylase digests starch, and an enzyme (not covered by this exercise) digests protein; produces enzymes to finalize digestion to nutrient molecules that enter the blood

Large intestine
absorbs water and salt to form feces

Rectum
stores and regulates elimination of feces

Anus

What Is a Control?

The experiments in today's laboratory have both a positive control and a negative control, *which should be saved for comparison purposes until the experiment is complete.* The **positive control** goes through all the steps of the experiment and does contain the substance being tested. Therefore, positive results are expected. The **negative control** goes through all the steps of the experiment, except it does not contain the substance being tested. Therefore, negative results are expected.

For example, if a test tube contains glucose (the substance being tested) and Benedict's reagent (blue) is added, a red color develops upon heating. This test tube is the positive control; it tests positive for glucose. If a test tube does not contain glucose and Benedict's reagent is added, Benedict's is expected to remain blue. This test tube is the negative control; it tests negative for glucose.

What benefit is a positive control? Positive controls give you a standard by which to tell if the substance being tested is present (or acting properly) in an unknown sample. Negative controls ensure that the experiment is giving reliable results; after all, if a negative control should happen to give a positive result, then the entire experiment may be faulty and unreliable.

13.1 Protein Digestion by Pepsin

Certain foods, such as meat and egg whites, are rich in protein. Egg whites contain albumin, which is the protein used in this Experimental Procedure. Protein is digested by **pepsin** in the stomach (Fig. 13.2), a process described by the following reaction:

$$\text{protein} + \text{water} \xrightarrow{\text{pepsin (enzyme)}} \text{peptides}$$

The stomach has a very low pH. Does this indicate that pepsin works effectively in an acidic or a

basic environment? _____ This is the pH that allows the enzyme to maintain its normal shape so that it will combine with the substrate. A warm temperature causes molecules to move about more rapidly and increases the encounters between enzyme and substrate. Therefore you would hypothesize that the yield from this enzymatic reaction will be higher if the

pH is _____ and the temperature is _____ (body temperature 37°C).

Test for Protein Digestion

Biuret reagent is used to test for protein digestion. If digestion has not occurred, biuret reagent turns purple, indicating that protein is present. If digestion has occurred, biuret reagent turns pinkish-purple, indicating that peptides are present.

> ⚠️ **Biuret reagent** is highly corrosive. Exercise care in using this chemical. If any should spill on your skin, wash the area with mild soap and water. Follow your instructor's directions for its disposal.

Experimental Procedure: Protein Digestion

1. Label four clean test tubes (1 to 4). Using the designated graduated pipet, add 2 ml of the albumin solution to all tubes. Albumin is a protein.
2. Add 2 ml of the pepsin solution to tubes 1 to 3, as listed in Table 13.1.
3. Add 2 ml of 0.2% HCl to tubes 1 and 2. HCl simulates the acidic conditions of the stomach.
4. Add 2 ml of water to tube 3 and 4 ml of water to tube 4, as listed in Table 13.1.
5. Swirl to mix the tubes. Tube 2 remains at room temperature, but the other three are incubated for 1½ hours. Record the temperature for each tube in Table 13.1.
6. Remove the tubes from the incubator and place all four tubes in a tube rack. Add 2 ml of biuret reagent to all tubes and observe. Record your results in Table 13.1 as + or − to indicate digestion or no digestion.

gastric pit

esophagus

muscles of the stomach wall

cells that secrete mucus

cells that secrete HCl and enzymes

gastric gland

400×

Figure 13.2   Digestion of protein.
Pepsin, produced by the gastric glands of the stomach, helps digest protein.   © Ed Reschke

Table 13.1 Protein Digestion by Pepsin

Tube	Contents	Temperature	Digestion (+ or –)	Explanation
1	Albumin Pepsin HCl Biuret reagent			
2	Albumin Pepsin HCl Biuret reagent			
3	Albumin Pepsin Water Biuret reagent			
4	Albumin Water Biuret reagent			

Conclusions: Protein Digestion

- Explain your results in Table 13.1 by giving an explanation why digestion did or did not occur. To be complete, consider all the requirements for an enzymatic reaction as listed in Table 13.4. Now show here that tube 1 met all the requirements for digestion:

  Pepsin is the correct _____.

  Albumin is the correct _____.

  37°C is the optimum _____.

  HCL provides the optimum _____.

  1½ hours provides _____ for the reaction to occur.

- Review "What Is a Control?" on page 168. Which tube was the negative control? _____

  Explain. _____

  _____

- If this control tube had given a positive result for protein digestion, what could you conclude about

  this experiment? _____

  _____

13.2 Fat Digestion by Pancreatic Lipase

Lipids include fats (e.g., butterfat) and oils (e.g., sunflower, corn, olive, and canola). Lipids are digested by **pancreatic lipase** in the small intestine (Fig. 13.3).

Figure 13.3 Emulsification and digestion of fat.
Bile from the liver (stored in the gallbladder) enters the small intestine, where lipase in pancreatic juice from the pancreas digests fat.

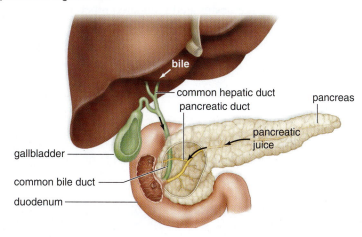

The following two reactions describe fat digestion:

1. $$\text{fat} \xrightarrow{\text{bile (emulsifier)}} \text{fat droplets}$$

2. $$\text{fat droplets} + \text{water} \xrightarrow{\text{lipase (enzyme)}} \text{glycerol} + \text{fatty acids}$$

With regard to the first step, consider that fat is not soluble in water; yet, lipase makes use of water when it digests fat. Therefore, bile is needed to emulsify fat—cause it to break up into fat droplets that disperse in water. The reason for dispersal is that bile contains molecules with two ends. One end (the nonpolar end) is soluble in fat, and the other end (the polar end) is soluble in water. Bile can emulsify fat because of this.

With regard to the second step, would the pH of the solution be lower before or after the enzymatic reaction? (*Hint:* Remember that an acid decreases pH and a base increases pH.) _____

_____

Test for Fat Digestion

In the test for fat digestion, you will be using a pH indicator, which changes color as the solution in the test tube goes from basic conditions to acidic conditions. Phenol red is a pH indicator that is red in basic solutions and yellow in acidic solutions.

Experimental Procedure: Fat Digestion

1. Label three clean test tubes (1 to 3). Using the designated graduated pipet, add 1 ml of vegetable oil to all tubes.
2. Add 2 ml of phenol red solution to each tube. What role does phenol red play? _____
3. Add 2 ml of pancreatic lipase (pancreatin) to tubes 1 and 2 and 2 ml of water to tube 3, as listed in Table 13.2. What role does lipase play? _____
4. Add a pinch of bile salts to tube 1.
5. Record the initial color of all tubes in Table 13.2.
6. Incubate all three tubes at 37°C and check every 20 minutes.
7. Record any color change and how long it took to see this color change in Table 13.2.

Table 13.2   Fat Digestion by Pancreatic Lipase

Tube	Contents	Color		Time Taken	Explanation
		Initial	*Final*		
1	Vegetable oil Phenol red Pancreatin Bile salts				
2	Vegetable oil Phenol red Pancreatin				
3	Vegetable oil Phenol red Water				

Conclusions: Fat Digestion

- Explain your results in Table 13.2 by giving an explanation why digestion did or did not occur.
- What role did bile salts play in this experiment? _____
  _____
- What role did phenol red play in this experiment? _____
- Review "What Is a Control?" on page 168. Which test tube in this experiment could be considered a negative control? _____
  _____

13.3 Starch Digestion by Pancreatic Amylase

Starch is present in bakery products and in potatoes, rice, and corn. Starch is digested by **pancreatic amylase** in the small intestine, a process described by the following reaction:

$$\text{starch} + \text{water} \xrightarrow{\text{amylase (enzyme)}} \text{maltose}$$

1. If digestion *does not* occur, which will be present—starch or maltose? _____

2. If digestion *does* occur, which will be present—starch or maltose? _____

Tests for Starch Digestion

You will be using two tests for starch digestion:

1. If digestion has not taken place, the iodine test for starch will be positive (+) and a blue-black color will be observed. If digestion has occurred, the iodine test for starch will be negative (–) and the iodine will remain yellowish-brown.

2. If digestion has taken place, the Benedict's test for sugar (maltose) will be positive (+) and a color change ranging from green to red will be observed. If digestion has not taken place, the Benedict's test for sugar will be negative (–) and the solution will remain blue.

> ⚠️ **Benedict's reagent** is highly corrosive. Use protective eyewear when performing this experiment. Exercise care in using this chemical. If any should spill on your skin, wash the area with mild soap and water. Follow your instructor's directions for disposal of this chemical.

To test for sugar, add five drops of Benedict's reagent to each test tube. Place the tube in a boiling water bath for a few minutes, and note any color changes (see Table 3.5 on page 32). Boiling the test tube is necessary for the Benedict's reagent to react.

Experimental Procedure: Starch Digestion

1. Label six clean test tubes (1 to 6).
2. Using the designated graduated transfer pipet, add 1 ml of pancreatic amylase solution to tubes 1 to 4 and 1 ml of water to tubes 5 and 6.
3. Test tubes 1 and 2 immediately

   Tube 1 Shake the starch solution and add 1 ml of starch solution. Immediately add five drops of iodine to test for starch. Put this tube in a test tube rack and record your results in Table 13.3.

   Tube 2 Shake the starch solution and add 1 ml of starch solution. Immediately test for sugar with the Benedict's test following the preceding directions. Put this tube in a test tube rack and record your results in Table 13.3.

4. Shake the starch solution and add 1 ml of starch solution to tubes 3 to 6. Allow the tubes to stand for 30 minutes.

   Tubes 3 and 5 After the 30 minutes have passed, test for starch using the iodine test. Place these tubes in the test tube rack and record your results in Table 13.3.

   Tubes 4 and 6 After the 30 minutes have passed, test for sugar with the Benedict's test following the preceding directions. Place these tubes in the test tube rack and record your results in Table 13.3.

5. Examine all your tubes in the test tube rack and decide whether digestion occurred (+) or did not occur (–). Complete Table 13.3.

Table 13.3   Starch Digestion by Amylase

Tube	Contents	Time*	Type of Test	Results	Explanation
1	Pancreatic amylase Starch	0	Iodine	+	
2	Pancreatic amylase Starch				
3	Pancreatic amylase Starch				
4	Pancreatic amylase Starch				
5	Water Starch				
6	Water Starch				

* Enter either 0 for immediately or T for after 30 minutes.

Conclusions: Starch Digestion

- Considering tubes 1 and 2, this experimental procedure showed that _____ must pass for digestion to occur.

- Considering tubes 5 and 6, this experimental procedure showed that an active _____ must be present for digestion to occur.

- Why would you not recommend doing the test for starch and the test for sugar on the same tube? _____

  _____

  _____

- Which test tubes served as a negative control in this experiment? _____

  Explain your answer. _____

  _____

Absorption of Sugars and Other Nutrients

Figure 13.4 shows that the folded lining of the small intestine has many fingerlike projections called villi. The small intestine not only digests food; it also absorbs the products of digestion, such as sugars from carbohydrate digestion, amino acids from protein digestion, and glycerol and fatty acids from fat digestion at the villi.

Figure 13.4   Anatomy of the small intestine.
Nutrients enter the bloodstream across the much-convoluted walls of the small intestine.

13.4   Requirements for Digestion

Explain in Table 13.4 how each of the requirements listed influences effective digestion.

Table 13.4 Requirements for Digestion	
Requirement	**Explanation**
Specific enzyme	
Specific substrate	
Warm temperature	
Specific pH	
Time	
Fat emulsifier	

To lose weight, some obese individuals undergo an operation in which (1) the stomach is reduced to the size of a golf ball, and (2) food bypasses the duodenum (first 2 feet) of the intestine. Answer these questions to explain how this operation would affect the requirements for digestion.

1. How is the amount of substrate reduced? _____

2. How is the amount of digestive enzymes reduced? _____

3. How is time reduced? _____

4. What makes the pH of the small intestine higher than before? _____

5. How is fat emulsification reduced? _____

6. How does surgery to reduce obesity sometimes result in malnutrition? _____

_____ 1. What part of an enzyme is specific for and accommodates its substrate?

_____ 2. What two environmental conditions affect the ability of an enzyme to function properly?

_____ 3. Where are carbohydrates first digested by an enzyme?

_____ 4. What organ produces the lipase and amylase that perform digestion in the small intestine?

_____ 5. What is the name given to a sample that contains the factor being tested and that goes through all the steps of the experiment?

_____ 6. Does pepsin function in an acidic or a basic pH?

_____ 7. What results from the digestion of proteins by pepsin?

_____ 8. What reagent was used to test for the digestion of proteins?

_____ 9. What molecule emulsifies fats in the small intestine?

_____ 10. What organ stores bile?

_____ 11. What component of foods is digested by pancreatic amylase?

_____ 12. What molecule is present after pancreatic amylase digests starch?

_____ 13. What two tests were done to test for starch digestion?

_____ 14. What structures absorb the products of digestion in the small intestine?

Thought Questions

15. Salivary amylase digests starch to maltose in the mouth while chewing takes place. Why is it beneficial to have the enzyme that digests starch, amylase, in two different locations?

16. How would the digestion of fat be impacted by the removal of the gallbladder? Explain.

17. Explain why consuming excess antacids might affect protein digestion. How could you investigate the impact of consuming antacids on protein digestion?

14

Cardiovascular System

Learning Outcomes

14.1 The Heart
- Name the four chambers of the heart and the blood vessels connected to these chambers   178–180
- Name and locate the valves of the heart, and trace the path of blood through the heart.   180–181
- Distinguish the role of the pulmonary circuit from that of the systemic circuit.   182
- Use a diagram to trace the path of blood from the heart to and from various organs.   183

14.2 Vessels of the Pulmonary Circuit
- Identify the pulmonary blood vessels and state whether they carry O_2-rich or O_2-poor blood.   184

14.3 Vessels of the Systemic Circuit
- Identify the coronary arteries and cardiac veins.   185–187
- Identify the arteries and veins in the thoracic cavity.   185–187
- Identify the arteries and veins in the abdominal cavity.   188–189

14.4 Blood Vessel Comparison
- Describe each of the layers of a blood vessel wall.   190–191
- Contrast the wall of an artery with that of a vein.   190–191

Introduction

Blood must circulate to serve the body. The **heart,** located in the thoracic cavity, pumps the blood, which moves away from the heart in **arteries** and **arterioles** and returns to the heart in **venules** and **veins.** When the heart contracts, blood leaves under high pressure. Therefore, arteries have thicker walls than veins, which return blood under low pressure to the heart. Skeletal muscle contractions in the limbs add pressure to venous blood by pushing on the thin walls of veins, which have valves to keep blood moving toward the heart. Capillaries, which connect arterioles to venules, have thin walls that allow exchange of molecules with tissue fluid.

lung

heart   diaphragm

© Professor David F. Cox/McGraw-Hill Education

In today's laboratory, you will first learn the anatomy of the heart and how to trace the path of blood through the heart in two major circuits. In the **pulmonary circuit,** blood moves to and from the lungs and in the **systemic circuit,** the blood moves to and from the other organs of the body. Once you have traced the path of blood in both circuits, major blood vessels in the fetal pig will be dissected. Humans and pigs are both mammals. Therefore, the work done today will give you an appreciation of your own heart and blood vessels and of how the blood moves through these vessels to service your organs.

14.1 The Heart

The heart has right and left sides divided by the **septum.** To tell the right from the left side, position Figure 14.1 so it corresponds to your body. There are four **chambers:** two upper, thin-walled atria and two lower, thick-walled ventricles. The heart has valves that keep the blood flowing in one direction; special muscles and tendons secure the valves to prevent backflow from the ventricles to the atria. The right side of the heart sends blood through the lungs, and the left side sends blood into the body. Therefore, the heart is called a double pump.

Observation: External Anatomy of the Heart

1. In a heart model and/or sheep heart, identify the **right atrium** and its attached blood vessels, the superior and inferior **venae cavae** (Fig. 14.1). The superior vena cava and the inferior vena cava return blood from the head and body, respectively, to the right atrium. Is the blood that enters the right atrium O_2-poor or

   O_2-rich? _____

   Explain. _____

   _____

2. Identify the **right ventricle** and its attached blood vessel, the **pulmonary trunk.** The pulmonary trunk leaves the ventral side of the heart from the top of the right ventricle and then passes diagonally forward before branching into the **right** and **left pulmonary arteries.**

3. Identify the **left atrium** and its attached blood vessels, the left and right **pulmonary veins.** The pulmonary veins return blood from the lungs to the left atrium. Is the blood that enters the left atrium O_2-poor or O_2-

   rich? _____ Explain. _____

   _____

4. Identify the **left ventricle** and its attached blood vessel, the **aorta,** which arises from the anterior end of the left ventricle, just dorsal to the origin of the pulmonary trunk. The aorta soon bends to the animal's left as the aortic arch. The aorta carries blood to the body proper.

5. Identify the **coronary arteries** and the **cardiac veins,** which service the needs of the heart wall. The coronary arteries branch off the aorta as soon as it leaves the heart and appear on the surface of the heart. The cardiac veins, also on the surface of the heart, join and then enter the right atrium through the coronary sinus on the dorsal side of the heart.

a. Photograph of human heart, external view

aortic arch

pulmonary trunk

right atrium

left atrium

coronary blood vessels

right ventricle

left ventricle

apex

left subclavian artery

left common carotid artery

brachiocephalic artery

superior vena cava

aortic arch

aorta

left pulmonary artery

right pulmonary artery

pulmonary trunk

left pulmonary veins

right pulmonary veins

left atrium

right atrium

left cardiac vein

right coronary artery

right ventricle

left ventricle

inferior vena cava

apex

b. Drawing of human heart, external view

Figure 14.1 External view of heart.

Externally, notice the coronary arteries and cardiac veins that serve the heart. (a) © A. & F. Michler/Getty Images

Remove the ventral half of the human heart model (Fig. 14.2) and/or sheep heart. To use a sheep heart that requires dissection, position the heart as it is shown in Figure 14.2*a*. Using a sharp scalpel, cut through the left atrium and the left ventricle to the apex of the heart. Be sure the cut is deep enough to reach the septum. Then position the heart so the apex is uppermost, and cut from the apex of the heart through the right atrium, being sure the cut is deep enough to reach the septum of the heart.

1. Identify the four chambers of the heart in longitudinal section: right atrium, right ventricle, left atrium, and left ventricle.

2. Which ventricle is more muscular? _____

   Why is this appropriate? _____

   _____

3. Find the **right atrioventricular** (tricuspid) valve, located between the right atrium and the right ventricle.
4. Find the **left atrioventricular** (bicuspid or mitral) valve, located between the left atrium and the left ventricle.
5. Find the **pulmonary semilunar** valve, located in the base of the pulmonary trunk.
6. Find the **aortic semilunar** valve, located in the base of the aorta.

   What is the function of the atrioventricular valves? _____

   _____

   What is the function of the semilunar valves? _____

   _____

7. Note the **chordae tendineae** ("heartstrings") that hold the atrioventricular valves in place while the heart contracts. These extend from the papillary muscles. The chordae tendineae prevent the atrioventricular valves from inverting into the atria when the ventricles contract.

Path of Blood Through the Heart

To demonstrate that O_2-poor blood is kept separate from O_2-rich blood, trace the path of blood from the right side of the heart to the aorta by filling in the following blanks. The blood passes through the lungs to go from the right to the left sides of the heart. Which side of the heart pumps O_2-poor blood? _____ Which side pumps O_2-rich blood? _____

Venae Cavae

_____

_____ valve

_____

_____ valve

_____

Lungs

Lungs

_____

_____

_____ valve

_____

_____ valve

Aorta

aorta
left atrium
right atrium
right ventricle
septum
left ventricle
apex

a. Photograph of sheep heart, internal view

left common carotid artery
brachiocephalic artery
superior vena cava
pulmonary trunk
right pulmonary artery
right pulmonary veins
pulmonary semilunar valve
right atrium
atrioventricular (tricuspid) valve
chordae tendineae
papillary muscles
right ventricle
inferior vena cava

left subclavian artery
aorta
left pulmonary artery
left pulmonary veins
left atrium
atrioventricular (bicuspid) valve
septum
left ventricle

b. Drawing of human heart, internal view

Figure 14.2   Internal view of heart.
Internally, the heart has four chambers, and there is a septum that separates the left side from the right side.   © Eric Wise

Path of Blood in the Body

In adult mammals, including humans, the heart is a double pump. The right ventricle pumps blood into the **pulmonary circuit**—to the lungs and back to the heart (Fig. 14.3). While the blood is in the lungs, it gives up carbon dioxide and gains oxygen. The left ventricle pumps blood to the **systemic circuit**—throughout the whole body except to the lungs. Blood in the systemic circuit gives up oxygen and gains carbon dioxide.

Figure 14.3 shows how to trace the path of blood in both the pulmonary and systemic circuits in adult humans. It also helps you to learn the names of some of the major blood vessels.

Pulmonary Circuit

1. Trace the path of blood in the pulmonary circuit from the heart to the lungs and then from the lungs to the heart. Follow the arrows in Figure 14.3, and use the label names provided there.

   right ventricle of heart ⟶ _____ ⟶ lungs ⟶ _____ ⟶

   _____ of heart

Systemic Circuit

2. Trace the path of blood in the systemic circuit from the heart to the kidneys, and then from the kidneys to the heart.

   left ventricle of heart ⟶ _____ ⟶ _____ ⟶ kidneys ⟶

   _____ ⟶ _____ ⟶ _____ of heart

Names of Blood Vessels

3. Use Figure 14.3 to complete Table 14.1.

Table 14.1	Major Blood Vessels in the Systemic Circuit	
Body Part	**Artery**	**Vein**
Heart		
Head		
Arms		
Kidney		
Legs		
Intestines		

What vessel lies between the digestive tract and the liver? _____

A portal system lies between two sets of capillaries.

Figure 14.3   Diagram of the human cardiovascular system.

In the pulmonary circuit, the pulmonary arteries take O_2-poor blood to the lungs (i.e., the capillaries in the lungs), and the pulmonary veins return O_2-rich blood to the heart. In the systemic circuit, the aorta branches into the various arteries that go to all other parts of the body. After blood passes through arterioles, capillaries, and venules, it enters various veins and then the superior and inferior venae cavae, which return it to the heart.

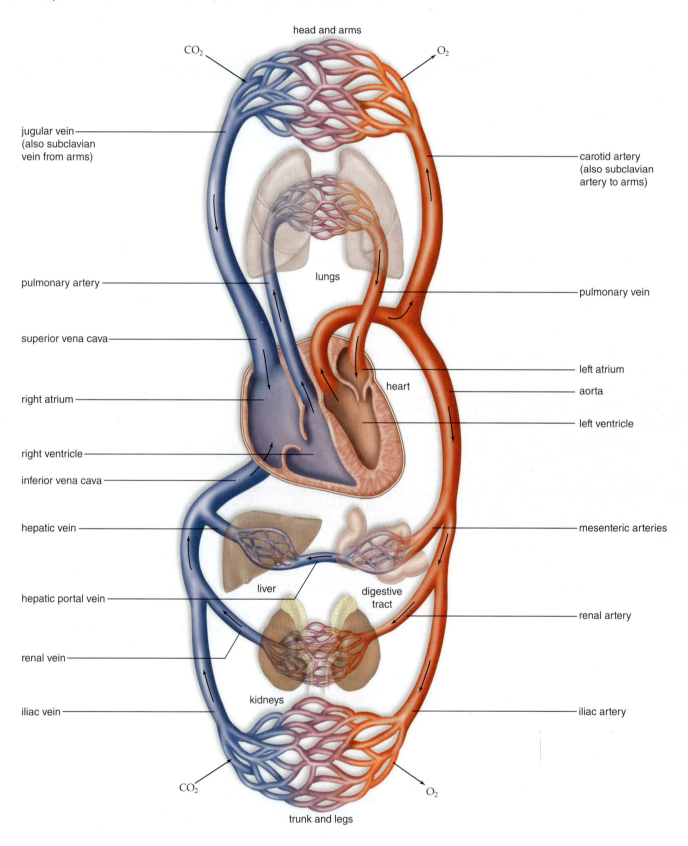

14.2 Vessels of the Pulmonary Circuit

In this section, we will use the fetal pig to examine the pulmonary arteries and veins that occur in mammals, including humans. The fetal pig will have pulmonary arteries and veins, even though they are not functional until after the pig is born. Fetal circulation has two features, numbered 1 and 2 in Table 14.2, that keep blood from entering the pulmonary arteries.

Table 14.2 Unique Features of Fetal Circulation

1. Oval opening (foramen ovale): an opening between the atria of the heart that allows blood to bypass the lungs
2. Arterial duct (ductus arteriosus): a short, stout vessel leading directly from the pulmonary trunk to the aorta; provides another way to bypass the lungs
3. Umbilical arteries: take blood from the iliac arteries to the placenta, the organ that nourishes the fetus
4. Umbilical vein: returns blood from the placenta to the liver
5. Venous duct (ductus venosus): a continuation of the umbilical vein that takes blood to the inferior vena cava

Observation: Vessels of the Pulmonary Circuit

In the pulmonary circuit of adult mammals, pulmonary arteries take blood away from the heart to the lungs, and pulmonary veins take blood from the lungs to the heart. Remember that in your pig, *all arteries* have been injected with red latex, and *all veins* have been injected with blue latex.

> ⚠ **Wear protective latex gloves and eyewear** when handling preserved animal organs. Exercise caution when using sharp instruments during this laboratory. Wash hands thoroughly upon completion of this laboratory.

Pulmonary Trunk and Pulmonary Arteries

1. Locate the pulmonary trunk, which arises from the ventral side of the heart. It may appear white because the thick wall prevents the color of the red latex from showing through.
2. If necessary, remove the pericardial sac from the heart to reveal the blood vessels entering the heart. Veins take blood to the heart.
3. Trace the pulmonary trunk, and notice that it seems to connect directly with the aorta, the major artery. This is the **arterial duct,** a connection found only in a fetus (see Table 14.2).
4. In addition to this duct, look closely and you will find the pulmonary arteries, which leave the pulmonary trunk and go to the lungs.

Pulmonary Veins

1. The pulmonary veins are hard to find. To view them, clean away the membrane dorsal to the heart, and carefully note the vessels (pulmonary veins) that leave the lungs. Trace these to the left atrium of the heart.
2. In the adult, which blood vessels—pulmonary arteries or pulmonary veins—carry O_2-rich blood?

14.3 Vessels of the Systemic Circuit

In adult mammals, the systemic circuit serves all parts of the body except where gas exchange occurs in the lungs. Arteries take O_2-rich blood from the heart to the organs, and veins take O_2-poor blood from the organs to the heart. The aorta is the major artery, and the venae cavae are the major veins. It will be possible for you to identify the arteries branching from the aorta and the corresponding veins branching from the venae cavae. In the fetal pig, the venae cavae are called the *anterior* vena cava and the *posterior* vena cava because the normal

position of the body is horizontal rather than vertical. Use the figures to trace blood vessels, but **do not remove any organs. You will need them in Laboratory 15.**

Observation: Aorta and Venae Cavae

1. Follow the aorta as it extends through the thoracic cavity (Fig. 14.4). To do this, gently move the lungs and heart to the right side of the thoracic cavity. Notice how the thoracic or dorsal aorta is a large, whitish vessel extending through the thoracic cavity and then through the diaphragm to become the abdominal aorta. Also notice the esophagus, a smaller, flattened tube more toward the midline. The esophagus goes through the diaphragm to join with the stomach. Locate the anterior vena cava coming off the top of the heart and the posterior vena cava just to the right of the midline. The posterior vena cava also passes through the diaphragm.

2. Name three structures that pass through the diaphragm. 1. _____

   2. _____ 3. _____

Blood Vessels of the Upper Body

The **coronary arteries** and **cardiac veins** lie on the surface of the heart. The **carotid arteries** and **jugular veins** serve the neck and head regions of the body.

Observation: Blood Vessels of the Upper Body

Coronary Arteries and Cardiac Veins

1. Locate the coronary arteries and cardiac veins, which are easily visible on the heart's surface (see Fig. 14.1).
2. The coronary arteries arise from the aorta just as it leaves the heart, and the cardiac veins go directly into the right atrium.

Carotid Arteries and Jugular Veins

1. Find the aorta as it leaves the left ventricle and curves downward. As it arches, it gives off branches. Its first branch, the **brachiocephalic arterial trunk,** divides almost immediately into the **right subclavian** (to the pig's right shoulder) **artery** and the **right** and **left common carotid arteries** (Figs. 14.4 and 14.5).
2. Find the second branch of the aorta, which is the **left subclavian artery.**
3. Both the right and left subclavian arteries have branches called the right and left brachial arteries, which serve the upper limbs.
4. Find the jugular veins alongside the carotid arteries (Figs. 14.4 and 14.6). Trace the carotid arteries and jugular veins as far as possible toward the head. What part of the body is serviced by the carotid arteries

   and the jugular veins? _____

Subclavian Arteries and Veins

1. Locate the **subclavian arteries** (Fig. 14.4 and 14.5) and **subclavian veins** (Fig. 14.4 and 14.6), which serve the forelimbs and are easily identified.
2. Note that while the left subclavian artery branches from the aorta, the right subclavian artery branches from the brachiocephalic arterial trunk.
3. Note that the jugular veins and the subclavian veins join to form the brachiocephalic venous trunk, which enters the anterior vena cava.

Figure 14.4   Ventral view of fetal pig arteries and veins.

Use this diagram to trace blood vessels, *but do not remove any organs. You will need them in Laboratory 15.*
(a. = artery; v. = vein.)

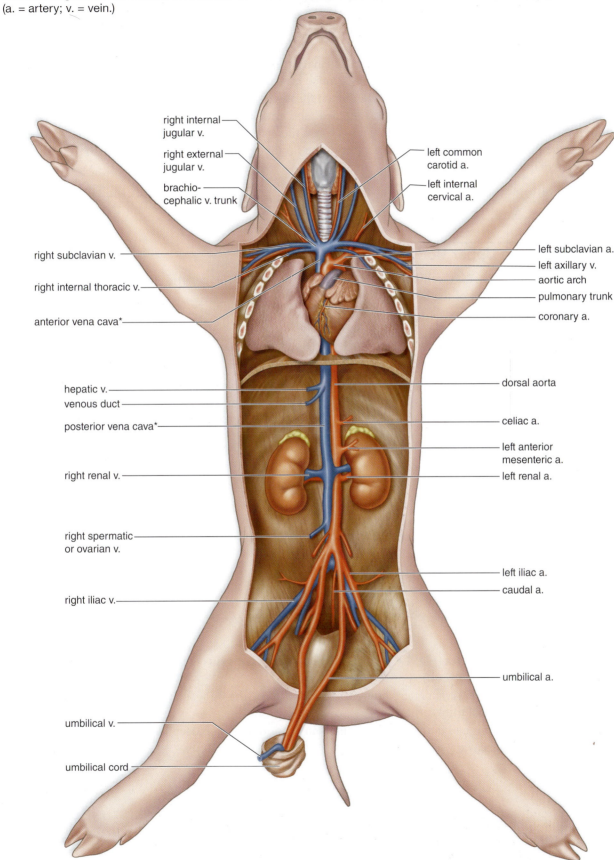

right internal jugular v.

right external jugular v.

brachio-cephalic v. trunk

right subclavian v.

right internal thoracic v.

anterior vena cava*

hepatic v.

venous duct

posterior vena cava*

right renal v.

right spermatic or ovarian v.

right iliac v.

umbilical v.

umbilical cord

left common carotid a.

left internal cervical a.

left subclavian a.

left axillary v.

aortic arch

pulmonary trunk

coronary a.

dorsal aorta

celiac a.

left anterior mesenteric a.

left renal a.

left iliac a.

caudal a.

umbilical a.

* Because the pig walks on all four limbs, the superior vena cava in humans is called the anterior vena cava in pigs and the inferior vena cava in humans is called the posterior vena cava in pigs.

Figure 14.5 Arteries of the upper body of the fetal pig.

The subscapular artery supplies the shoulder muscles. (a. = artery.)

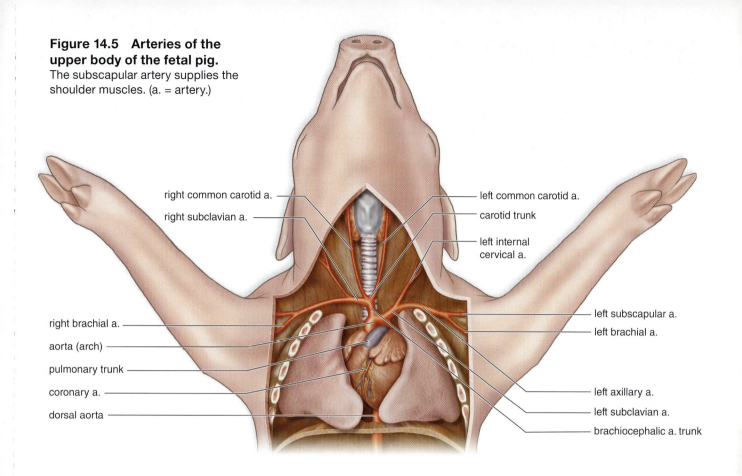

right common carotid a.

right subclavian a.

left common carotid a.

carotid trunk

left internal cervical a.

right brachial a.

aorta (arch)

pulmonary trunk

coronary a.

dorsal aorta

left subscapular a.

left brachial a.

left axillary a.

left subclavian a.

brachiocephalic a. trunk

Figure 14.6 Veins in the thoracic cavity of the fetal pig.

The subclavian veins enter a brachiocephalic vein. (v. = vein.)

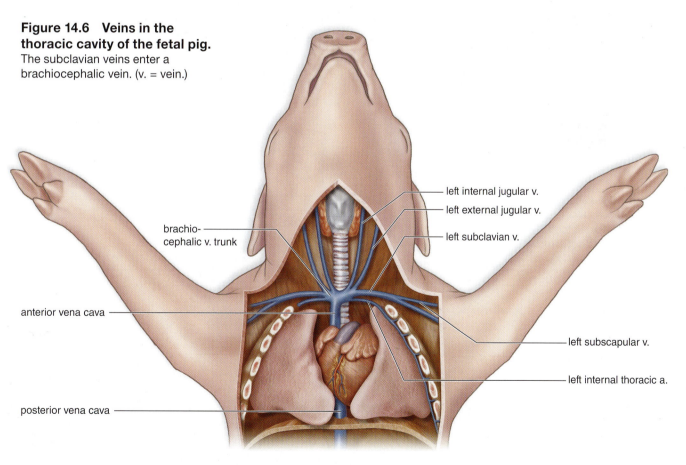

brachio-cephalic v. trunk

left internal jugular v.

left external jugular v.

left subclavian v.

anterior vena cava

left subscapular v.

left internal thoracic a.

posterior vena cava

Vessels of the Abdominal Cavity

Several major blood vessels are in the abdominal cavity. Use Figures 14.7 and 14.8 to trace the blood vessels, but **do not remove any organs. You will need them in Laboratory 15.**

Observation: Vessels of the Abdominal Cavity

Celiac and Mesenteric Arteries

1. Carefully lift up the liver and stomach, and put them aside to your left. Dissect the dorsal mesentery to see the **celiac artery** as it leaves the aorta. Tributaries from this vessel eventually reach the stomach, duodenum, liver, and spleen.
2. Branching from the aorta just posterior to the celiac artery is a long, unpaired trunk, called the **anterior mesenteric artery,** which has tributaries to the pancreas and small intestine (see Fig. 14.7).
3. The celiac and mesenteric arteries take blood to the intestines. Thereafter, the hepatic portal vein takes blood from the intestinal capillaries to capillaries in the liver. A portal system is defined as a circulatory unit that goes from one capillary bed to another without passing through the heart (see Fig. 14.3).

Renal Arteries and Veins

1. Locate the **renal arteries** as they branch from the aorta, and trace these arteries as they go into the kidneys.
2. Locate the renal veins as they leave the kidneys (Fig. 14.8), and trace these veins as they join the posterior vena cava.

Figure 14.7 Arteries in the abdominal cavity of the fetal pig. The celiac and mesenteric artery serve the digestive and associated organs. The external iliac arteries proceed into the hindlimbs from the caudal end of the dorsal aorta. They give rise to the umbilical arteries and continue as much smaller vessels. (aa. = arteries.) Use this diagram to trace blood vessels, *but do not remove any organs. You will need them in Laboratory 15.*

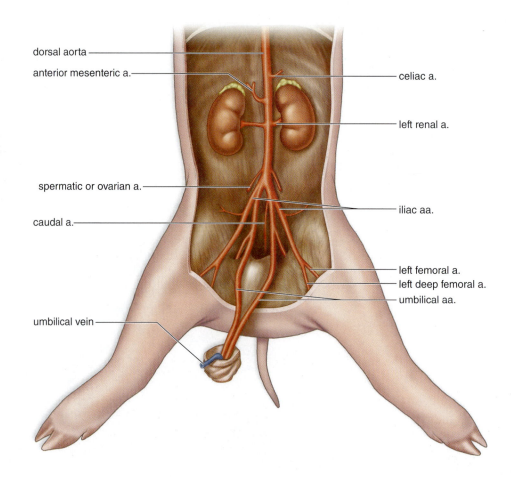

dorsal aorta
anterior mesenteric a.
celiac a.
left renal a.
spermatic or ovarian a.
iliac aa.
caudal a.
left femoral a.
left deep femoral a.
umbilical aa.
umbilical vein

Iliac Arteries and Veins

1. At its posterior end, the aorta branches into the paired **iliac arteries.** Locate the iliac arteries at the posterior end of the aorta, and trace these arteries into the hindlimbs (see Fig. 14.7).
2. Find the **iliac veins** alongside the iliac arteries, and trace these veins as they join the posterior vena cava (Fig. 14.8).

Umbilical Arteries and Veins

1. Locate the **umbilical arteries** on either side of the bladder. Trace these arteries as they branch from the iliac arteries and as they pass into the **umbilical cord** (see Fig. 14.7).
2. When you were exposing the abdominal cavity, you cut the **umbilical vein.** Trace the umbilical vein from the umbilical cord to the liver. The umbilical vein is joined to the posterior vena cava by the **venous duct** (ductus venosus), which passes through the posterior portion of the liver.

Posterior Vena Cava

1. Locate the **posterior vena cava,** which is easily seen as a large, blue vessel just ventral to the dorsal aorta (Fig. 14.8).
2. Note that this vessel seems to disappear in the region of the liver. Here the posterior vena cava receives the hepatic veins coming from the liver. Scrape away some of the liver tissue in order to see these veins.
3. Locate the posterior vena cava as it passes through the diaphragm into the thoracic cavity and as it enters the right atrium.
4. Trace the posterior vena cava from the iliac veins to the right atrium of the heart (see Fig. 14.4).

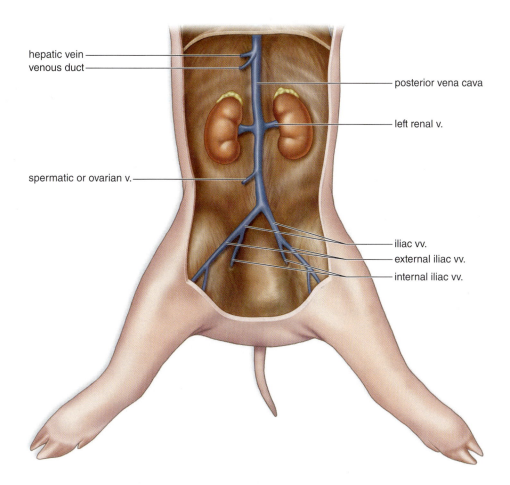

hepatic vein
venous duct
posterior vena cava
left renal v.
spermatic or ovarian v.
iliac vv.
external iliac vv.
internal iliac vv.

Figure 14.8 Veins in the abdominal cavity of the fetal pig.

The posterior vena cava divides into the common iliac veins. The common iliac veins branch into the external and internal iliac veins. (vv. = veins.) Use this diagram to trace blood vessels, but *do not remove any organs. You will need them in Laboratory 15.*

Storage of Pigs

1. Before leaving the laboratory, place your pig in the plastic bag provided.
2. Expel excess air from the bag, and tie it shut.
3. Write your name and section on the tag provided, and attach it to the bag. Your instructor will indicate where the bags are to be stored until the next laboratory period.
4. Clean the dissecting tray and tools, and return them to their proper location.
5. Wipe off your goggles.
6. Wash your hands.

14.4 Blood Vessel Comparison

Blood pressure keeps the blood in arteries moving away from the heart. Skeletal muscle contraction pushing in on veins keeps the blood in veins moving toward the heart. Do you predict that arteries or veins are generally more superficial in the body? _____

_____

Wall of an Artery Compared with Wall of a Vein

Both arteries and veins have three distinct layers, or **tunicas,** that form a wall around the lumen, the space through which blood flows. The three tunicas are called the inner layer, the middle layer, and the outer layer. Figure 14.9 shows that arteries have a thicker wall than veins because the middle layer consisting of smooth muscle and elastic fibers is much thicker. The elastic fibers in the wall of an artery allows it to expand when the blood pours into it with each heartbeat. The smooth muscle allows the arterial wall to constrict when needed to keep blood pressure normal.

(a) Artery   (b) Vein

Figure 14.9   Blood vessel comparison.

1. Obtain a microscope slide that shows an artery and a vein in cross section.
2. View the slide, under both low and high power, and with the help of Figure 14.10, determine which is the artery and which is the vein.
3. Identify the **outer layer,** which contains many collagen and elastic fibers and often appears white in specimens.
4. Identify the **middle layer,** the thickest layer, which is composed of smooth muscle and elastic tissue. Does

   this layer appear thicker in arteries than in veins? _____
5. Identify the **inner layer,** a smooth lining of simple squamous epithelial cells called the endothelium. In veins, the endothelium forms valves that keep the blood moving toward the heart. Arteries do not have valves. Considering the relationship of arteries and veins to the heart, why do veins have valves, while arteries do

   not? _____

   _____

Figure 14.10   Photomicrograph of an artery and a vein.

© Ed Reschke

100×

Conclusions

- Which type of blood vessel (arteries or veins) has thicker walls? What makes the wall thicker? _____

  _____

  _____

- Which type of blood vessel has thinner walls? _____

  _____

- Which type of blood vessel is more apt to lose its elasticity, leading to a discoloration that can be

  externally observed? _____

  _____

  What is this condition called? _____

_____ 1. What blood vessels connect arterioles to venules and exchange molecules with tissue fluid?

_____ 2. Which blood vessels return blood to the right atrium?

_____ 3. Does the pulmonary artery carry blood back to or away from the heart?

_____ 4. Is the blood entering the left ventricle O_2-poor or O_2-rich?

_____ 5. What valve prevents the backflow of blood when the left ventricle relaxes?

_____ 6. Which circuit delivers blood to the spleen, the pulmonary or the systemic?

_____ 7. What vessel transports blood to the liver from the digestive tract?

_____ 8. Do pulmonary veins carry O_2-poor or O_2-rich blood?

_____ 9. In addition to the esophagus and the posterior vena cava, what structure passes through the diaphragm?

_____ 10. What is the name of the veins that serve the neck and head regions of the body?

_____ 11. What organs are served by the anterior mesenteric artery?

_____ 12. What arteries form from the branching of the posterior end of the aorta?

_____ 13. What is the name of the vein that is joined to the posterior vena cava by the venous duct (ductus venosus)? (Hint: it was cut when the abdominal cavity was exposed.)

_____ 14. Which type of blood vessel (artery or vein) has a thick wall with a lot of smooth muscle that helps regulate blood pressure?

Thought Questions

15. Explain why narrowing or blockages of the coronary arteries can be life threatening, and identify which arteries, when blocked, increase the risk of having a stroke.

16. Explain why an airline recently introduced an exercise video to combat deep vein thrombosis (clots that form in deep veins after prolonged inactivity). Consider the pressure of blood in veins and venous structure in your reply.

17. In figures illustrating arteries and veins, arteries usually appear red and veins usually appear blue. Explain why the pulmonary arteries seen in Figure 14.2 appear blue and the pulmonary veins in Figure 14.2 appear red.

15

Basic Mammalian Anatomy II

Introduction

The **urinary system** and the **reproductive system** are so closely associated in mammals that they are often considered together as the **urogenital system.** They are particularly associated in males, where certain structures function in both systems. In this laboratory, we will focus first on dissecting the urinary and reproductive systems in the fetal pig.

The kidneys of the urinary system produce urine, which is stored in the bladder before being released to the exterior. As the kidneys produce urine they also regulate the volume and the composition of the blood so that the water and salt balance and the acid-base balance of the blood stays within normal limits.

In mammalian reproductive systems, the testes are the male gonads, and the ovaries are the female gonads. The testes produce sperm, and the ovaries produce oocytes that become eggs. We will compare the anatomy of the reproductive systems in pigs with those in humans.

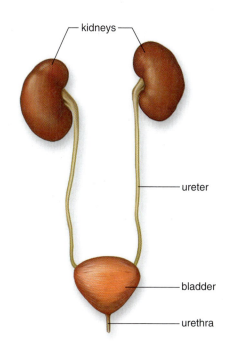

kidneys

ureter

bladder

urethra

Urinary system

Figure 15.1 Urinary system of the fetal pig.

In (a) females and (b) males, urine is made by the kidneys, transported to the bladder by the ureters, stored in the bladder, and then excreted from the body through the urethra.

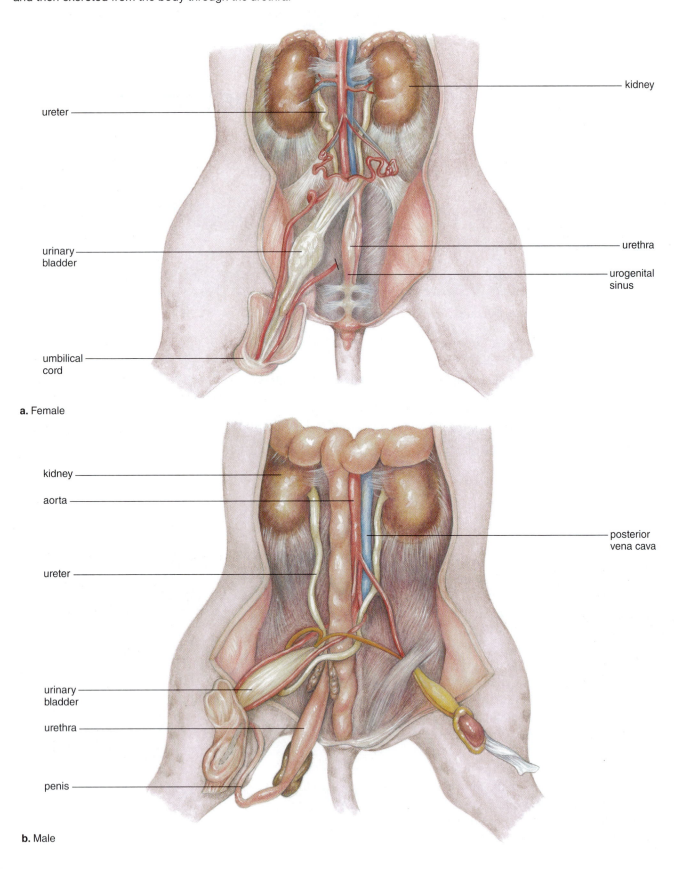

ureter

kidney

urinary
bladder

urethra

urogenital
sinus

umbilical
cord

a. Female

kidney

aorta

posterior
vena cava

ureter

urinary
bladder

urethra

penis

b. Male

15.1 Urinary System

The urinary system consists of the **kidneys,** which produce urine; the **ureters,** which transport urine to the **urinary bladder,** where urine is stored; and the **urethra,** which transports urine to the outside. In males, the urethra also transports sperm during ejaculation.

Observation: Urinary System in Pigs

During this dissection, compare the urinary system structures of male and female fetal pigs. Later in this laboratory period, exchange specimens with a neighboring team for a more thorough inspection.

1. The large, paired kidneys (Fig. 15.1) are reddish organs covered by **peritoneum,** a membrane that anchors them to the dorsal wall of the abdominal cavity, sometimes called the **peritoneal cavity.** Clean the peritoneum away from one of the kidneys, and study it more closely.

2. Locate the ureters, which leave the kidneys and run posteriorly under the peritoneum (Fig. 15.1).

> ⚠️ **Wear protective latex gloves and eyewear** when handling preserved animal organs. Exercise caution when using sharp instruments during this laboratory. Wash hands thoroughly upon completion of this laboratory.

3. Clean the peritoneum away, and follow a ureter to the urinary bladder, which normally lies in the ventral portion of the abdominal cavity. The urinary bladder is on the inner surface of the flap of tissue to which the umbilical cord was attached.

4. The urethra arises from the bladder posteriorly and joins the urogenital sinus. Follow the urethra until it passes from view into the ring formed by the pelvic girdle.

5. Sequence the organs in the urinary system to trace the path of urine from its production to its exit. _____

_____

6. Using a scalpel, section one of the kidneys in place, cutting it lengthwise (Fig. 15.2). At the center of the medial portion of the kidney is an irregular, cavity-like reservoir, the **renal pelvis.** The outermost portion of the kidney (the **renal cortex**) shows many small striations perpendicular to the outer surface. This region and the more even-textured **renal medulla** contain **nephrons** (excretory tubules), microscopic organs that produce urine.

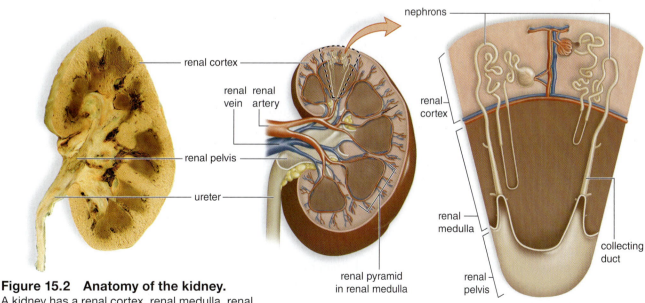

Figure 15.2 Anatomy of the kidney.
A kidney has a renal cortex, renal medulla, renal pelvis, and microscopic tubules called nephrons.

© MedImage/Science Source

15.2 Male Reproductive System

The **male reproductive system** consists of the **testes** (sing., testis), which produce sperm, and the **epididymides** (sing., epididymis), which store sperm before they enter the **vasa deferentia** (sing., vas deferens). Just prior to ejaculation, sperm leave the vasa deferentia and enter the urethra, which eventually passes into the penis. The **penis** is the male organ of sexual intercourse. **Seminal vesicles,** the **prostate gland,** and the **bulbourethral glands** (Cowper's glands) add fluid to semen after sperm reach the urethra. Table 15.1 summarizes the male reproductive organs.

The testes begin their development in the abdominal cavity, just anterior and dorsal to the kidneys. Before birth, however, they gradually descend into paired **scrotal sacs** within the scrotum, suspended anterior to the anus. Each scrotal sac is connected to the body cavity by an **inguinal canal,** the opening of which can be found in the pig. The passage of the testes from the body cavity into the scrotal sacs is called the descent of the testes. The testes in most of the male fetal pigs being dissected will probably be partially or fully descended.

Table 15.1   Male Reproductive Organs and Functions

Organ	Function
Testis	Produces sperm and sex hormones
Epididymis	Stores sperm as they mature
Vas deferens	Conducts and stores sperm
Seminal vesicle	Contributes secretions to semen
Prostate gland	Contributes secretions to semen
Urethra	Conducts sperm
Bulbourethral glands	Contribute secretions to semen
Penis	Organ of copulation

Observation: Male Reproductive System in Pigs

While doing this dissection, consult Table 15.1 for the function of the male reproductive organs.

Inguinal Canal, Testis, Epididymis, and Vas Deferens

1. Locate the opening of the left inguinal canal, which leads to the left scrotal sac (Fig. 15.3).
2. Expose the canal and sac by making an incision through the skin and muscle layers from a point over this opening back to the left scrotal sac.
3. Open the sac, and find the testis. Note the much-coiled tubule—the epididymis—that lies alongside the testis. An epididymis is continuous with a vas deferens, which runs toward the abdominal cavity.
4. Each vas deferens loops over an umbilical artery and ureter and unites with the urethra as it leaves the urinary bladder. We will dissect this juncture below.

Penis, Urethra, and Accessory Glands

1. Cut through the ventral skin surface just posterior to the umbilical cord. This will expose the rather undeveloped penis, which contains a long portion of the urethra.
2. Lay the penis to one side, and then cut down through the ventral midline, laying the legs wide apart in the process (Fig. 15.4). The cut will pass between muscles and through pelvic cartilage (bone has not developed yet). Do not cut any of the ducts or tracts in the region.
3. You will now see the urethra ventral to the rectum. It is somewhat heavier in the male due to certain accessory glands:
   a. Bulbourethral glands (Cowper's glands), about 1 cm in diameter, are further along the urethra and are more prominent than the other accessory glands.
   b. The prostate gland, about 4 mm across and 3 mm thick, is located on the dorsal surface of the urethra, just posterior to the juncture of the urinary bladder with the urethra. It is often difficult to locate and is not shown in Figures 15.3 and 15.4.
   c. Small, paired seminal vesicles may be seen on either side of the prostate gland.

Figure 15.3 Male reproductive system of the fetal pig.

In males, the urinary system and the reproductive system are joined. The vasa deferentia (sing., vas deferens) enter the urethra, which also carries urine.

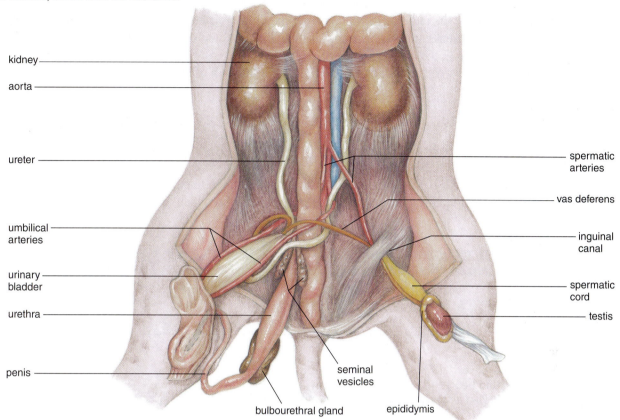

- kidney
- aorta
- ureter
- umbilical arteries
- urinary bladder
- urethra
- penis
- bulbourethral gland
- seminal vesicles
- epididymis
- spermatic arteries
- vas deferens
- inguinal canal
- spermatic cord
- testis

Figure 15.4 Photograph of the male reproductive system of the fetal pig.

Compare Figure 15.3 to this photograph to help identify the structures of the male urogenital system.

© Carlyn Iverson/McGraw-Hill Education

- kidney
- ureter
- seminal vesicle
- inguinal canal
- bulbourethral gland
- umbilical cord
- urogenital orifice
- umbilical artery
- urinary bladder
- penis
- urethra
- vas deferens
- scrotal sac
- epididymis
- testis

4. Trace the urethra as it leaves the bladder. When it nears the end of the abdominal cavity, it turns rather abruptly and runs anteriorly just under the skin where you have just dissected it. This portion of the urethra is within the penis.

5. Now you should be able to see the vasa deferentia enter the urethra. If necessary, free these structures from surrounding tissue to see them enter the urethra near the location of the prostate gland. In males, the urethra transports sperm and urine to the urogenital opening (Fig 12.1*c*).

6. Sequence the organs in the male reproductive system to trace the path of sperm from the organ of production to the penis. _____

_____

Comparison of Male Fetal Pig and Human Male

Use Figure 15.5 to help compare the male pig reproductive system with the human male reproductive system. Complete Table 15.2, which compares the location of the penis in these two mammals.

Table 15.2 Location of Penis in Male Fetal Pig and Human Male	
Fetal Pig	**Human**
Penis	

Figure 15.5   Human male urogenital system.
In the fetal pig, but not in the human male, the penis lies beneath the skin and exits at the urogenital opening.

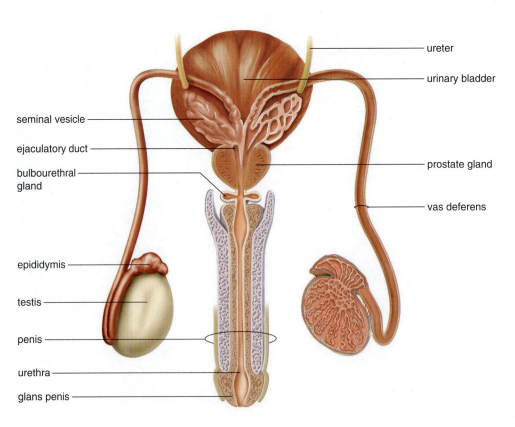

15.3 Female Reproductive System

The **female reproductive system** (Table 15.3) consists of the **ovaries** (sing., ovary), which produce eggs, and the **oviducts,** which transport eggs to the **uterus,** where development occurs. In the fetal pig, the uterus does not form a single organ, as in humans, but is partially divided into structures called **uterine horns,** which connect with the oviduct. The **vagina** is the birth canal and the female organ of sexual intercourse.

Table 15.3 Female Reproductive Organs and Functions	
Organ	**Function**
Ovary	Produces egg and sex hormones
Oviduct (fallopian tube)	Conducts egg toward uterus
Uterus	Houses developing fetus
Vagina	Receives penis during copulation and serves as birth canal

Observation: Female Reproductive System in Pigs

While doing this dissection, consult Table 15.3 for the function of the female reproductive organs.

Ovaries and Oviducts

1. Locate the paired ovaries, small bodies suspended from the peritoneal wall in mesenteries, posterior to the kidneys (Figs. 15.6 and 15.7).
2. Closely examine one ovary. Note the small, short, coiled oviduct, sometimes called the **fallopian tube.** The oviduct does not attach directly to the ovary but ends in a funnel-shaped structure with fingerlike processes (fimbriae) that partially encloses the ovary. The egg produced by an ovary enters an oviduct, where it is fertilized by a sperm, if reproduction will occur. Any resulting embryo passes to the uterus.

Uterine Horns

1. Locate the **uterine horns.** (Do not confuse the uterine horns with the oviducts; the latter are much smaller and are found very close to the ovaries.)
2. Find the body of the uterus, located where the uterine horns join.

Vagina

1. Separate the hindlimbs of your specimen, and cut down along the midventral line. The cut will pass through muscle and the cartilaginous pelvic girdle. With your fingers, spread the cut edges apart, and use blunt dissecting instruments to separate connective tissue.
2. Now find the vagina, which passes from the uterus to the urogenital sinus. The vagina is dorsal to the urethra, which also enters the urogenital sinus, and ventral to the rectum, which exits at the anus. The urogenital sinus opens at the urogenital papilla (Figs. 15.6 and 15.7).
3. The vagina is the organ of copulation and is the birth canal. The receptacle for the vagina in a pig, the urogenital sinus, is absent in adult humans and several other adult female mammals in which both the urethra and vagina have their own openings.
4. The vagina plays a critical role in reproduction, even though development of the offspring occurs in the uterus. Explain. _____

_____

Figure 15.6 Female reproductive system of the fetal pig.
In the fetal pig, both the vagina and the urethra enter the urogenital sinus, which opens at the urogenital papilla.

ureter

uterine
horn

urinary
bladder

umbilical
arteries

umbilical
cord

kidney

ovarian
vein

ovary

body of
uterus

vagina

urethra

urogenital
sinus

urogenital
papilla

Figure 15.7 Photograph of the female reproductive system of the fetal pig.
Compare Figure 15.6 with this photograph to help identify the structures of the female urinary and reproductive systems.

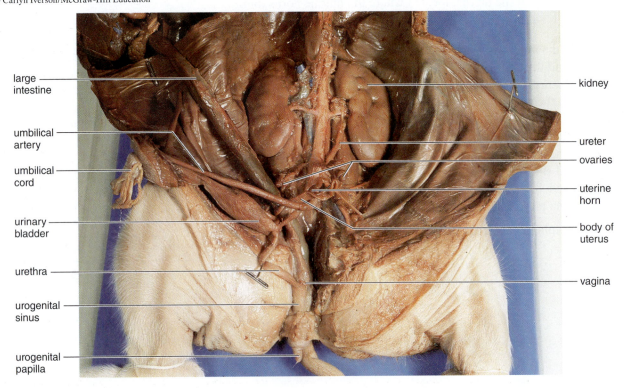

large
intestine

umbilical
artery

umbilical
cord

urinary
bladder

urethra

urogenital
sinus

urogenital
papilla

kidney

ureter

ovaries

uterine
horn

body of
uterus

vagina

Comparison of Female Fetal Pig with Human Female

Use Figure 15.8 to compare the female pig reproductive system with the human female reproductive system. Complete Table 15.4, which compares the appearance of the oviducts and the uterus, as well as the presence or absence of a urogenital sinus, in these two mammals.

Figure 15.8 Human female reproductive system.
Especially compare the anatomy of the oviducts in humans with that of the uterine horns in a pig. In a pig, the fetuses develop in the uterine horns; in a human female, the fetus develops in the body of the uterus.

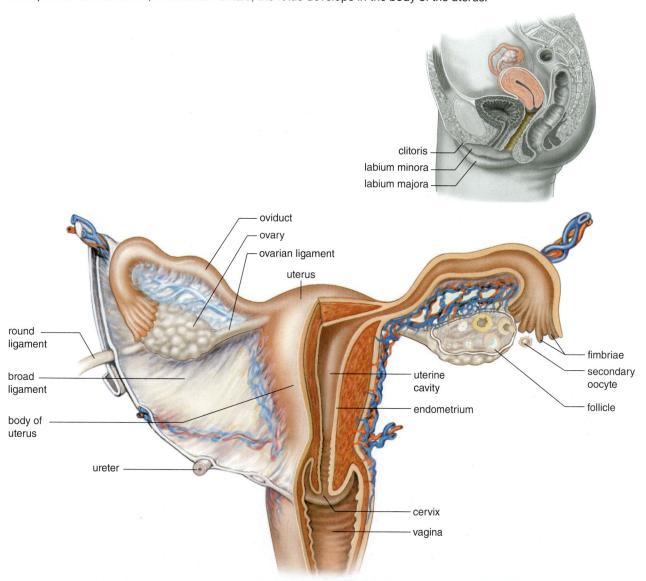

Table 15.4	Comparison of Female Fetal Pig with Human Female	
	Fetal Pig	**Human**
Oviducts		
Uterus		
Urogenital sinus		

Figure 15.9 Internal anatomy of the fetal pig.

Most of the major organs are shown in this photograph. The stomach has been removed. The spleen, gallbladder, and pancreas are not visible.   © Ken Taylor/Wildlife Images

15.4 Review of the Respiratory, Digestive, and Cardiovascular Systems

In Laboratory 12, you dissected the respiratory, digestive, and cardiovascular systems of the fetal pig. Review your knowledge of these systems by reexamining your dissection of the fetal pig and *label Figure 15.9*. In this portion of today's lab, you will review each system and examine some organs in more detail. **Do not remove any organs** unless told to do so by your instructor.

Observation: Respiratory System in Pigs

1. Using these terms (bronchiole, bronchus, glottis, larynx, pharynx, trachea), trace the path of air from the nasal passages to the lungs. List the first three organs in the left column and the last three organs in the right column.

   nasal passages

   _____   _____

   _____   _____

   _____   lungs _____

2. Make sure you have cut the corners of the mouth as directed in Lab 12, page 156. In the **pharynx,**

   you should be able to locate the **glottis,** an opening to the _____.

3. If necessary, make a midventral incision in the neck to expose the **larynx.**
4. Clear away the "straplike" muscles covering the **trachea.** Now you should be able to feel the cartilaginous rings that hold the trachea open. Locate the esophagus, which lies below (dorsal to) the trachea.
5. If available, observe a slide on display showing a section through the trachea and esophagus. Notice in Figure 15.10 that the air and food pathways cross in the pharynx.
6. Open the pig's mouth, insert a blunt probe into the glottis, and carefully work the probe down through the larynx to the level of the **bronchi.**
7. Observe the **lungs,** and if available, observe a prepared slide of lung tissue.
8. If so directed by your instructor, remove a portion of the trachea, the bronchi, and the lungs, keeping them all in one piece. Place this specimen in a small container of water. Holding the trachea with your forceps, gently but firmly stroke the lung repeatedly with the blunt wooden base of one of your probes. If you work carefully, the alveolar tissue will be fragmented and rubbed away, leaving the branching system of air tubes and blood vessels.

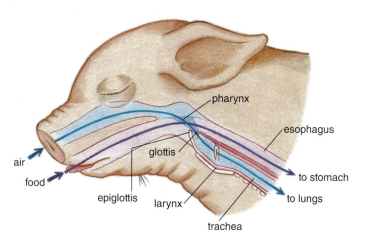

Figure 15.10   Air and food passages in the fetal pig.
A probe can pass from the mouth to the larynx and to the esophagus.

1. Using these terms (esophagus, large intestine, small intestine, stomach) trace the path of food from the mouth to the anus:

   mouth _____   _____

   _____   _____

   _____   anus _____

2. Open the **mouth,** and insert a blunt probe into the esophagus (Fig. 15.10). Then trace the **esophagus** to the stomach.

3. Open one side of the **stomach,** and examine its interior surface. Does it appear smooth or

   rough? _____

4. Find the pyloric sphincter, the muscle that surrounds the entrance to the duodenum, the first part of the

   small intestine. Record the length of the small intestine. _____

   If you have not done so before, find the bile duct that empties into the duodenum. The bile duct comes

   from the _____.

5. Find the **cecum,** a projection where the small intestine enters the large intestine. How does the appearance

   of the pig's large intestine differ from that of a human? _____

   _____

6. Carefully cut the mesenteries holding the colon of the **large intestine** in place, and uncoil the large

   intestine. Record the length of the large intestine. _____ How does the length of the large intestine

   compare with that of the small intestine? _____

7. Locate again the liver, pancreas, and gallbladder, three accessory organs of digestion.

Observation: Cardiovascular System in Pigs

Heart

1. Trace the path of blood through the heart, starting with the vena cava and ending with the aorta. Mention all the chambers of the heart and the valves (see Fig. 14.2).

   To the heart: From the lungs:

   vena cava   _____

   _____   _____

   _____ valve   _____ valve

   _____   _____

   _____ valve   _____ valve

   _____   aorta

   _____

2. Keeping the heart inside the pig, cut the pericardial sac (the tissue that surrounds the heart).

3. Look for and identify the vessels attached to the heart.

4. Section the heart, and look for its four chambers. Remnants of the atrioventricular valves can be seen as thin sheets of whitish tissue attached to fine, white, tendinous strands.

5. With your blunt probe, find the oval opening in the wall between the two atrial chambers. Recall that this is a shunt that allows blood to bypass lung circulation prior to birth.

Blood Vessels

1. In general, arteries take blood _____ the heart, and veins take blood _____ the heart.
2. Locate the following blood vessels in your pig. State their origin and destination.

   a. coronary artery _____

   b. cardiac vein _____

   c. carotid artery _____

   d. jugular vein _____

   e. subclavian artery _____

   f. subclavian vein _____

   g. renal artery _____

   h. renal vein _____

   i. iliac artery _____

   j. iliac vein _____

Hepatic Portal System and Associated Vessels

A **portal system** goes from one capillary bed to another without passing through the heart. For example, the hepatic portal vein takes blood from the intestinal capillaries to capillaries in the liver. The liver plays an important role in processing and storing materials absorbed from the intestine. The hepatic veins take blood from the liver to the inferior (posterior) vena cava.

1. In your pig, the hepatic portal vein is dorsal to the bile duct and will not be blue if the latex did not enter it. To find the hepatic portal vein, break the mesenteries in the region of the bile duct (Fig. 15.11).
2. The **hepatic veins** consist of three or four vessels from the liver to the inferior vena cava. To see them, scrape away the liver with the blunt side of the scalpel until liver material has been removed and only a mass of cords remains.
3. Identify the **umbilical vein** leading into the liver and, on its posterior surface, the **venous duct** which is the main channel through the liver in the fetus (see Table 14.2).

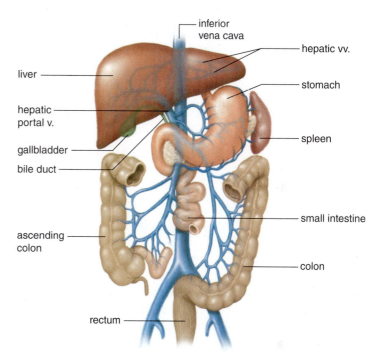

Figure 15.11 Hepatic portal system.
The hepatic portal vein lies between the digestive tract and the liver. The hepatic veins enter the inferior vena cava. (v. = vein; vv. = veins)

Storage of Pigs

1. Before leaving the laboratory, place your pig in the plastic bag provided.
2. Expel excess air from the bag, and tie it shut.
3. Write your name and section on the tag provided, and attach it to the bag. Your instructor will indicate where the bags are to be stored until the next laboratory period.
4. Clean the dissecting tray and tools, and return them to their proper location.
5. Wipe off your goggles and wash your hands.

Laboratory Review 15

_____ 1. What two individual systems make up the urogenital system?

_____ 2. What is the cavity-like space in a kidney called?

_____ 3. What organ stores urine before it is transported to the outside?

_____ 4. What structure associated with the urinary system transports urine and sperm?

_____ 5. Where are sperm stored while they mature?

_____ 6. When descending, what do the testes travel through to get to the scrotal sacs?

_____ 7. What accessory gland for reproduction in human males is inferior (posterior) to the urinary bladder?

_____ 8. What female reproductive structures transport eggs to the location where embryos/fetuses develop?

_____ 9. Where, in a pig, are the embryos/fetuses located during development?

_____ 10. What female reproductive structure produces sex hormones?

_____ 11. What respiratory structure has cartilaginous rings that keep the airway open?

_____ 12. What does food travel through to enter the duodenum when leaving the stomach?

_____ 13. What is the tissue surrounding the heart called?

_____ 14. What does blood pass through when it goes from one capillary bed to another without passing through the heart?

Thought Questions

15. Why do women experience more urinary tract infections than men? It may help to observe Figures 15.5 and 15.8 (the lateral view at top right) and think about where the bladder and urethra are in the drawings. You may also refer to figures of the human female reproductive system in your textbook.

16. What male reproductive structure is severed during a vasectomy? Do the testes still produce sperm following a vasectomy? Why does sterility result from a vasectomy?

17. The hormone insulin lowers blood glucose. How will insulin affect what happens to glucose once blood travels through the hepatic portal system?

16

Homeostasis

Introduction

Homeostasis refers to the dynamic equilibrium of the body's internal environment. The **internal environment** consists of blood and tissue fluid. The body's cells take nutrients from tissue fluid and return their waste molecules to it. Tissue fluid, in turn, exchanges molecules with the blood. This is called capillary exchange. All internal organs contribute to homeostasis, but this laboratory specifically examines the contributions of the blood, lungs, and kidneys (Fig. 16.1).

Figure 16.1   Contributions of organs to homeostasis.
The lungs exchange gases with blood; the kidneys remove nitrogenous wastes from blood; and the intestinal tract adds nutrients as regulated by the liver to blood.

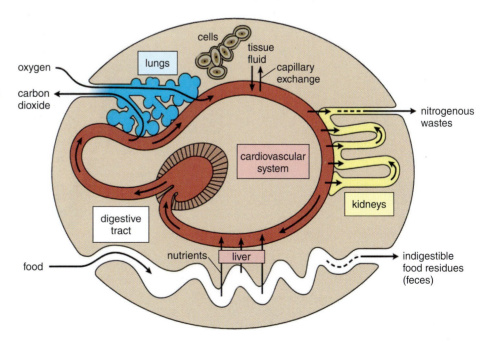

16.1 Heartbeat and Blood Flow

Recall that the cardiovascular system consists of the heart, blood vessels, and blood (Fig. 16.2). Arteries carry blood away from the heart while veins transport blood toward the heart. Arteries branch into smaller vessels called arterioles that enter capillary beds. Capillary beds are present throughout the organs and tissues of the body. An exchange of gases takes place across the thin walls of **pulmonary capillaries.** In the lungs, CO_2 leaves the blood and O_2 enters the blood. An exchange of gases and nutrients for metabolic wastes takes place across the thin walls of **systemic capillaries.** In the body tissues, O_2 and nutrients exit the blood, while CO_2 and metabolic wastes enter the blood.

We will see that the heart is vital to homeostasis because its contraction (called the **heartbeat**) keeps the blood moving in the arteries and arterioles which take blood to the capillaries. The exchanges that take place across capillaries help maintain homeostasis.

Liver

The hepatic portal vein lies between the

_____ and

the _____.

This placement allows the liver to regulate what molecules enter the blood from the digestive tract. For example, if the hormone insulin is present, the liver removes excess glucose and stores it as glycogen. Later, the liver breaks down glycogen to glucose to keep the blood glucose concentration constant.

Figure 16.2 The circulatory system.
The heart provides the pumping action that transports the blood through the arteries, capillary beds, and veins.

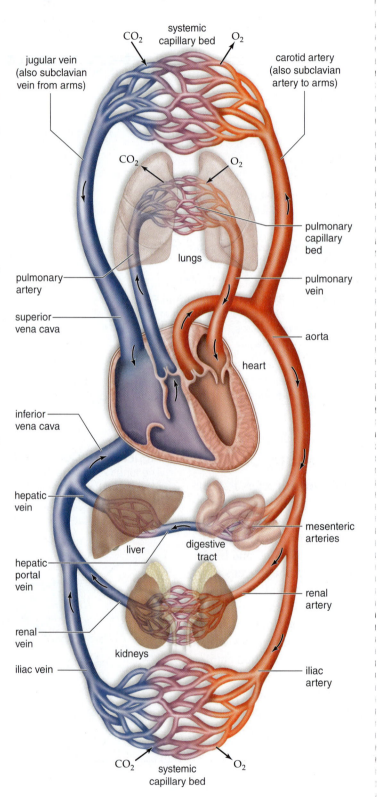

Heartbeat

During a heartbeat, first the atria contract and then the ventricles contract. When a chamber contracts, it is called **systole,** and when a chamber relaxes, it is called **diastole.** The atria and ventricles take turns being in systole.

Time	Atria	Ventricles
0.15 sec	Systole	Diastole
0.30 sec	Diastole	Systole
0.40 sec	Diastole	Diastole

Usually, there are two heart sounds with each heartbeat (Fig. 16.3). The first sound (*lub*) is low and dull and lasts longer than the second sound. It is caused by the closure of valves following atrial systole. The second sound (*dub*) follows the first sound after a brief pause. The sound has a snapping quality of higher pitch and shorter duration. The *dub* sound is caused by the closure of valves following ventricle systole.

Figure 16.3   The heartbeat sounds.
a. When the atria contract (are in systole), the ventricles fill with blood. **b.** Closure of the valves between atria and ventricles results in a **lub** sound. When the ventricles contract, blood enters the attached arteries (aorta and pulmonary trunk). **c.** Closure of valves between ventricles and arteries results in a **dub** sound. Blood enters the heart from the attached veins (venae cavae and pulmonary veins) once more.

In the following procedure, you will work with a partner and use a stethoscope to listen to the heartbeat. It will not be necessary for you to count the number of beats per minute.

1. Obtain a stethoscope, and properly position the earpieces. They should point forward. Place the bell of the stethoscope on the left side of your partner's chest between the fourth and fifth ribs. This is where the apex (tip) of the heart is closest to the body wall.

2. Which of the two sounds (**lub** or **dub**) is louder? _____

3. Now switch, and your partner will determine your heartbeat.

Blood Pressure

Blood pressure is highest just after ventricular systole (contraction), and it is lowest during ventricular diastole (relaxation). Why? _____

We would expect a person to have lower blood pressure readings at rest than after exercise. Why?

_____

A number of different types of digital blood pressure monitors are available, and your instructor will instruct you on how to use the type you will be using for this Experimental Procedure. The resting blood pressure readings for an individual are displayed on the monitor shown in Figure 16.4. A blood pressure reading at or below 120/80 (systolic/diastolic) is considered normal.

Figure 16.4 Measurement of blood pressure and pulse.
There are many different types of digital blood pressure/pulse monitors now available. The one shown here uses a cuff to be placed on the arm. Others use a cuff for the wrist.

© Ilene MacDonald/Alamy RF

During this experimental procedure you may work with a partner or by yourself. If working with a partner, each of you will assist the other in taking blood pressure readings. When you note the blood pressure readings, also note the pulse reading.

Blood Pressure and Pulse at Rest

1. Reduce your activity as much as possible.
2. Use the blood pressure monitor to obtain several blood pressure readings, average them, and record your results in Table 16.1. Also note the pulse rates and average. Record in Table 16.1.

Blood Pressure and Pulse After Exercise

1. Run in place for 1 minute.
2. Immediately use the blood pressure monitor to obtain a blood pressure reading, and record it in Table 16.1. Also note the pulse rate and record in Table 16.1.

Table 16.1 Blood Pressure				
	Rates at Rest		**Rates After Exercise**	
	Blood Pressure	**Pulse**	**Blood Pressure**	**Pulse**
Partner				
Yourself				

Conclusions: Blood Pressure

- Knowing that exercise increases the heart rate, offer an explanation for your results. _____
_____

- Under what conditions in everyday life would you expect the heart rate and the blood pressure to increase, even though you were not exercising? _____
_____

When might this be an advantage? _____

A disadvantage? _____

16.2   Blood Flow and Systemic Capillary Exchange

We associate death with lack of a heartbeat, but the real problem is lack of blood flow to the capillaries.

Blood Flow

The beat of the heart moves blood into the aorta, from which other arteries branch off to specific locations or organs. These arteries divide into arterioles and then arterioles divide into capillaries. Venules, which receive blood from capillaries, combine to form veins, which take blood back to the heart.

Experimental Procedure: Blood Flow

1. Observe blood flow through arterioles, capillaries, and venules, either in the tail of a goldfish or in the webbed skin between the toes of a frog, as prepared by your instructor.
2. Examine under low and high power of the microscope.
3. Watch the pulse and the swiftly moving blood in the arterioles.
4. Contrast this with the more slowly moving blood that circulates in the opposite direction in the venules. Many criss-crossing capillaries are visible.
5. Look for blood cells floating in the bloodstream. Don't confuse blood cells with chromatophores, irregular black patches of pigment that may be visible in the skin.

Systemic Capillary Exchange

The beat of the heart creates blood pressure, and blood pressure is necessary to capillary exchange. Blood pressure acts to move water out of a capillary, while osmotic pressure (created by the presence of proteins in the blood) acts to move water into a capillary (Fig. 16.5). Blood pressure is higher than osmotic pressure at the arteriole end of a capillary, and water moves out of a capillary. But blood pressure lessens as the blood moves through a capillary bed. This means that osmotic pressure is higher than blood pressure at the venule end of a capillary, and water moves back into the capillary.

Figure 16.5   Systemic capillary exchange.
At a systemic capillary, an exchange takes place across the capillary wall. Between the arterial end and the venule end, molecules follow their concentration gradient. Oxygen and nutrients move out of a capillary while carbon dioxide and wastes move into the capillary.

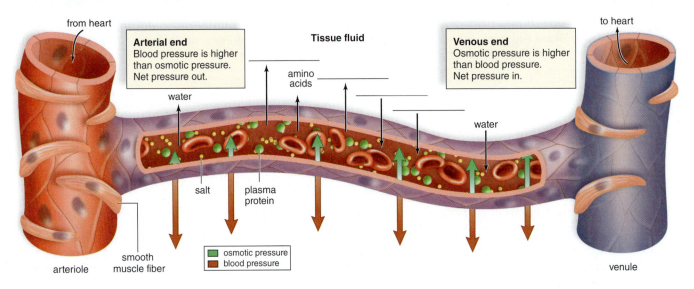

Conclusions: Systemic Capillary Exchange

- What generates blood pressure? _____
- Why are tissue cells always in need of glucose and oxygen? _____

  Add glucose at the end of an appropriate arrow in Figure 16.5. Do the same for oxygen.

- Why are tissue cells always producing carbon dioxide? _____

  Add carbon dioxide at the start of an appropriate arrow in Figure 16.5. Do the same for metabolic wastes.

16.3   Lung Structure and Human Respiratory Volumes

The right and left lungs lie in the thoracic cavity on either side of the heart. Air moves from the nasal passages to the trachea, bronchi, bronchioles, and finally, lungs.

Lung Structure

A **lung** is a spongy organ consisting of irregularly shaped air spaces called **alveoli** (sing., alveolus) (Fig. 16.6*a*). The alveoli are surrounded by a rich network of tiny blood vessels called pulmonary capillaries. *In Figure. 16.6a, use a labeled arrow to show oxygen entering blood from an alveolus and another labeled arrow to show carbon dioxide entering an alveolus from the blood.*

1. Observe a prepared slide of a stained section of a lung (Fig. 16.6*b*). In stained slides, the nuclei of the cells forming the thin alveolar walls appear purple or dark blue.
2. Look for areas that show red or orange disc-shaped **erythrocytes.** These are the red blood cells that contain hemoglobin, which takes up oxygen and transports it to the tissues. When these appear in strings, you are looking at capillary vessels in side view.
3. In some part of the slide, you may even observe an artery. Thicker circular or oval structures with a lumen (cavity) are cross sections of **bronchioles,** tubular pathways through which air reaches the air spaces.
4. In Figure 16.6*c*, note that in emphysema, alveoli have burst. In smokers, small bronchioles collapse, and trapped air in alveoli causes them to burst. Now gas exchange is minimal.

Figure 16.6   Healthy lung tissue versus emphysema.
a. The lungs normally contain many air sacs called alveoli where gas exchange occurs. **b.** Micrograph of normal lung tissue. **c.** In smokers, emphysema can occur; the alveoli burst and gas exchange is inadequate.

(b) © Dr. Keith Wheeler/Science Source; (c) © CMSP/Getty Images

bronchus

bronchiole

capillary network

alveoli

a. Normal lung

red blood cells

alveolus

artery

b. Micrograph of normal lung   36×

c. Emphysema

Human Respiratory Volumes

Breathing in, called **inspiration** or inhalation, is the active part of breathing because that's when contraction of rib cage muscles causes the rib cage to move up and out, and contraction of the diaphragm causes the diaphragm to lower. Due to an enlarged thoracic cavity, the lungs expand and air is drawn into them. Breathing out, called **expiration** or exhalation, occurs when relaxation of these same muscles causes the thoracic cavity to resume its original capacity. Now air is pushed out of the lungs (Fig. 16.7).

Figure 16.7 Inspiration and expiration.

a. Inspiration occurs after the rib cage moves up and out and the diaphragm moves down. Air rushes into lungs because they expand as the thoracic cavity expands. **b.** Expiration occurs as the rib cage moves down and in and the diaphragm moves up. As the thoracic cavity and lungs get smaller, air is pushed out.

a. Inspiration

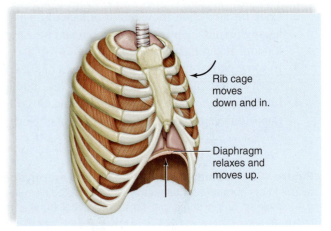

b. Expiration

Experimental Procedure: Human Respiratory Volumes

During this Experimental Procedure, you will be working with a spirometer, an instrument that measures the amount of exhaled air (Fig. 16.8). Normally, about 500–600 ml of air move into and out of the lungs with each breath. This is called the **tidal volume** (**TV**). You can inhale deeply after a normal breath and more air will enter the lungs; this is the **inspiratory reserve volume** (**IRV**). You can also force more air out of your lungs after a normal breath; this is the **expiratory reserve volume** (**ERV**). **Vital capacity** is the volume of air that can be forcibly exhaled after forcibly inhaling.

Figure 16.8   Nine-liter student wet spirometer.
© Phipps & Bird, Inc., Richmond, VA

Tidal Volume (TV)

1. When it's your turn to use the spirometer, install a new disposable mouthpiece and set the spirometer to zero.
2. Inhale normally, then exhale normally (with *no* extra effort) through the mouthpiece of the spirometer. Record your measurement in Table 16.2.
3. Three readings are needed, so twice more set the spirometer to zero and repeat the same procedure. Record your measurements in Table 16.2.
4. If the spirometer being used indicates measurements in liters (l), change your readings to milliliters (ml) by multiplying your values by 1,000. Calculate your average TV in ml by adding your three readings together and dividing by 3.

   In your own words, what is tidal volume? _____

   _____

Expiratory Reserve Volume (ERV)

1. Make sure the spirometer is set to zero.
2. Inhale and exhale normally and then force as much air out as possible into the spirometer. Record your measurement in Table 16.2.

3. Three readings are needed, so twice more, set the spirometer to zero and repeat the same procedure. Record your measurements in Table 16.2.
4. Later, if necessary, change your readings to ml, and calculate your average ERV.

In your own words, what is expiratory reserve volume? _____

_____

Vital Capacity (VC)

1. Make sure the spirometer is set to zero.
2. Inhale as much as possible and then exhale as much as possible into the spirometer.
3. Three readings are needed, so twice more, set the spirometer to zero and repeat the same procedure. Record your measurements in Table 16.2.
4. Later, if necessary, change your readings to ml, and calculate your average VC.

In your own words, what is vital capacity? _____

_____

Inspiratory Reserve Volume (IRV)

It will be necessary for us to calculate IRV because a spirometer measures only exhaled air, not inhaled air.

Explain. _____

From having measured vital capacity (VC) you can see that VC = TV + IRV + ERV. To calculate IRV, simply subtract the average TV + the average ERV from the value you recorded for the average VC:

IRV = VC − (TV + ERV) = _____ ml. Record your IRV in Table 16.2.

Table 16.2 Measurements of Lung Volumes			
Tidal Volume (TV)	**Expiratory Reserve Volume (ERV)**	**Vital Capacity (VC)**	**Inspiratory Reserve Volume (IRV)**
1st	1st	1st	_____
2nd	2nd	2nd	_____
3rd	3rd	3rd	_____
Average ml	Average ml	Average ml	Calculated value = ml

Conclusions: Human Respiratory Volumes

- Vital capacity varies with age, gender, and height; however, typically for men, vital capacity is about 5,200 ml and for women, it is about 4,000 ml. How does your vital capacity compare to the typical values for your gender? _____ If smaller than normal, are you a smoker or is there any health reason why it would be smaller? If larger than normal, do you participate in endurance sports of some kind or do you play a musical instrument that involves inhaling and exhaling deeply? _____

- Diffusion alone accounts for pulmonary gas exchange. Therefore, how does good lung ventilation assist gas exchange?_____

_____

16.4 Kidneys

The **kidneys** are bean-shaped organs that lie along the dorsal wall of the abdominal cavity.

Kidney Structure

Figure 16.9 shows the structure of a kidney, macroscopic and microscopic. The macroscopic structure of a kidney is due to the placement of over 1 million **nephrons.** Nephrons are tubules that do the work of producing urine.

Figure 16.9 Longitudinal section of a kidney.
a. The kidneys are served by the renal artery and renal vein. **b.** Macroscopically, a kidney has three parts: renal cortex, renal medulla, and renal pelvis. **c.** Microscopically, each kidney contains over a million nephrons.

Observation: Kidney Model

Study a model of a kidney, and with the help of Figure 16.9, locate the following:

1. **Renal cortex:** a granular region
2. **Renal medulla:** contains the renal pyramids
3. **Renal pelvis:** where urine collects

Study a nephron model and, with the help of Figure 16.10, identify the following parts of a nephron:

1. **Glomerular capsule** (Bowman's capsule): closed end of the nephron pushed in on itself to form a cuplike structure; the inner layer has pores that allow **glomerular filtration** to occur; substances move from the blood to inside the nephron.
2. **Proximal convoluted tubule:** the inner layer of this region has many microvilli that allow **tubular reabsorption** to occur; substances move from inside the nephron to the blood.
3. **Loop of the nephron:** nephron narrows to form a U-shaped portion which functions in water reabsorption.
4. **Distal convoluted tubule:** second convoluted section that lacks microvilli and functions in **tubular secretion;** substances move from blood to inside nephron.

Several nephrons enter one collecting duct. The **collecting ducts** also function in water reabsorption, and they conduct urine to the pelvis of a kidney.

Observation: Circulation About a Nephron

Study a nephron model and, with the help of Figure 16.10 and Table 16.3, trace the path of blood from the renal artery to the renal vein:

1. **Afferent arteriole:** small vessel that conducts blood from the renal artery to a nephron.
2. **Glomerulus:** capillary network that exists inside the glomerular capsule; small molecules move from inside the capillary to the inside of the glomerulus during glomerular filtration.
3. **Efferent arteriole:** small vessel that conducts blood from the glomerulus to the peritubular capillary network.
4. **Peritubular capillary network:** surrounds the proximal convoluted tubule, the loop of the nephron, and the distal convoluted tubule.
5. **Venule:** takes blood from the peritubular capillary network to the renal vein.

Table 16.3 Blood Vessels Serving the Nephron

Name of Structure	Significance
Afferent arteriole	Brings arteriolar blood to the glomerulus
Glomerulus	Capillary tuft enveloped by glomerular capsule
Efferent arteriole	Takes arteriolar blood away from the glomerulus
Peritubular capillary network	Capillary bed that envelops the rest of the nephron
Venule	Takes venous blood away from the peritubular capillary network

Kidney Function

The kidneys produce urine and in doing so help maintain homeostasis in several ways. Urine formation requires three steps: **glomerular filtration, tubular reabsorption,** and **tubular secretion** (Fig. 16.10).

Figure 16.10 Nephron structure and blood supply.

The three main processes in urine formation are described in boxes and color coded to arrows that show the movement of molecules out of or into the nephron at specific locations. In the end, urine is composed of the substances within the collecting duct (see brown arrow).

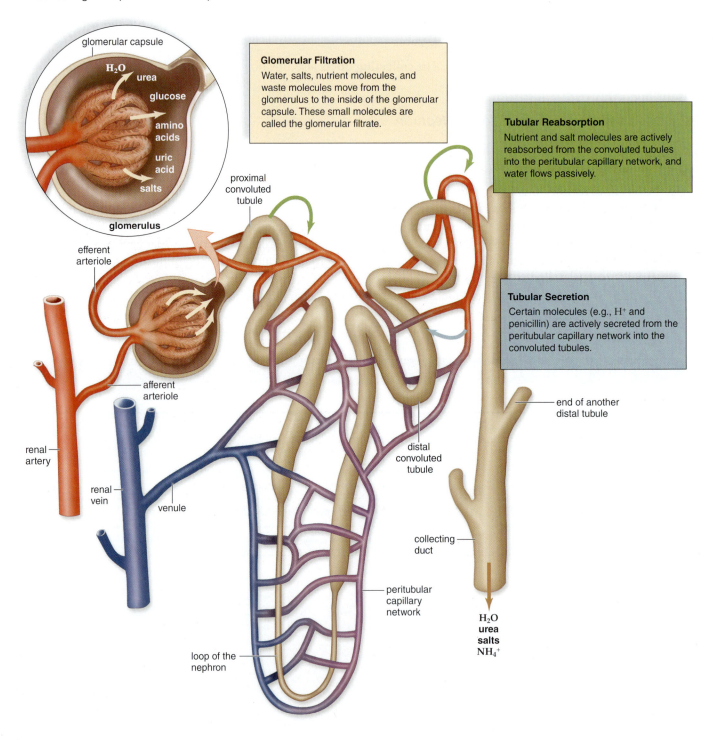

Glomerular Filtration
Water, salts, nutrient molecules, and waste molecules move from the glomerulus to the inside of the glomerular capsule. These small molecules are called the glomerular filtrate.

Tubular Reabsorption
Nutrient and salt molecules are actively reabsorbed from the convoluted tubules into the peritubular capillary network, and water flows passively.

Tubular Secretion
Certain molecules (e.g., H^+ and penicillin) are actively secreted from the peritubular capillary network into the convoluted tubules.

Glomerular Filtration

1. Blood entering the glomerulus contains blood cells, proteins, glucose, amino acids, salts, urea, and water. Blood cells and proteins are too large to pass through the glomerular wall and enter the filtrate.

2. Blood pressure causes small molecules of glucose, amino acids, salts, urea, and water to exit the blood and enter the glomerular capsule. The fluid in the glomerular capsule is called the **filtrate.**

3. In the list that follows, draw an arrow from left to right for the small molecules that leave the glomerulus and become part of the filtrate.

   Glomerulus   **Glomerular (Filtrate)**
   Cells
   Proteins
   Glucose
   Amino acids
   Urea
   Water and salts

4. Complete the second column in Table 16.4, page 220. Use an X to indicate that the substance is at the locations noted.

Tubular Reabsorption

1. When the filtrate enters the proximal convoluted tubule, it contains glucose, amino acids, urea, water, and salts. Some water and salts remain in the nephron, but enough are *passively* reabsorbed into the peritubular capillary to maintain blood volume and blood pressure. Use this information to state a way kidneys help

   maintain homeostasis. _____

   _____

2. The cells that line the proximal convoluted tubule are also engaged in active transport and usually completely reabsorb nutrients (glucose and amino acids) into the peritubular capillary. What would happen

   to cells if the body lost all its nutrients by way of the kidneys? _____

   _____

3. Which of the filtrate substances is reabsorbed the least and will become a part of urine? _____

   Urea is a nitrogenous waste. State here another way kidneys contribute to homeostasis. _____

   _____

4. In the list that follows, draw an arrow from left to right for all those molecules passively reabsorbed into the blood of the peritubular capillary. Use darker arrows for those that are normally reabsorbed completely by active transport.

   Proximal Convoluted Tubule   **Peritubular Capillary**
   Water and salts
   Glucose
   Amino acids
   Urea

Tubular Secretion

1. During tubular secretion, certain substances—for example, penicillin and histamine—are actively secreted from the peritubular capillary into the fluid of the tubule. Also, hydrogen ions (H^+) and ammonia (NH_3) are secreted as NH_4^+ as necessary. Complete the last column of Table 16.4. Check your entries against Figure 16.10.
2. The blood is buffered, but only the kidneys can excrete H^+. The excretion of H^+ by the kidneys raises the pH of the blood. Use this information to state a third way the kidneys contribute to homeostasis. _____

_____

Table 16.4   Urine Constituents

In Glomerulus	In Filtrate	In Urine
Blood cells		
Proteins		
Glucose		
Amino acids		
Urea		
Water and salts		
NH_4^+		

Urinalysis: A Diagnostic Tool

Urinalysis can indicate whether the kidneys are functioning properly or whether an illness such as diabetes mellitus is present. The procedure is easily performed with a Chemstrip test strip, which has indicator spots that produce specific color reactions when certain substances are present in urine.

Experimental Procedure: Urinalysis

A urinalysis has been ordered, and you are to test the urine for a possible illness. (In this laboratory, you will be testing simulated urine.)

Assemble Supplies

1. Obtain three Chemstrip urine test strips each of which tests for leukocytes, pH, protein, glucose, ketones, and blood, as noted in Figure 16.11.
2. The color key on the diagnostic color chart or on the Chemstrip vial label will explain what any color changes mean in terms of the pH level and amount of each substance present in the urine sample. You will use these color blocks to read the results of your test.
3. Obtain three "specimen containers of urine" marked 1 through 3. Among them are a normal specimen and two that indicate the patient has an illness. Have a piece of absorbent paper ready to use.

Test the Specimen

1. Briefly (no longer than 1 second) dip a test strip into the first specimen of urine. Be sure the chemically treated patches on the test strip are totally immersed.
2. Draw the edge of the strip along the rim of the specimen container to remove excess urine.
3. Turn the test strip on its side, and tap once on a piece of absorbent paper to remove any remaining urine and to prevent the possible mixing of chemicals.
4. After 60 seconds, read the results as follows: Hold the strip close to the color blocks on the diagnostic color chart (Fig. 16.11) or vial label, and match carefully, ensuring that the strip is properly oriented to the color chart. Enter the test results in Figure 16.11. Use a negative symbol (−) for items that are not present in the urine, a plus symbol (+) for those that are present, and a number for the pH.
5. Test the other two specimens and complete Figure 16.11.

Figure 16.11 Urinalysis test.
A Chemstrip test strip can help determine illness in a patient by detecting substances in the urine. If leukocytes (white blood cells), protein, or blood are in the urine, the kidneys are not functioning properly. If glucose and ketones are in the urine, the patient has diabetes mellitus (type 1 or type 2).

Tests For:	Normal	Results		
		Test 1	Test 2	Test 3
leukocytes	−			
pH	pH 5			
protein	−			
glucose	−			
ketones	−			
blood	−			

strip handle

Chemstrip before urine test

Conclusion: Urinalysis

- State below if the urinalysis is normal or indicates a urinary tract infection (leukocytes, blood, and possibly protein in the urine) or that the patient has diabetes mellitus.

Test strip 1 _____

Test strip 2 _____

Test strip 3 _____

- The hormone insulin promotes the uptake of glucose by cells. When glucose is in the urine, either the pancreas is not producing insulin (diabetes mellitus type 1) or cells are resistant to insulin (diabetes mellitus type 2). Ketones (acids) are also in the urine because the cells are metabolizing fat instead of glucose.

Explain why. _____

_____

Why is the pH of urine lower than normal? _____

_____

- If urinalysis shows that proteins are excreted instead of retained in the blood, would capillary exchange in the tissues (see Fig. 16.5) be normal? _____ Why or why not? _____

_____

Laboratory Review 16

_____ 1. What organs contribute to homeostasis by removing nitrogenous waste from the blood?

_____ 2. What blood vessels are involved in the exchange of gases that takes place in the lungs?

_____ 3. Does the liver store or release glucose if the hormone insulin is present?

_____ 4. When a heart chamber contracts, is it called systole or diastole?

_____ 5. If blood pressure is measured to be 112/72, which number corresponds to the relaxation of the ventricles?

_____ 6. What lung structures have burst in someone who has emphysema?

_____ 7. During inspiration, does the diaphragm contract or relax, and does that make it move up or down?

_____ 8. Which respiratory volume is obtained by measuring the amount of air moved in and out of the lungs with normal breathing?

_____ 9. What is the name of the instrument used to measure lung volumes?

_____ 10. Besides gender and age, what characteristic affects lung volumes?

_____ 11. What is the outermost granular region of the kidney called?

_____ 12. During what stage of urine formation is glucose transferred from the tubule to the peritubular capillary network?

_____ 13. What blood contents are too big to pass through the glomerular wall and enter the filtrate?

_____ 14. What appears in the urine of someone with diabetes mellitus besides glucose?

Thought Questions

15. If someone has high blood pressure, they might be prescribed a beta blocker medication to lower the heart rate or a medication that dilates blood vessels. Explain why lower blood pressure results from treatment with either kind of medication.

16. Why would someone with emphysema complain of shortness of breath and have difficulty climbing stairs? Why would regular exercise be recommended for someone with emphysema?

17. Predict the results (pH and substances present) of a urinalysis on an individual following a high protein, low carbohydrate diet. Explain.

LABORATORY

17

Nervous System and Senses

Learning Outcomes

17.1 Central Nervous System
- Identify the parts of the brain studied, and state the functions of each part.   224–227
- Label a diagram of the human brain, and give examples to show that the parts of the brain work together.   226–227
- Describe the anatomy of the spinal cord, and tell how the cord functions as a relay station.   228

17.2 Peripheral Nervous System
- Distinguish between cranial nerves and spinal nerves on the basis of location and function.   228
- Describe the anatomy and physiology of a spinal reflex arc.   228–230

17.3 The Human Eye
- Identify the parts of the eye and state a function for each part.   230–231
- Explain how to test for the blind spot and accommodation.   232–233

17.4 The Human Ear
- Identify the parts of the ear and state a function for each part.   233–235

17.5 Sensory Receptors in Human Skin
- Describe the anatomy of the human skin, and explain the distribution and function of sensory receptors.   235–236
- Relate the abundance of touch receptors to the ability to distinguish between two different touch points.   236

17.6 Human Chemoreceptors
- Relate the ability to distinguish foods to the senses of smell and taste.   237

Introduction

The nervous system has two major divisions. One is the central nervous system (CNS), which consists of the brain and spinal cord. The other is the peripheral nervous system (PNS), which contains cranial nerves and spinal nerves (Fig. 17.1). Sensory receptors detect changes in environmental stimuli, and nerve impulses move along sensory nerve fibers to the brain and the spinal cord. The brain and spinal cord sum up the data before sending impulses via motor nerve fibers to effectors (muscles and glands) so a response to stimuli is possible. The brain and spinal cord contain all parts of neurons—dendrites, cell body, axon, and axon terminal—but nerves contain only axons.

Figure 17.1   The nervous system.
The central nervous system (CNS) is in the midline of the body, and the peripheral nervous system (PNS) is outside the CNS.

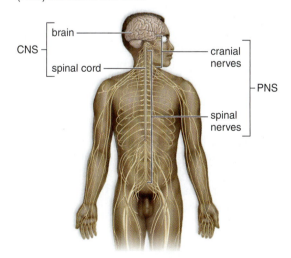

17.1 Central Nervous System

The brain is the enlarged anterior end of the spinal cord. It contains centers that receive input from and can command other regions of the nervous system.

> ⚠️ **Latex gloves** Wear protective latex gloves when handling preserved animal organs. Use protective eyewear and exercise caution when using sharp instruments during this laboratory. Wash hands thoroughly upon completion of this laboratory.

Preserved Sheep Brain

The sheep brain (Fig. 17.2) is often used to study the brain. It is easily available and large enough that individual parts can be identified.

Observation: Preserved Sheep Brain

Examine the exterior and a midsagittal (longitudinal) section of a preserved sheep brain or a model of the human brain, and with the help of Figure 17.2, identify the following:

1. **Ventricles:** interconnecting spaces that produce and serve as a reservoir for cerebrospinal fluid, which cushions the brain. Toward the anterior, note the lateral ventricle on one longitudinal section and similarly a lateral ventricle on the other longitudinal section.

2. **Cerebrum:** most developed area of the brain; responsible for higher mental capabilities. The cerebrum is divided into the right and left **cerebral hemispheres,** joined by the **corpus callosum,** a broad sheet of white matter. The outer portion of the cerebrum is highly convoluted and divided into the following surface lobes (see Fig. 17.4):

   a. **Frontal lobe:** controls motor functions and permits voluntary muscle control; it also is responsible for abilities to think, problem solve, speak, and smell.

   b. **Parietal lobe:** receives information from sensory receptors located in the skin and also the taste receptors in the mouth. A groove called the **central sulcus** separates the frontal lobe from the parietal lobe.

   c. **Occipital lobe:** interprets visual input and combines visual images with other sensory experiences. The optic nerves split and enter opposite sides of the brain at the optic chiasma, located in the diencephalon.

   d. **Temporal lobe:** has sensory areas for hearing and smelling. The olfactory bulb contains nerve fibers that communicate with the olfactory cells in the nasal passages and take nerve impulses to the temporal lobe.

3. **Diencephalon:** portion of the brain where the third ventricle is located. The hypothalamus and thalamus are also located here.

   a. **Thalamus:** two connected lobes located in the roof of the third ventricle. The thalamus is the highest portion of the brain to receive sensory impulses before the cerebrum. It is believed to control which received impulses are passed on to the cerebrum. For this reason, the thalamus sometimes is called the "gatekeeper to the cerebrum."

   b. **Hypothalamus:** forms the floor of the third ventricle and contains control centers for appetite, body temperature, blood pressure, and water balance. Its primary function is homeostasis. The hypothalamus also has centers for pleasure, reproductive behavior, hostility, and pain.

4. **Cerebellum:** located just posterior to the cerebrum as you observe the brain dorsally, the cerebellum's two lobes make it appear rather like a butterfly. In cross section, the cerebellum has an internal pattern that looks like a tree. The cerebellum coordinates equilibrium and motor activity to produce smooth movements.

Figure 17.2   The sheep brain.

(a) © Dr. J. Timothy Cannon; (b) © Dr. J. Timothy Cannon; (c) © Dr. J. Timothy Cannon

olfactory bulb

right cerebral
hemisphere

temporal lobe
of cerebrum

pons

medulla
oblongata

longitudinal
fissure

left cerebral
hemisphere

optic
chiasma

midbrain

cranial
nerve

cerebellum

spinal cord

a. Ventral view

occipital
lobe

parietal
lobe

cerebellum

central sulcus

spinal cord

frontal
lobe

medulla
oblongata

pons

temporal
lobe

olfactory
bulb

b. Lateral view

pineal body

hypothalamus

thalamus

diencephalon

cerebellum

cerebrum

fourth
ventricle

corpus
callosum

spinal cord

lateral
ventricle

medulla
oblongata

pons

midbrain

third
ventricle

optic
chiasma

brain stem

c. Longitudinal section

5. **Brain stem:** part of the brain that connects with the spinal cord. Because it includes the pons and medulla oblongata, it contains centers for the functioning of internal organs. Based on its location, it serves as a relay station for nerve impulses passing from the cord to the brain. Therefore, it helps keep the rest of the brain alert and functioning.

   a. **Midbrain:** anterior to the pons, the midbrain serves as a relay station for sensory input and motor output. It also contains a reflex center for eye muscles.

   b. **Pons:** the ventral, bulblike enlargement on the brain stem. It serves as a passageway for nerve impulses running between the medulla and the higher brain regions.

   c. **Medulla oblongata** (or simply **medulla**): the most posterior portion of the brain stem. It controls internal organs; for example, blood pressure, cardiac, and breathing control centers are present in the medulla. Nerve impulses pass from the spinal cord through the medulla to and from higher brain regions.

The Human Brain

Based on your knowledge of the sheep brain, complete Table 17.1 by stating the major functions of each part of the brain listed. *Also label Figure 17.3.*

Table 17.1 Summary of Brain Functions	
Part	**Major Functions**
Cerebrum	
Cerebellum	
Diencephalon Thalamus	
Hypothalamus	
Brain stem Midbrain	
Pons	
Medulla oblongata	

Which parts of the brain work together to achieve the following?

1. Good eye–hand coordination _____

2. Concentrating on homework when TV is playing _____

3. Avoiding dark alleys while walking home at night _____

4. Keeping the blood pressure within the normal range. _____

Figure 17.3 The human brain (longitudinal section).

The cerebrum is larger in humans than in sheep. Label where indicated using: brain stem, cerebellum, cerebrum, diencephalon, hypothalamus, medulla oblongata, midbrain, pons, thalamus.

Cerebral Lobes

As stated previously, the outer portion of the cerebrum is highly convoluted and divided into lobes as illustrated in Figure 17.4. The various sense organs send nerve impulses to a particular lobe where the nerve impulses are integrated to give us our senses of vision, hearing, smell, taste, and touch. Although not stated in Figure 17.4, the frontal lobe helps us remember smells of some significance to us. *Write the name of a sense next to the appropriate lobe in Figure 17.4.*

Figure 17.4 The cerebral lobes.

Each lobe has centers for integrating nerve impulses received from a particular type of sense organ. Our five senses result from this activity of the brain.

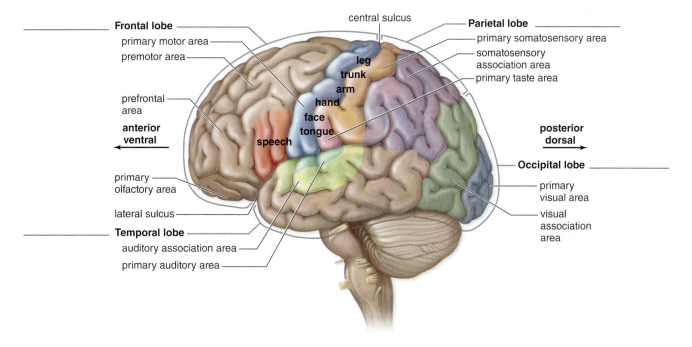

The Spinal Cord

The spinal cord is a part of the central nervous system. It lies in the middorsal region of the body and is protected by vertebrae.

Observation: The Spinal Cord

1. Examine a prepared slide of a cross section of the spinal cord under the lowest magnification possible. For example, some microscopes are equipped with a short scanning objective that enlarges about 3.5×, with a total magnification of 35×. If a scanning objective is not available, observe the slide against a white background with the naked eye.
2. Identify the following with the help of Figure 17.5:

   a. **Gray matter:** a central, butterfly-shaped area composed of axons of sensory neurons, interneurons, and motor neuron cell bodies.

   b. **White matter:** masses of long fibers that lie outside the gray matter and carry impulses up and down the spinal cord. In living animals, white matter appears white because an insulating myelin sheath surrounds long fibers.

Figure 17.5   The spinal cord.
Photomicrograph of spinal cord cross section.
© Kage Mikrofotografie/Phototake

central canal

gray matter

white matter

17.2 Peripheral Nervous System

The peripheral nervous system contains the cranial nerves and the spinal nerves. Twelve pairs of cranial nerves project from the inferior surface of the brain. The cranial nerves are largely concerned with nervous communication between the head, neck, and facial regions of the body and the brain. The 31 pairs of spinal nerves emerge from either side of the spinal cord (Fig. 17.6).

Spinal Nerves

Each spinal nerve contains long fibers of sensory neurons and long fibers of motor neurons. In Figure 17.6, identify the following:

1. **Sensory neuron:** takes nerve impulses from a sensory receptor to the spinal cord. The cell body of a sensory neuron is in the dorsal root ganglion.
2. **Interneuron:** lies completely within the spinal cord. Some interneurons have long fibers and take nerve impulses to and from the brain. The interneuron in Figure 17.6 transmits nerve impulses from the sensory neuron to the motor neuron.
3. **Motor neuron:** takes nerve impulses from the spinal cord to an effector—in this case, a muscle. Muscle contraction is one type of response to stimuli.

Suppose you were walking barefoot and stepped on a prickly sandbur. Describe the pathway of information, starting with the pain receptor in your foot, that would allow you to both feel and respond to this unwelcome

stimulus. _____

_____

_____

Spinal Reflexes

A **reflex** is an involuntary and predictable response to a given stimulus that allows a quick response to environmental stimuli without communicating with the brain. When you touch a sharp tack, you immediately

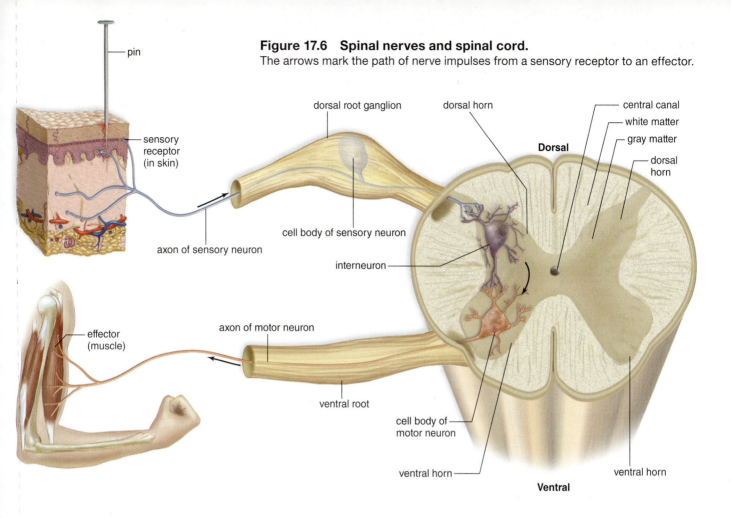

Figure 17.6 Spinal nerves and spinal cord.
The arrows mark the path of nerve impulses from a sensory receptor to an effector.

pin

sensory receptor (in skin)

dorsal root ganglion

dorsal horn

central canal

white matter

gray matter

Dorsal

dorsal horn

axon of sensory neuron

cell body of sensory neuron

interneuron

effector (muscle)

axon of motor neuron

ventral root

cell body of motor neuron

ventral horn

ventral horn

Ventral

withdraw your hand (Fig. 17.6). When a spinal reflex occurs, a sensory receptor is stimulated and generates nerve impulses that pass along the three neurons mentioned earlier—the sensory neuron, interneuron, and motor neuron—until the effector responds. During the investigation of the knee-jerk reflex (Fig. 17.7), a receptor detects the tap, and sensory neurons conduct nerve impulses to interneurons in the spinal cord. The interneurons send a message via motor neurons to the effectors, muscles in the leg or foot. These reflexes are involuntary because the brain is not involved in formulating the response. *Consciousness* of the stimulus lags behind the response because information must be sent up the spinal cord to the brain before you can become aware of the tap.

Experimental Procedure: Spinal Reflex

Although many reflexes occur in the body, only a tendon reflex is investigated in this Experimental Procedure. One easily tested tendon reflex involves the **patellar tendon.** When this tendon is tapped with a reflex hammer (Fig. 17.7) or meter stick, the attached muscle is stretched. This causes a receptor to generate

Figure 17.7 Knee-jerk reflex.
The quick response when the patellar tendon is stimulated by tapping with a reflex hammer indicates that a reflex has occurred. © Ph. Gerbier/Science Source

Knee-jerk (patellar) reflex

nerve impulses which are transmitted along sensory neurons to the spinal cord. Nerve impulses from the cord then pass along motor neurons and stimulate the muscle, causing it to contract. As the muscle contracts, it tugs on the tendon, causing movement of a bone opposite the joint. Receptors in other tendons, such as the Achilles tendon, respond similarly. Such reflexes help the body automatically maintain balance and posture.

Knee-Jerk (Patellar) Reflex

1. Have the subject sit on a stool or table so that his or her legs hang freely.
2. Sharply tap one of the patellar tendons just below the patella (kneecap) with a meter stick.
3. In this relaxed state, does the leg flex (move toward the buttocks) or extend (move away from the

   buttocks)? _____

17.3 The Human Eye

The human eye is responsible for sight. Light rays enter the eye and strike the **rod cells** and **cone cells,** the photoreceptors for sight. The rods and cones generate nerve impulses that go to the brain via the optic nerve.

Observation: The Human Eye

1. Examine a human eye model, and identify the structures listed in Table 17.2 and depicted in Figure 17.8.
2. Trace the path of light from outside the eye to the retina. _____

   _____

   _____

   _____

3. During **accommodation,** the lens rounds up to aid in viewing near objects or flattens to aid in viewing

   distant objects. Which structure holds the lens and is involved in accommodation? _____

   _____

   _____

4. **Refraction** is the bending of light rays so that they can be brought to a single focus. Which of the structures

   listed in Table 17.2 aid in refracting and focusing light rays? _____

   _____

5. Specifically, what are the sensory receptors for sight, and where are they located in the eye? _____

   _____

   _____

6. What structure takes nerve impulses to the brain from the rod cells and cone cells? _____

   _____

7. Which cerebral lobe processes nerve impulses from an eye? _____

Table 17.2 Parts of the Human Eye

Part	Location	Function
Sclera	Outer layer of eye	Protects and supports eyeball
Cornea	Transparent portion of sclera	Refracts light rays
Choroid	Middle layer of eye	Absorbs stray light rays
Retina	Inner layer of eye	Contains receptors for sight
Rod cells	In retina	Make black-and-white vision possible
Cone cells	Concentrated in fovea centralis	Make color vision possible
Fovea centralis	Special region of retina	Makes acute vision possible
Lens	Interior of eye between cavities	Refracts and focuses light rays
Ciliary body	Extension from choroid	Holds lens in place; functions in accommodation
Iris	More anterior extension of choroid	Regulates light entrance
Pupil	Opening in middle of iris	Admits light
Humors (aqueous and vitreous)	Fluid media in anterior and posterior compartments, respectively, of eye	Transmit and refract light rays; support eyeball
Optic nerve	Extension from posterior of eye	Transmits impulses to occipital lobe of brain

Figure 17.8 Anatomy of the human eye.
The sensory receptors for vision are the rod cells and cone cells present in the retina of the eye.

The Blind Spot of the Eye

The **blind spot** occurs where the optic nerve fibers exit the retina. No vision is possible at this location because of the absence of rod cells and cone cells.

Experimental Procedure: Blind Spot of the Eye

This Experimental Procedure requires a laboratory partner. Figure 17.9 shows a small circle and a cross several centimeters apart.

Figure 17.9 Blind spot.
This dark circle (or cross) will disappear at one location because there are no rod cells or cone cells at each eye's blind spot, where vision does not occur.

Left Eye

1. Hold Figure 17.9 approximately 30 cm from your eyes. The cross should be directly in front of your left eye. If you wear glasses, keep them on.
2. Close your right eye.
3. Stare only at the cross with your left eye. You should also be able to see the circle in the same field of vision. Slowly move the paper toward you until the circle disappears.
4. Repeat the procedure as many times as needed to find the blind spot.
5. Then slowly move the paper closer to your eyes until the circle reappears. Because only your left eye is open, you have found the blind spot of your left eye.
6. With your partner's help, measure the distance from your eye to the paper when the circle first

   disappeared. Left eye: _____ cm

Right Eye

1. Hold Figure 17.9 approximately 30 cm from your eyes. The circle should be directly in front of your right eye. If you wear glasses, keep them on.
2. Close your left eye.
3. Stare only at the circle with your right eye. You should also be able to see the cross in the same field of vision. Slowly move the paper toward you until the cross disappears.
4. Repeat the procedure as many times as needed to find the blind spot.
5. Then slowly move the paper closer to your eyes until the cross reappears. Because only your right eye is open, you have found the blind spot of your right eye.
6. With your partner's help, measure the distance from your eye to the paper when the cross first

   disappeared. Right eye: _____ cm

Why are you unaware of a blind spot under normal conditions? Although the eye detects patterns of light and color, it is the brain that determines what we visually perceive. The brain interprets the visual input based in part on past experiences. In this exercise, you created an artificial situation in which you became aware of how your perception of the world is constrained by the eye's anatomy.

Accommodation of the Eye

When the eye accommodates to see objects at different distances, the shape of the lens changes. The lens shape is controlled by the ciliary muscles attached to it. When you are looking at a distant object, the lens is in a flattened state. When you are looking at a closer object, the lens becomes more rounded. The elasticity of the lens determines how well the eye can accommodate. Lens elasticity decreases with increasing age, a condition called **presbyopia.** Presbyopia is the reason many older people need bifocals or reading glasses to see near objects.

Experimental Procedure: Accommodation of the Eye

This Experimental Procedure requires a laboratory partner. It tests accommodation of either your left or your right eye.

1. Hold a pencil upright by the eraser and at arm's length in front of whichever of your eyes you are testing (Fig. 17.10).
2. Close the opposite eye.
3. Move the pencil from arm's length toward your eye.
4. Focus on the end of the pencil.
5. Move the pencil toward you until the end is out of focus. Measure the distance (in centimeters)

   between the pencil and your eye: _____ cm. This is the distance at which your eye can no longer accommodate for distance.

6. If you wear glasses, repeat this experiment without your glasses, and note the accommodation distance of your eye without glasses:

   _____ cm. (Contact lens wearers need not make these determinations, and they should write the words *contact lens* in this blank.)

7. The "younger" lens can easily accommodate for closer distances. The nearest point at which the end of the pencil can be clearly seen is called the **near point.** The more elastic the lens, the "younger" the eye

   (Table 17.3). How "old" is the eye you tested? _____

Figure 17.10 Accommodation.
When testing the ability of your eyes to accommodate to see a near object, always keep the pencil in this position.

Table 17.3 Near Point and Age Correlation						
Age (years)	10	20	30	40	50	60
Near point (cm)	9	10	13	18	50	83

17.4 The Human Ear

The human ear, whose parts are listed and depicted in Table 17.4 and Figure 17.11, serves two functions: hearing and balance.

Observation: The Human Ear

Examine a human ear model, and find the structures depicted in Figure 17.11 based on the information given in Table 17.4.

Table 17.4 Parts of the Human Ear

Part	Medium	Function	Mechanoreceptor
Outer ear	Air		
Pinna		Collects sound waves	—
Auditory canal		Filters air	—
Middle ear	Air		
Tympanic membrane and ossicles		Amplify sound waves	—
Auditory tube		Equalizes air pressure	—
Inner ear	Fluid		
Semicircular canals		Rotational equilibrium	Stereocilia embedded in cupula
Vestibule (contains utricle and saccule)		Gravitational equilibrium	Stereocilia embedded in otolithic membrane
Cochlea (spiral organ)		Hearing	Stereocilia embedded in tectorial membrane

Figure 17.11 Anatomy of the human ear.
The outer ear extends from the pinna to the tympanic membrane. The middle ear extends from the tympanic membrane to the oval window. The inner ear encompasses the semicircular canals, the vestibule, and the cochlea.

When you hear, sound waves are picked up by the tympanic membrane and amplified by the malleus, incus, and stapes. This creates pressure waves in the canals of the cochlea that lead to stimulation of hair cells, the receptors for hearing. Hair cells in the utricle and saccule of the vestibule and in semicircular canals are receptors for equilibrium (i.e., balance). Nerve impulses from the cochlea travel by way of the cochlear nerve and the vestibular nerve to the brain and eventually are interpreted by the _____ lobe of the brain as sound.

Humans locate the direction of sound according to how fast it is detected by either or both ears. A difference in the hearing ability of the two ears can lead to a mistaken judgment about the direction of sound. You and a laboratory partner should perform this Experimental Procedure on each other. Enter the data for *your* ears, not your partner's ears, in the spaces provided.

1. Ask the subject to be seated, with eyes closed. Then strike a tuning fork or rap two spoons together at the five locations listed in number 2. *Use a random order.*

2. Ask the subject to give the exact location of the sound in relation to his or her head. Record the subject's perceptions when the sound is

   a. directly below and behind the head. _____

   b. directly behind the head. _____

   c. directly above the head. _____

   d. directly in front of the face. _____

   e. to the side of the head. _____

3. Is there an apparent difference in hearing between your two ears? _____

17.5 Sensory Receptors in Human Skin

The sensory receptors in human skin respond to touch, pain, temperature, and pressure (Fig. 17.12). There are individual sensory receptors for each of these stimuli, as well as free nerve endings able to respond to pressure, pain, and temperature.

Figure 17.12   Sensory receptors in the skin.
Each type of receptor shown responds primarily to a particular stimulus.

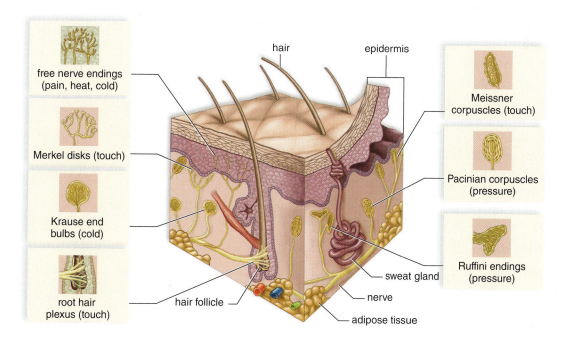

Sense of Touch

The dermis of the skin contains touch receptors, whose concentration differs in various parts of the body.

Experimental Procedure: Sense of Touch

You will need a laboratory partner to perform this Experimental Procedure. Enter *your* data, not the data of your partner, in the spaces provided.

1. Ask the subject to be seated, with eyes closed.
2. Then test the subject's ability to discriminate between the two points of a hairpin, a pair of scissors, or calipers at the four locations listed in number 7.
3. Gently touch the given skin area with the points of the hairpin, scissors, or calipers closed (making one point).
4. Ask the subject whether the experience involves one or two touch sensations.
5. If the subject reports one touch sensation, separate the points of the test item slightly. Gently touch the given skin area again and repeat number 4.
6. Repeat number 5 until the subject reports feeling two touch sensations.
7. Record the shortest distance between the hairpin, scissors, or caliper points for a two-point discrimination.

   a. Forearm: _____ mm   **c.** Index finger: _____ mm

   b. Back of the neck: _____ mm   **d.** Back of the hand: _____ mm

8. Which of these areas apparently contains the greatest density of touch receptors? _____

   Why is this useful? _____

9. Do you have a sense of touch at every point in your skin? _____ Explain. _____

   _____

10. What specific part of the brain processes nerve impulses from touch and pain receptors? _____

   _____

Sense of Heat and Cold

Temperature receptors respond to a change in temperature. If the temperature remains constant, the temperature receptors will experience accommodation to that temperature.

Experimental Procedure: Sense of Heat and Cold

1. Obtain three 1,000 ml beakers and add 500 ml *cold* (tap with *a bit* of ice added to it) water to one, *room temperature* water to the second, and *warm water* (45°–50°C) to the third.
2. Immerse your left hand in the cold-water beaker and your right hand in the warm-water beaker for 30 seconds.
3. Then place both hands in the beaker with room-temperature tap water.
4. Record the sensation in the right and left hands.

   a. Right hand: _____

   b. Left hand: _____

5. Explain your results: _____

   _____

17.6 Human Chemoreceptors

The taste receptors, called *taste buds,* located in the mouth, and the smell receptors, called *olfactory cells,* located in the nasal cavities, are the chemoreceptors that respond to molecules in the air and water. Nerve impulses from taste receptors go to the _____ lobe of the brain, while those from smell receptors go to the _____ lobe of the brain.

Experimental Procedure: Sense of Taste and Smell

You will need a laboratory partner to perform the following procedures. It will not be necessary for all tests to be performed on both partners. You should take turns being either the subject or the experimenter.

Taste and Smell

1. Students work in groups. Each group has one experimenter and several subjects.
2. The experimenter should obtain a LifeSavers® candy from the various flavors available, without letting the subject know what flavor it is.
3. The subject closes both eyes and holds his or her nose.
4. The experimenter gives the LifeSavers® candy to the subject, who places it on his or her tongue.
5. The subject, while still holding his or her nose, guesses the flavor of the candy. The experimenter records the guess in Table 17.5.
6. The subject releases his or her nose and guesses the flavor again. The experimenter records the guess and the actual flavor in Table 17.5.

Table 17.5   Taste and Smell Experiment

Subject	Actual Flavor	Flavor While Holding Nose	Flavor After Releasing Nose
1			
2			
3			
4			
5			

Conclusions: Sense of Taste and Smell

- From your results, how would you say smell affects the taste of LifeSavers® candy? _____

_____

- What do you conclude about the effect of smell on your sense of taste? _____

_____

_____

_____

_____ 1. Is the sciatic nerve part of the central nervous system or the peripheral nervous system?

_____ 2. What part of the brain is divided into right and left hemispheres?

_____ 3. What portion of the brain looks like a tree in cross section?

_____ 4. What is the most posterior portion of the brain stem?

_____ 5. What cerebral lobe is associated with the sense of vision?

_____ 6. What structures protect the spinal cord?

_____ 7. What makes the white matter of the spinal cord appear white?

_____ 8. What type of neuron is found completely within the central nervous system?

_____ 9. What type of neuron is responsible for transmitting nerve impulses from the spinal cord to an effector?

_____ 10. What part of the eye refracts and focuses light rays?

_____ 11. What part of the ear contains the sensory receptors for hearing?

_____ 12. What is the anatomical name for the eardrum that picks up sound waves?

_____ 13. What layer of the skin contains Meissner and Pacinian corpuscles?

_____ 14. What are the names of the chemoreceptors that respond to molecules in air and water?

Thought Questions

15. Explain why people who develop cataracts (cloudy eye lenses) require surgery to implant new lenses.

16. When you stared at the dark cross with your left eye, why did the dark circle "disappear" from view as you moved the illustration toward you?

17. In a drag race, drivers must wait until the green light is illuminated before they can move their vehicle. Explain why a time delay exists, based on the information presented in this exercise.

18

Musculoskeletal System

Learning Outcomes

Introduction

The term **musculoskeletal** system recognizes that contraction of muscles causes the bones to move. The skeletal system consists of the bones and joints, along with the cartilage and ligaments that occur at the joints. The muscular system contains three types of muscles: smooth, cardiac, and skeletal. The skeletal muscles are most often attached to bones after crossing a joint (Fig. 18.1). In humans, the biceps brachii muscle has two **origins,** and the triceps brachii has three origins on the humerus and scapula. Find the biceps brachii **insertion** on the anterior surface of your elbow after contracting your biceps muscle. Locate the ulna at your posterior elbow. The triceps brachii tendon inserts on the ulna.

Muscles work in antagonistic pairs. For example, when the biceps brachii contracts, the bones of the forearm are pulled upward, and the triceps brachii relaxes; and when the triceps brachii contracts, the bones of the forearm are pulled downward, and the biceps brachii relaxes.

Figure 18.1   Muscular action.
Muscles, such as these muscles of the arm (which have their origin on the scapula and their insertion on the bones of the forearm), cause bones to move.

scapula

humerus

origin

biceps
brachii

triceps
brachii

radius

ulna

insertion

18.1 Anatomy of a Long Bone

Although the bones of the skeletal system vary considerably in shape as well as in size, a long bone, such as the human femur, illustrates the general principles of bone anatomy (Fig. 18.2).

Figure 18.2 Anatomy of a long bone from the macroscopic to the microscopic level.
A long bone is encased by the periosteum except at the ends, where it is covered by hyaline (articular) cartilage (see micrograph, *top left*). Spongy bone located at the ends may contain red bone marrow. The medullary cavity contains yellow bone marrow and is bordered by compact bone, which is shown in the enlargement and micrograph (*right*).

(hyaline) © Ed Reschke; (bone) © Dennis Strete/McGraw-Hill Education; (osteocyte) © Biophoto Associates/Science Source

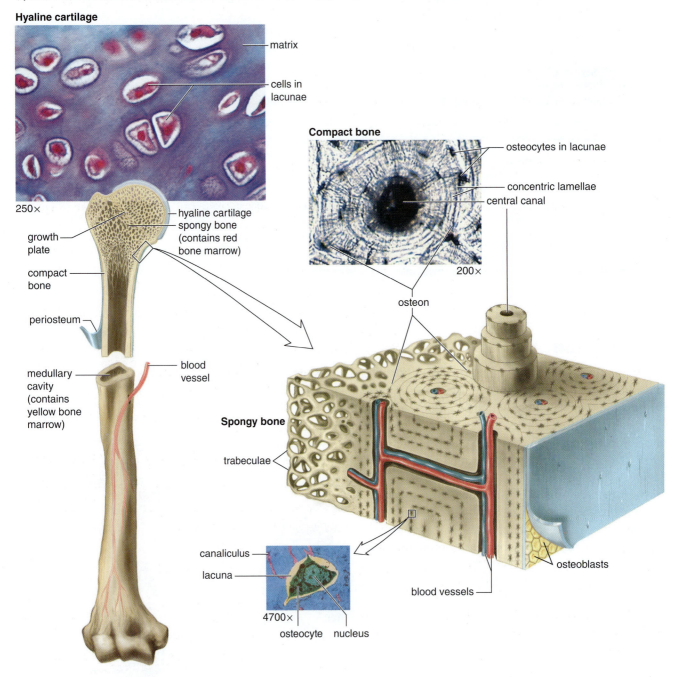

Hyaline cartilage
- matrix
- cells in lacunae
- 250×

Compact bone
- osteocytes in lacunae
- concentric lamellae
- central canal
- 200×
- osteon

- growth plate
- compact bone
- periosteum
- hyaline cartilage
- spongy bone (contains red bone marrow)
- blood vessel
- medullary cavity (contains yellow bone marrow)

Spongy bone
- trabeculae
- canaliculus
- lacuna
- 4700×
- osteocyte
- nucleus
- blood vessels
- osteoblasts

Observation: Anatomy of a Long Bone

Examine the exterior and a longitudinal section of a long bone or a model of a long bone, and with the help of Figure 18.2, identify the following:

1. **Periosteum:** tough, fibrous connective tissue covering continuous with the ligament and tendons that anchor bones; the periosteum allows blood vessels to enter the bone and service its cells.
2. Expanded portions at each end of the bone (*epiphysis*) that contain spongy bone.
3. Extended portion, or shaft (*diaphysis*) of a long bone that lies between the epiphyses; walls of diaphysis are compact bone.
4. **Hyaline (articular) cartilage:** layer of cartilage where the bone articulates with (meets) another bone; decreases friction between bones during movement.
5. **Medullary cavity:** cavity located in the diaphysis that stores yellow marrow, which contains a large amount of fat.

   Label the diaphysis and the epiphysis (twice) in Figure 18.2. Which of these contains the growth line where a long bone can grow in length? _____

Observation: Tissues of a Long Bone

The medullary cavity is bounded at the sides by **compact bone** and at the ends by **spongy bone.** Beyond a thin shell of compact bone is the layer of articular cartilage. **Red marrow,** a specialized tissue that produces all types of blood cells, occurs in the spongy bone of the skull, ribs, sternum, and vertebrae and in the ends of the long bones.

1. Examine a prepared slide of compact bone, and with the help of Figure 18.2, identify
   a. **Osteons:** cylindrical structural units
   b. **Lamellae:** concentric rings of matrix
   c. **Matrix:** nonliving material maintained by osteocytes; contains mineral salts (notably calcium salts) and protein
   d. **Lacunae:** cavities between the lamellae that contain osteocytes (bone cells)
   e. **Central canal:** canal in the center of each osteon; Figure 18.2 shows that there are _____ in a central canal.
   f. **Canaliculi:** tiny channels that contain the processes of cells; these processes allow nutrients to pass between the osteocytes. (sing., *canaliculus*)

   Describe how an osteocyte located near a central canal can pass nutrients to osteocytes located far from the central canal. _____

2. Examine a prepared slide of spongy bone, and with the help of Figure 18.2, identify
   a. **Trabeculae:** bony bars and plates made of mineral salts and protein
   b. **Lacunae:** cavities scattered throughout the trabeculae that contain osteocytes
   c. **Red bone marrow** within large spaces separated by the trabeculae
   What activity occurs in red bone marrow? _____
3. Examine a prepared slide of hyaline cartilage, and with the help of Figure 18.2, identify
   a. **Lacunae:** cavities in twos and threes scattered throughout the matrix, which contain chondrocytes (cells that maintain cartilage)
   b. **Matrix:** material more flexible than bone because it consists primarily of protein
   Seniors tend to have joints that creak. What might be the matter? _____

18.2 The Skeleton

The human skeleton is divided into axial and appendicular components. The **axial skeleton** is the main longitudinal portion and includes the skull (Fig. 18.3), the vertebral column, the sternum, and the ribs. The **appendicular skeleton** includes the bones of the appendages and their supportive pectoral and pelvic (shoulder and hip) girdles.

Figure 18.3   Three views of skull.

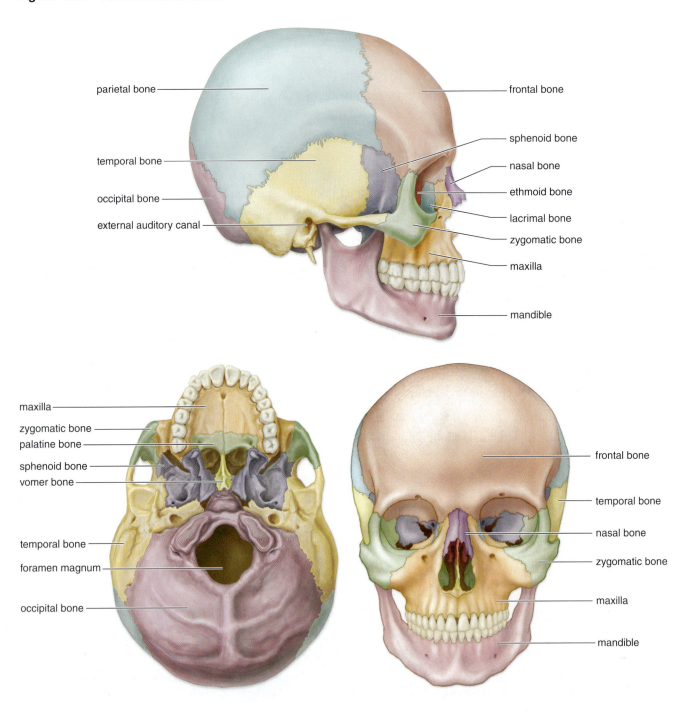

Examine a human skeleton, and with the help of Figures 18.3 and 18.4, identify the **foramen magnum,** a large opening through which the spinal cord passes, and the following bones.

1. The **skull** is composed of many bones fused together at fibrous joints called **sutures.** Note the following in the cranium (brain case):

   a. **Frontal bone:** forms forehead

   b. **Parietal bones:** extend to sides of skull

   c. **Occipital bone:** curves to form base of skull

Figure 18.4   Human skeletal system.

a. Anterior view. **b.** Posterior view. The axial skeleton appears in blue, while the appendicular skeleton is shown in tan.

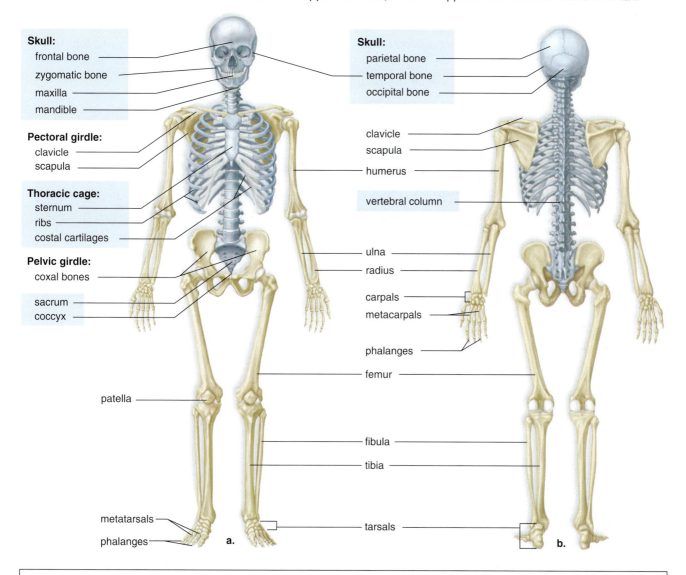

Skull:
frontal bone
zygomatic bone
maxilla
mandible

Pectoral girdle:
clavicle
scapula

Thoracic cage:
sternum
ribs
costal cartilages

Pelvic girdle:
coxal bones
sacrum
coccyx

patella

metatarsals
phalanges

a.

Skull:
parietal bone
temporal bone
occipital bone

clavicle
scapula
humerus

vertebral column

ulna
radius

carpals
metacarpals

phalanges

femur

fibula
tibia

tarsals

b.

NOTE: According to accepted terminology, the upper limb is composed of the humerus in the **arm** and the radius and ulna in the **forearm.** The lower limb is composed of the femur in the **thigh** and the tibia and fibula in the **leg.**

d. **Temporal bones:** located on sides of skull

e. **Sphenoid bone:** helps form base and sides of skull, as well as part of the orbits

Which of these bones contribute to forming the face? _____

Which could best be associated with wearing glasses? _____

2. Note the facial bones:

   a. **Mandible:** the lower jaw

   b. **Maxillae:** the upper jaw and anterior portion of the hard palate

   c. **Palatine bones:** posterior portion of hard palate and floor of nasal cavity

   d. **Zygomatic bones:** cheekbones

   e. **Nasal bones:** bridge of nose

   Which of these is movable and allows you to chew your food? _____

3. The **vertebral column** provides support and houses the **spinal cord.** It is composed of many vertebrae separated from one another by intervertebral disks. The vertebral column customarily is divided into five series:

   a. Seven **cervical vertebrae** (forming the neck region)

   b. Twelve **thoracic vertebrae** (with which the ribs articulate)

   c. Five **lumbar vertebrae** (in the abdominal region)

   d. Five fused sacral vertebrae, called the **sacrum**

   e. Four fused caudal vertebrae forming the **coccyx** in humans

   Which of the vertebrae can be associated with the chest? _____

   Which of the vertebrae could be the cause of your aching back? _____

4. The twelve pairs of **ribs** and their associated muscles form a bony case that supports the thoracic cavity wall. The ribs connect posteriorly with the thoracic vertebrae, and some are also attached by cartilage directly or indirectly to the sternum. Those ribs without any anterior attachment are called **floating ribs.**

   Which of the ribs help form the protective part of the rib cage? _____

After studying Figure 18.4, give several reasons why the axial skeleton is critical to life.

Observation: Appendicular Skeleton

Examine a human skeleton, and with the help of Figure 18.4, identify the following bones.

1. The **pectoral girdles,** which support the upper limbs, are composed of the **clavicle** (collarbone) and **scapula** (shoulder bone).

   Why is the shoulder apt to become dislocated? _____

2. The upper limb (arm plus the forearm) is composed of the following:

   a. **Humerus:** the large long bone of the arm

   b. **Radius:** the long bone of the forearm, with a pivot joint at the elbow that allows rotational motion

   c. **Ulna:** the other long bone of the forearm, with a hinge joint at the elbow that allows motion in only one plane. Take hold of your elbow, and twist the forearm to show that the radius rotates over the ulna but the ulna doesn't move during this action.

   d. **Carpals:** a group of small bones forming the wrist

   e. **Metacarpals:** slender bones forming the palm

   f. **Phalanges:** the bones of the fingers

   Which of these bones would you use to pick up a teacup? _____

3. The **pelvic girdle** forms the basal support for the lower limbs and is composed of two **coxal (hip) bones.** Each coxal bone consists of the *ilium* (which is superior to the other two), the *pubis,* and the *ischium.* (The pubis is ventral to the ischium.)

ilium
coxal bone
pubis
ischium

pelvic girdle

The female pelvis is much broader and shallower than that of the male. The angle between the pubic bones looks like a U in females and a V in males.

How is it advantageous for the female pelvis to be broader and more shallow than that of a

male? _____

4. The lower limb (the thigh plus the leg) is composed of a series of loosely articulated bones, including the following:

   a. **Femur:** the long bone of the thigh

   b. **Patella:** kneecap

   c. **Tibia:** the larger of the two long bones of the leg; feel for the bump on the inside of the ankle

   d. **Fibula:** the smaller of the two long bones of the leg; feel for the bump on the outside of the ankle

   e. **Tarsals:** a group of small bones forming the ankle

   f. **Metatarsals:** slender anterior bones of the foot

   g. **Phalanges:** the bones of the toes

   Which of these bones would you use to kick a soccer ball? _____

Hip Replacement

Find the hip joint in Figure 18.4. The terminology "hip replacement" is a misnomer because no part of the coxal bone is replaced. Instead, the hip joint is replaced. The head of the femur is replaced with a metal ball attached to a metal stem. The stem is inserted into the femur. The inside of the socket on the coxal bone is bolstered with a plastic (or plastic and metal) cup. After the replacement, no part of the joint is living tissue.

18.3 The Skeletal Muscles

This laboratory is concerned with skeletal muscles—those muscles that make up the bulk of the human body. With the help of Figure 18.5 and Tables 18.1 and 18.2, identify the major muscles of the body.

Naming Muscles

Muscles are named for various characteristics, as shown in the following list:

1. **Size:** The *gluteus maximus* is the largest muscle, and it forms the buttocks.
2. **Shape:** The *deltoid* is shaped like a Greek letter delta, or triangle.
3. **Direction of fibers:** The *rectus abdominis* is a longitudinal muscle of the abdomen (*rectus* means "straight").
4. **Location:** The *frontalis* overlies the frontal bone.
5. **Number of attachments:** The *biceps brachii* has two attachments, or origins.
6. **Action:** The *extensor digitorum* extends the fingers, or digits.

Figure 18.5   Human superficial skeletal muscles.

All the muscles of the human body are named in accordance with the structure and/or function of the underlying bone. The muscles highlighted here are those noted in the observation on page 249.

orbicularis oculi

zygomaticus

orbicularis oris

sternocleidomastoid

trapezius

deltoid

pectoralis major

latissimus dorsi

biceps brachii

rectus abdominis

external oblique

flexor carpi group

flexor digitorum

iliopsoas

adductor longus

quadriceps femoris group

peroneus longus
tibialis anterior

extensor digitorum longus

frontalis

masseter

sartorius

Anterior view

occipitalis

sternocleido-mastoid

trapezius

deltoid

latissimus dorsi

external oblique

gluteus maximus

gastrocnemius

triceps brachii

gluteus medius

extensor carpi group

extensor digitorum

biceps femoris

peroneus longus

Posterior view

Limbs
arm: above the elbow
forearm: below the elbow
thigh: above the knee
leg: below the knee

Frontalis

Occipitalis

Masseter

Sternocleidomastoid

Temporalis

Orbicularis oculi

Zygomaticus major

Zygomaticus minor

Buccinator

Orbicularis oris

Platysma

Table 18.1   Muscles (Anterior View)

Name	Action
Head and Neck	
Frontalis	Wrinkles forehead and lifts eyebrows
Orbicularis oculi	Closes eye (winking)
Zygomaticus	Raises corner of mouth (smiling)
Masseter	Closes jaw (chewing)
Orbicularis oris	Closes and protrudes lips (kissing)
Upper Limb and Trunk	
External oblique	Compresses abdomen; rotates trunk
Rectus abdominis	Flexes spine
Pectoralis major (pulls arm across chest)	Flexes and adducts shoulder and arm ventrally
Deltoid	Abducts and raises arm at shoulder joint
Biceps brachii	Flexes forearm and supinates hand
Flexor carpi group	Flexes wrist
Flexor digitorum	Flexes fingers
Lower Limb	
Adductor longus	Adducts and flexes thigh
Iliopsoas	Flexes thigh at hip joint
Sartorius	Rotates thigh (sitting cross-legged)
Quadriceps femoris group	Extends leg
Peroneus longus	Everts foot
Tibialis anterior	Dorsiflexes and inverts foot
Flexor digitorum longus	Flexes toes (not shown)
Extensor digitorum longus	Extends toes

Table 18.2   Muscles (Posterior View)

Name	Action
Head and Neck	
Occipitalis	Moves scalp backward
Sternocleidomastoid	Turns head to side; flexes neck and head
Trapezius	Extends head; raises and adducts shoulders dorsally (shrugging shoulders)
Upper Limb and Trunk	
Latissimus dorsi	Extends and adducts shoulder and arm dorsally (pulls arm across back)
Deltoid	Abducts and raises arm at shoulder joint
External oblique	Rotates trunk
Triceps brachii	Extends forearm
Extensor carpi group	Extends wrist
Extensor digitorum	Extends fingers
Buttocks and Lower Limb	
Gluteus medius	Abducts thigh
Gluteus maximus	Extends thigh (forms buttock)
Biceps femoris	Flexes leg and extends thigh at hip joint
Gastrocnemius	Flexes leg and foot (tiptoeing)

Match these muscles to these functions:

Doing a pelvic tilt and curving the spine _____

Swinging movements of arms during walking and swimming _____

Pulling forearm toward you when rowing _____

Helping maintain the trunk in an erect posture _____

Extending and separating the fingers _____

Wrinkling the skin of the forehead _____

Joint Movements

Figure 18.6 demonstrates the following types of joint movements:

Flexion	Moving jointed body parts toward each other
Extension	Moving jointed body parts away from each other
Adduction	Moving a part toward a vertical plane running through the longitudinal midline of the body
Abduction	Moving a part away from a vertical plane running through the longitudinal midline of the body
Rotation	Moving a body part around its own axis; **circumduction** is moving a body part in a wide circle
Inversion	A movement of the foot in which the sole is turned inward
Eversion	A movement of the foot in which the sole is turned outward

These terms describe the action of the muscles listed in Tables 18.1 and 18.2.

Antagonistic Pairs

Skeletal muscles are attached to the skeleton, and their contraction causes the movement of bones at a joint. Muscles shorten when they contract, so they can only pull; they cannot push. Therefore, muscles work in

Figure 18.6   Joint movements.
a. Flexion and extension. **b.** Adduction and abduction. **c.** Rotation and circumduction. **d.** Inversion and eversion. Red dots indicate pivot points.

antagonistic pairs. Usually, contraction of one member of the pair causes a bone to move in one direction, and contraction of the other member of the pair causes the same bone to move in an opposite direction.

Locate the following antagonistic pairs in Figure 18.5a and b. In each case, state their opposing actions by inserting one of these functions: *flexes, extends, adducts,* or *abducts.*

1. The biceps brachii _____ the forearm.

   The triceps brachii _____ the forearm.

2. The sternocleidomastoid _____ the head.

   The trapezius _____ the head.

3. The iliopsoas _____ the thigh.

   The gluteus maximus _____ the thigh.

4. The quadriceps femoris group _____ the leg.

   The biceps femoris _____ the leg.

Isotonic and Isometric Contractions

A muscle contains many muscle fibers. When a muscle contracts, usually some fibers undergo isotonic contraction, and others undergo isometric contraction. When the tension of muscle fibers is sufficient to lift a load, many fibers change length as they lift the load. The muscle contraction is said to be **isotonic** (same tension). In contrast, when the tension of muscle fibers is used only to support rather than to lift a load, the muscle contraction is said to be **isometric** (same length). The length of many fibers remains the same, but their tension still changes.

Note: The upper limb is composed of the arm plus the forearm.

Isotonic Contraction

1. Start with your left forearm resting on a table. Watch the anterior surface of your left arm while you slowly bend your elbow and bring your left forearm toward the arm. An isotonic contraction of the biceps brachii produces this movement.

2. If a muscle contraction produces movement, is this an isometric or isotonic contraction? _____

Isometric Contraction

1. Place the palm of your left hand underneath a tabletop. Push up against the table while you have your right hand cupped over the anterior surface of your left arm so that you can feel the muscle there undergo an isometric contraction.

2. Is the biceps brachii or the triceps brachii located on the anterior surface of the arm? _____

3. What change did you notice in the firmness of this muscle as it contracted? _____

   _____

4. Did your hand or forearm move as you pushed up against the table? _____

5. Given your answer to question 4, did this muscle's fibers shorten as you pushed up against

   the tabletop? _____

18.4 Mechanism of Muscle Fiber Contraction

A whole skeletal muscle is made up of many cells called **muscle fibers,** and muscle fibers contain many myofibrils (Fig. 18.7). Muscle fibers are striated—that is, they have alternating light and dark bands as seen in a light micrograph of muscle fibers in longitudinal section.

Electron microscopy has shown that striations are due to the placement of protein filaments of **myosin** and **actin.** During contraction, units of the muscle, called **sarcomeres,** shorten because myosin filaments attach to actin filaments by cross-bridges that pull the actin filaments inward. ATP serves as the immediate energy source for sarcomere contraction. Potassium (K^+) and magnesium (Mg^{2+}) ions are cofactors for the breakdown of ATP by myosin.

Observation: Skeletal Muscle

Examine a prepared slide of skeletal muscles, and identify long, multinucleated fibers arranged in a parallel fashion. Note that the muscle fibers are striated (Fig. 18.7). See also section 11.1, page 147 for a photomicrograph of skeletal muscle.

Experimental Procedure: Muscle Fiber Contraction

1. Label two slides "slide 1" and "slide 2." Mount a strand of glycerinated muscle fibers in a drop of glycerol on each slide. Place each slide on a millimeter ruler, and measure the length of the strand. Record these lengths in the first row in Table 18.3.
2. If there is more than a small drop of glycerol on the slides, soak up the excess on a piece of lens paper held at the edge of the glycerol farthest from the fiber strand.
3. To slide 1, add a few drops of a salt solution containing potassium (K^+) and magnesium (Mg^{2+}) ions, and measure any change in strand length. Record your results in Table 18.3.
4. To slide 2, add a few drops of ATP solution, and measure any change in strand length. Record your results in Table 18.3.
5. Now add ATP solution to slide 1. Measure any change in strand length, and record your results in Table 18.3.
6. To slide 2, add a few drops of the K^+/Mg^{2+} salt solution, and measure any change in strand length. Record your results in Table 18.3.

Table 18.3 Glycerinated Muscle Contraction

Solution	Length (mm)	
	Slide 1	**Slide 2**
Glycerol alone		
K^+/Mg^{2+} salt solution alone		—
ATP alone		
Both salt solution and ATP		

7. To demonstrate that you understand the requirements for contraction, state the function of each of the substances listed in Table 18.4.

Table 18.4 Summary of Muscle Fiber Contraction

Substance	Function
Myosin	
Actin	
K^+/Mg^{2+} salt solution	
ATP	

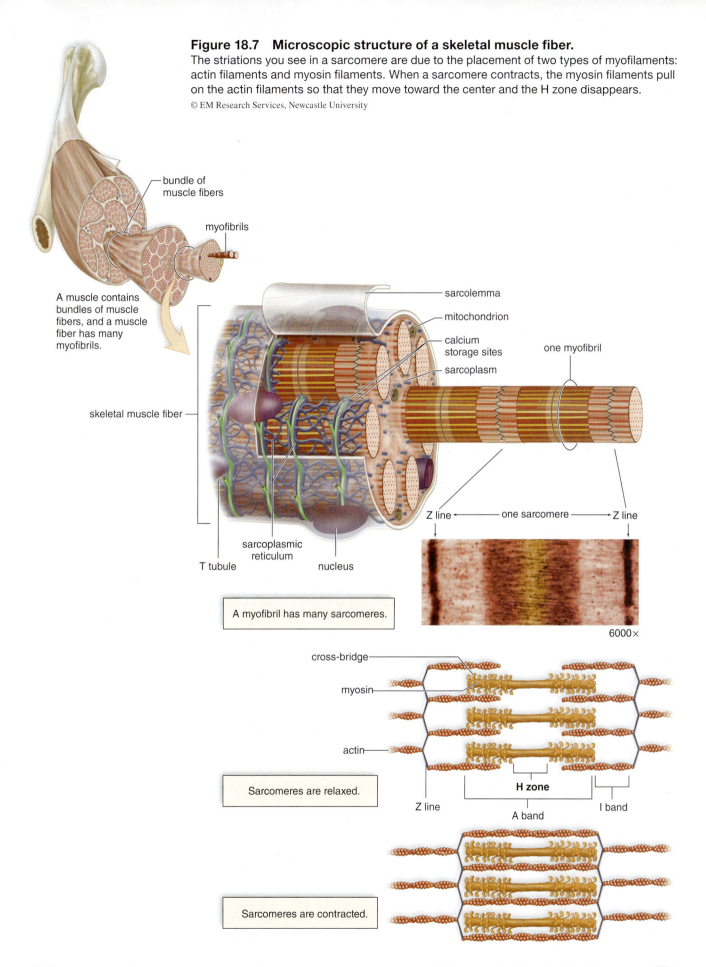

Figure 18.7 Microscopic structure of a skeletal muscle fiber.
The striations you see in a sarcomere are due to the placement of two types of myofilaments: actin filaments and myosin filaments. When a sarcomere contracts, the myosin filaments pull on the actin filaments so that they move toward the center and the H zone disappears.

© EM Research Services, Newcastle University

bundle of muscle fibers

myofibrils

A muscle contains bundles of muscle fibers, and a muscle fiber has many myofibrils.

skeletal muscle fiber

sarcolemma

mitochondrion

calcium storage sites

sarcoplasm

one myofibril

T tubule

sarcoplasmic reticulum

nucleus

Z line ← one sarcomere → Z line

A myofibril has many sarcomeres.

6000×

cross-bridge

myosin

actin

Sarcomeres are relaxed.

Z line

H zone

A band

I band

Sarcomeres are contracted.

_____ 1. Is the growth plate of a long bone associated with the epiphyses or the diaphysis?

_____ 2. What is produced by the red marrow of the spongy bone?

_____ 3. What are osteocytes?

_____ 4. Are the cervical vertebrae part of the appendicular or axial skeleton?

_____ 5. In which skull bone is the foramen magnum found?

_____ 6. What are the fibrous joints associated with the skull called?

_____ 7. Which skull bone encloses the middle and inner ear?

_____ 8. With which vertebrae do the ribs articulate?

_____ 9. What bones comprise the hip joint?

_____ 10. Which muscles contract and extend the phalanges?

_____ 11. Does contraction of the triceps brachii result in extension or flexion of the forearm?

_____ 12. What antagonistic pair of muscles causes adduction and abduction of the thigh?

_____ 13. What are the names of the protein filaments that slide across each other during muscle contractions?

_____ 14. What molecule supplies the energy needed for muscle contractions?

Thought Questions

15. Bone marrow transplants are used to treat aplastic anemia (bone marrow failure) and leukemia. What is the goal of transplanting bone marrow into someone with either disease?

16. What bones/areas of bones would a forensic anthropologist pay special attention to if he/she was trying to determine the age and gender of an unidentified person to whom the set of bones belonged?

17. Why would a potassium (K^+) deficiency cause muscle weakness and fatigue?

19
Development

Learning Outcomes

19.1 The First Stages of Development
- Identify the cellular stages of development with reference to slides of early sea star development.   254–255
- Identify the tissue stages of development with reference to slides of frog development.   256–257

19.2 The Organ Stages of Development
- Associate the germ layers with the development of various organs.   258
- Describe the significance of the process of induction during development.   258
- Identify which organs develop first in a vertebrate embryo (e.g., frog, chick, and human).   259–262
- Compare the later development of humans to that of chicks.   263

19.3 Extraembryonic Membranes, the Placenta, and the Umbilical Cord
- Distinguish between and give a function for the extraembryonic membranes, the placenta, and the umbilical cord.   264–266

19.4 Human Fetal Development
- Trace the main events of human fetal development.   267–268

Introduction

The early development of animals is quite similar, regardless of the species. The fertilized egg, or zygote, undergoes successive divisions by cleavage, forming a mulberry-shaped ball of cells called a morula and then a hollow ball of cells called a blastula. The fluid-filled cavity of the blastula is the blastocoel. Later, some of the surface cells fold inward, or invaginate, eventually forming a double-walled structure. The outer layer is called the ectoderm, and the inner layer is the endoderm. Between these layers, a middle layer, or mesoderm, arises. The embryo is now called a gastrula. In particular, the presence of yolk (nutrient material) influences how the gastrula comes about. All later development can be associated with the three germ layers (ectoderm, endoderm, and mesoderm) that give rise to different tissues and organ systems.

zygote   100×   embryo at one
                    week; implants in
                    uterine wall

embryo at eight weeks

fetus at three months

fetus at five months

Development occurs in stages.

(zygote): © Anatomical Travelogue/Science Source; (embryo, 1 week) © Bettman/Corbis; (embryo, 8 weeks) © Neil Harding/Getty Images; (fetus, 3 months) © Petit Format/Science Source; (fetus, 5 months) © John Watney/Science Source

19.1   The First Stages of Development

In this section, we will consider the cellular and tissue layer stages of development.

Cellular Stages of Development

The cellular stages of development include the following:

- **Zygote formation:** A single sperm fertilizes an egg and the result is a zygote, the first cell of the new individual.
- **Morula formation:** Zygote divides into a number of smaller cells until there is a cluster of 16–32 cells called a morula.
- **Blastula formation:** The morula becomes a blastula, a hollow ball of cells.

Observation: Cellular Stages of Development in the Sea Star

The cellular stages of development are remarkably similar in all animals. Therefore, we can view slides of sea star development to study the cellular stages of human development (Fig. 19.1). A sea star is an invertebrate that develops in the ocean and, therefore, will develop easily in the laboratory where it can be observed.

Obtain slides or view a model of sea star development and note the following:

1. **Zygote.** Both plants and animals begin life as a single cell, a zygote. A zygote contains chromosomes from each parent. Explain your answer. _____

   _____

2. **Cleavage.** View slides showing various numbers of cells due to the process of cleavage, cell division

   without growth until the morula stage. Is the morula about the same size as the zygote? _____

   Explain your answer. _____

   _____

3. **Blastula.** The cavity of a blastula is called the blastocoel. *Label the blastocoel in Figure 19.1.* The formation of a hollow cavity is important to the next stage of development.

Figure 19.1   Sea star development.
All animals, including sea stars and humans, go through the same cellular stages from cleavage to blastula. (Magnification: zygote is 400x; b-f are 75×)

(a) © Ed Reschke/Photolibrary/Getty Images; (b) © Carolina Biological Supply/Phototake; (c) © Carolina Biological Supply/Phototake; (d) © Carolina Biological Supply/Phototake; (e) © Carolina Biological Supply/Phototake; (f) © Carolina Biological Supply/Phototake

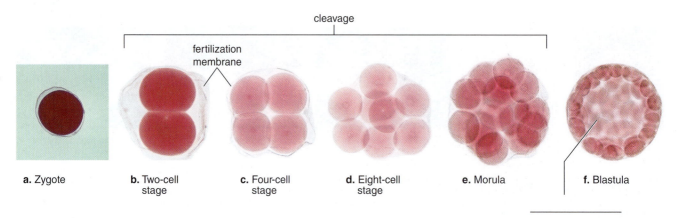

In Figure 19.2, or in a model of human development, observe the same stages of development already observed in sea star slides. Also, observe that fertilization in humans occurs in an oviduct following ovulation. As the embryo undergoes cleavage, it travels in the oviduct to the uterus.

If the embryo splits at the 2-cell stage, the result is identical twins. (Fraternal twins arise when two separate eggs are fertilized.) How might you account for the development of identical triplets? _____

_____

_____

The blastula in humans is called a blastocyst. The blastocyst contains an **inner cell mass** that becomes the embryo, and the outer group of cells (the trophoblast) will become membranes that nourish and protect it. At about day 6, the blastocyst has reached the uterus and implants into the uterine wall, where it will receive nourishment from the mother's bloodstream.

What's the main difference between the cellular stages in a sea star and in a human? _____

_____

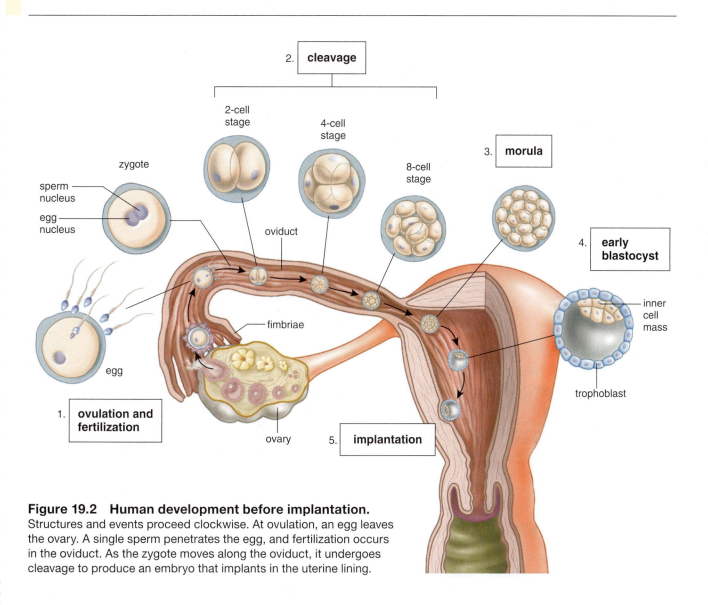

Figure 19.2   Human development before implantation.
Structures and events proceed clockwise. At ovulation, an egg leaves the ovary. A single sperm penetrates the egg, and fertilization occurs in the oviduct. As the zygote moves along the oviduct, it undergoes cleavage to produce an embryo that implants in the uterine lining.

Tissue Stages of Development

The tissue stages of development include the following:

- **Early gastrula stage.** This stage begins when certain cells begin to push or invaginate into the blastocoel, creating a double layer of cells. The outer tissue layer is called the **ectoderm,** and the inner tissue layer is called the **endoderm.**
- **Late gastrula stage.** Gastrulation is not complete until there are three layers of cells. The third tissue layer, called **mesoderm,** occurs between the other two layers already mentioned.

Observation: Tissue Stages of Development in Frogs

It is traditional to view gastrulation in a frog. A frog is a vertebrate, and so its development is expected to be closer to that of a human than is the development of a sea star. In Figure 19.3, note that the yellow (vegetal pole) cells are heavily laden with yolk, and the blue (animal pole) cells are the ones that invaginate into the blastocoel, forming the early gastrula.

1. **Early gastrula stage.** Obtain a cross section of a frog gastrula. Most likely, your slide is the equivalent of Figure 19.3*b,* number 3, in which case you will see two cavities, the old blastocoel and newly forming *archenteron,* which forms once the animal pole cells have invaginated. The archenteron will become the digestive tract, and the blastopore will become the anus.
2. **Late gastrula stage:** Invagination of cells occurs at the lateral and ventral lips of the blastopore only because the cells heavy laden with yolk (yellow cells) do not invaginate.

   Name the germ (tissue) layers that are now present in the embryo. _____

   _____

   _____

Figure 19.3   Drawings of frog developmental stages.

a. During cleavage, the number of cells increases but overall size remains the same. **b.** During gastrulation, three tissue layers form. Blue = ectoderm; yellow = endoderm; red = mesoderm.

animal pole

vegetal pole

Morula   **Blastula**

a. Cleavage

blastocoel   archenteron   ectoderm   mesoderm
endoderm

blastopore   yolk plug

b. Gastrulation   1.   2.   3.   4.   5.
Gastrula

Observation: Tissue Stages of Development in a Human

In a model of human development, observe the same stages of development already observed in frog slides. After implantation, gastrulation in humans turns the inner cell mass into the **embryonic disk.** Figure 19.4 shows the embryonic disk, which has the three layers of cells we have been discussing: the ectoderm, mesoderm, and endoderm. Figure 19.4 also shows the significance of these layers, often called the **germ layers.** The future organs of an individual can be traced back to one of the germ layers.

Figure 19.4 Embryonic disk.

The embryonic disk has three germ layers called ectoderm, mesoderm, and endoderm. Organs and tissues can be traced back to a particular germ layer as indicated in this illustration.

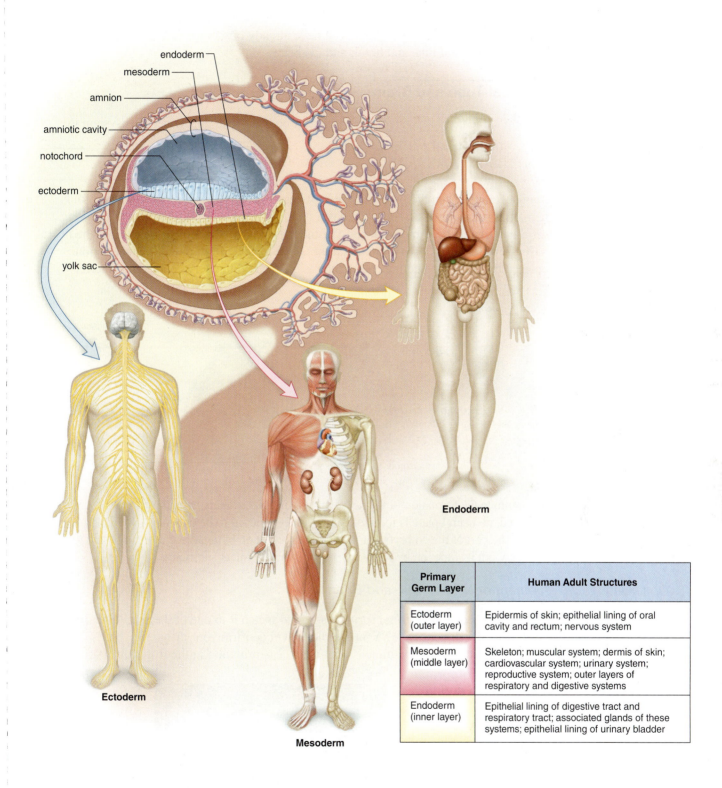

Primary Germ Layer	Human Adult Structures
Ectoderm (outer layer)	Epidermis of skin; epithelial lining of oral cavity and rectum; nervous system
Mesoderm (middle layer)	Skeleton; muscular system; dermis of skin; cardiovascular system; urinary system; reproductive system; outer layers of respiratory and digestive systems
Endoderm (inner layer)	Epithelial lining of digestive tract and respiratory tract; associated glands of these systems; epithelial lining of urinary bladder

19.2 The Organ Stages of Development

As soon as all three embryonic tissue layers (ectoderm, endoderm, and mesoderm) are established, the organ level of development begins. It continues until all organs have formed. Figure 19.4 pertains to the organ stages of development because it shows which germ layers give rise to which organs.

The first organs to develop are the (1) digestive tract (you have already observed the start of the archenteron during gastrulation); (2) neural tube and brain; and (3) heart.

Development of the Neural Tube and Brain

One of the first systems to form is the nervous system. Why might it be beneficial for the nervous system to begin development first? _____

_____

During nervous system development in the frog, two folds of ectoderm grow upward as the neural folds with a groove between them. The flat layer of ectoderm between them is the **neural plate.** The tube resulting from closure of the folds is the **neural tube.** An examination of the neurula in cross section shows that the neural tube develops directly above the **notochord,** a structure that arises from mesoderm in the middorsal region. Later, the notochord is replaced by vertebrae and the neural tube is then called the nerve cord. The anterior portion of the neural tube becomes the brain.

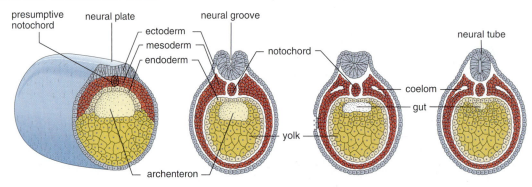

Neurulation

The notochord is said to **induce** the formation of the nervous system. Experiments have shown that if contact with notochord tissue is prevented, no neural plate is formed. **Induction** is believed to be one means by which development is usually orderly. The part of the embryo that induces the formation of an adjacent organ is said to be an **organizer** and is believed to carry out its function by releasing one or more chemical substances.

Observation: Development of Neural Tube and Brain

1. Obtain a slide showing a cross section of a frog neurula stage, and match it to one of the preceding drawings. Which drawing seems to best match your slide? Your instructor will confirm your match for you.
2. Obtain and examine frog embryos for a three-dimensional view of neurulation in the frog (Fig. 19.5).

Figure 19.5 Photograph of frog embryos during neurulation.

© Carolina Biological Supply Company/ Phototake

Development of the Heart

A chick embryo offers an opportunity to view a beating heart in an embryo. Your instructor may show you various stages. In particular you will want to observe the chick embryo from the 48-hour stage up to the 96-hour stage.

Observing Live Chick Embryos

Use the following procedure for selecting and opening the eggs of live chick embryos:

1. Choose an egg of the proper age to remove from the incubator, and put a penciled × on the uppermost side. The embryo is just below the shell.
2. Add warmed chicken Ringer solution to a finger bowl until the bowl is about half full. (Chicken Ringer solution is an isotonic salt solution for chick tissue that maintains the living state.) The chicken Ringer solution should not cover the yolk of the egg.
3. On the edge of the dish, gently crack the egg on the side opposite the ×.
4. With your thumbs placed over the ×, hold the egg in the chicken Ringer solution while you pry it open from below and allow its contents to enter the solution. If you open the egg too slowly or too quickly, the shell may damage the delicate membranes surrounding the embryo.

Observation: Forty-Eight-Hour Chick Embryo

1. Follow the standard procedure for selecting and opening an egg containing a 48-hour chick embryo.
2. The embryo has turned so that the head region is lying on its side. Refer to Figure 19.6, and identify the following:

   a. **Shape of the embryo,** which has started to bend. The head is now almost touching the heart.

   b. **Heart,** contracting and circulating blood. Can you make out a ventricle, an atrium, and the aortic arches in the region below the head? Later, only one aortic arch will remain.

   c. **Vitelline arteries** and **veins,** which extend over the yolk. The vitelline veins carry nutrients from the yolk sac to the embryo.

   d. **Brain** with several distinct regions.

   e. **Eye,** which has a developing lens.

   f. **Margin (edge) of the amnion,** which can be seen above the vitelline arteries (see next section for amnion).

   g. **Somites,** blocks of developing muscle tissue that differentiate from mesoderm, which now number 24 pairs.

   h. **Caudal fold** of the amnion. The embryo will be completely enveloped when the head fold and caudal fold meet the margin of the amnion.

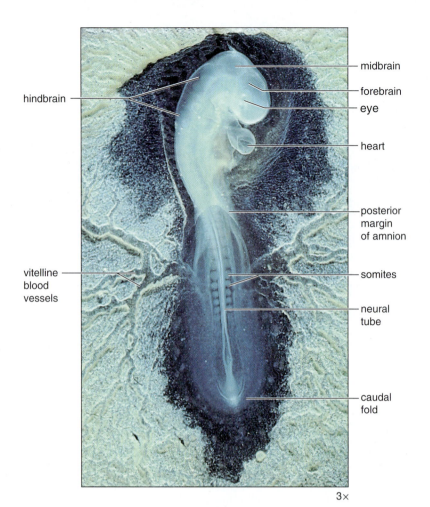

hindbrain

midbrain

forebrain

eye

heart

posterior
margin
of amnion

vitelline
blood
vessels

somites

neural
tube

caudal
fold

3×

Figure 19.6 Forty-eight-hour chick embryo.
The most prominent organs are labeled.
© Carolina Biological Supply/Phototake

Observation: Older Chick Embryos

As a chick embryo continues to grow, various organs differentiate further (Fig. 19.7). The neural tube closes along the entire length of the body and is now called the spinal cord. The allantois, an extraembryonic membrane, is seen as a sac extending from the ventral surface of the hindgut near the tail bud. The digestive system forms specialized regions, and there is both a mouth and an anus. The yolk sac, the extraembryonic membrane that encloses the yolk, is attached to the ventral wall, but when the yolk is used up, the ventral wall closes.

ear (otic vesicle)

aortic arches

somites

heart

hindbrain

midbrain

eye

forebrain

vitelline blood vessels

hindlimb bud

a. Photograph

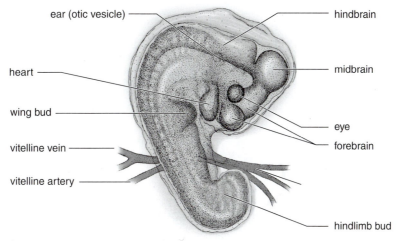

ear (otic vesicle)

heart

wing bud

vitelline vein

vitelline artery

hindbrain

midbrain

eye

forebrain

hindlimb bud

b. Drawing

Key: **Brain region**	**Becomes:**
forebrain	cerebrum diencephalon
midbrain	relay station
hindbrain	pons, cerebellum medulla oblongata

Figure 19.7   Ninety-six-hour chick embryo.
Brain regions listed in the key can now be seen.

(a) © Carolina Biological Supply/Phototake

Study models or other study aids available that show the development of the nervous system and the heart in humans and/or show models of human embryos of different ages. Also view Figure 19.8, which depicts the external appearance of the embryo from the fourth to the seventh week of development.

During the embryonic period of development, the growing baby is susceptible to environmental influences, including the following:

- Drugs, such as alcohol; certain prescriptions; and recreational drugs. These can cause birth defects.
- Infections such as rubella, also called German measles, and other viral infections.
- Nutritional deficiencies.
- X-rays or radiation therapy.

Figure 19.8 External appearance of the human embryo.
a. Weeks 4 and 5. **b.** Weeks 6 and 7.

a. Weeks 4 and 5
  • Head dominant, but body getting longer.
  • Limb buds are visible.
  • Eyes and ears begin to form.
  • Tissue for vertebrae extend into tail.

b. Weeks 6 and 7
  • Head still dominant, but tail has disappeared.
  • Facial features continue to develop.
  • Hands and feet have digits.
  • All organs are more developed.

As illustrated in Figure 19.9, the early stages of human development are quite similar to those of the chick. Differences become marked only as development proceeds.

Figure 19.9 Comparison of vertebrate embryos.
Successive stages in the development of chick and human. Early stages (*far left*) are similar; differences become apparent as development continues.

19.3 Extraembryonic Membranes, the Placenta, and the Umbilical Cord

- The **extraembryonic membranes** take their name from the observation that they are not part of the embryo proper. They are outside the embryo, and therefore they are "extra."
- In humans, the **placenta** is the structure that provides the embryo with nutrient molecules and oxygen and takes away its waste molecules, such as carbon dioxide. The fetal half of the placenta contains the fetal capillaries. The maternal half of the placenta is the uterine wall where maternal blood vessels meet the fetal capillaries.
- The **umbilical cord** is a tubular structure that contains two of the extraembryonic membranes (the allantois and the yolk sac) and also the **umbilical blood vessels.** The umbilical blood vessels bring fetal blood to and from the placenta. When a baby is born and begins to breathe on its own, the umbilical cord is cut and the remnants become the navel. *In this drawing, label the umbilical cord, the umbilical blood vessels, and the placenta, which contains the maternal blood vessels.*

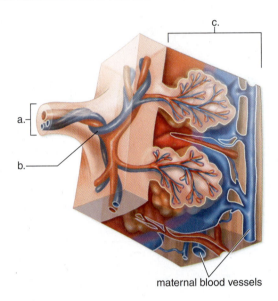

maternal blood vessels

Observation: The Extraembryonic Membranes

In a model, and in Figure 19.10, trace the development of the extraembryonic membranes. Also, note the development of the placenta and the umbilical cord. The extraembryonic membranes are as follows:

- **Chorion.** The chorion is the outermost membrane, and in chicks it lies just below the porous shell, where it functions in gas exchange. In humans, an outer layer of cells surrounding the inner cell mass at the blastocyst stage becomes the chorion. Notice in Figure 19.10 that the treelike **chorionic villi** are a part of the chorion.
- **Amnion.** The amnion forms the amniotic cavity, which envelops the embryo and contains the amniotic fluid that cushions and protects the developing offspring (Fig. 19.11). All animals, whether the sea star, the frog, the chick, or the human, develop in an aqueous environment. Birth of a human is imminent when "the water breaks," and the amniotic fluid is lost.
- **Allantois.** The allantois serves as a storage area for metabolic waste in the chick. In humans, the allantois extends into the umbilical cord. It accumulates the small amount of urine produced by the fetal kidneys and later contributes to urinary bladder formation. Its blood vessels become the umbilical blood vessels.
- **Yolk sac.** The yolk sac is the first embryonic membrane to appear. In the chick, the yolk sac does contain yolk, food for the developing embryo. In humans, the yolk sac contains plentiful blood vessels and is the first site of blood cell formation.

Figure 19.10 Development of extraembryonic membranes in humans.

a. At first, no organs are present in the embryo, only tissues. The amniotic cavity is above the embryonic disk, and the yolk sac is below. The chorionic villi are present. **b, c.** The allantois and yolk sac, two more extraembryonic membranes, are positioned inside the body stalk as it becomes the umbilical cord. **d.** At 35+ days, all membranes are present, and the umbilical cord takes blood vessels between the embryo and the chorion (placenta).

amniotic cavity

embryonic disk

yolk sac

chorionic villi

chorion

a. 18 days

placenta

umbilical cord

amnion

Figure 19.11 Fetus and amnion.
Photograph of a human fetus at 8 weeks. The scale bar is 3 cm.

© Martin Rotker/Phototake

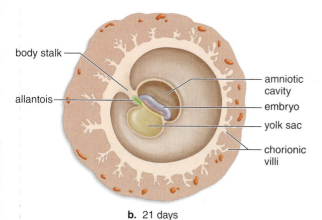

body stalk

allantois

amniotic cavity

embryo

yolk sac

chorionic villi

b. 21 days

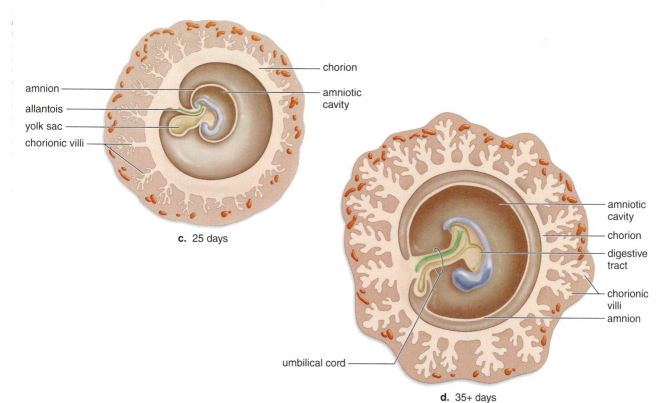

amnion

allantois

yolk sac

chorionic villi

chorion

amniotic cavity

c. 25 days

amniotic cavity

chorion

digestive tract

chorionic villi

amnion

umbilical cord

d. 35+ days

Comparison of Extraembryonic Membranes in Chick and Human

Consult Figure 19.12 and use the information on page 264 to complete Table 19.1, which compares the functions of the extraembryonic membranes in the chick and in the human. Reptiles, which we now know include birds, were the first animals to have extraembryonic membranes. These membranes allowed reptiles to develop on land. They also allow mammals, including humans, to develop inside the uterus of the mother.

Table 19.1 Functions of Extraembryonic Membranes in Chick and Human		
Membrane	Chick	Human
Amnion		
Allantois		
Chorion		
Yolk sac		

Chick

Human

Figure 19.12   Extraembryonic membranes.
The chick and a human have the same extraembryonic membranes, but except for the amnion, they have different functions.

19.4 Human Fetal Development

During fetal development (last seven months), the skeleton becomes ossified (bony), reproductive organs form, arms and legs develop fully, and the fetus enlarges in size and gains weight.

Three- to four-month-old fetus   Seven- to eight-month-old fetus

(left) © Petit format/Science Source; (right) © Petit format/Science Source

Observation: Fetal Development

1. Using Table 19.2 and Figure 19.13 to assist you, examine models of fetal development.
2. In Table 19.2, note the following.

   a. **External genitals:** About the third month, it is possible to tell male from female if an ultrasound is done.

   b. **Quickening:** Fetal movement is felt during the fourth or fifth months.

   c. **Vernix caseosa:** Beginning with the fifth month, the skin is covered with a cheesy coating called vernix caseosa.

   d. **Lanugo:** During the sixth and seventh months, the body is covered with fine, downy hair termed lanugo.

Table 19.2   Fetal Development

Month	Events for Mother	Events for Baby
Third month	Uterus is the size of a grapefruit.	Possible to distinguish sex. Fingernails appear.
Fourth month	Fetal movement is felt by those who have been previously pregnant. Heartbeat is heard by stethoscope.	Bony skeleton visible. Hair begins to appear. 150 mm (6 in.), 170 g (6 oz).
Fifth month	Fetal movement is felt by those who have not been previously pregnant. Uterus reaches up to level of umbilicus and pregnancy is obvious.	Protective cheesy coating, called vernix caseosa, begins to be deposited. Heartbeat can be heard.
Sixth month	Doctor can tell where baby's head, back, and limbs are. Breasts have enlarged, nipples and areolae are darkly pigmented, and colostrum is produced.	Body is covered with fine hair called lanugo. Skin is wrinkled and reddish.
Seventh month	Uterus reaches halfway between umbilicus and rib cage.	Testes descend into scrotum. Eyes are open. 300 mm (12 in.), 1,350 g (3 lb).
Eighth month	Weight gain is averaging about a pound a week. Difficulty in standing and walking because center of gravity is thrown forward.	Body hair begins to disappear. Subcutaneous fat begins to be deposited.
Ninth month	Uterus is up to rib cage, causing shortness of breath and heartburn. Sleeping becomes difficult.	Ready for birth. 530 mm (20½ in.), 3,400 g (7½ lb).

Figure 19.13 Human development.
Changes occurring from the fifth week to the eighth month.

(g) © Chris Downie/E+/Getty RF; (h) © James Stevenson/Science Source

lens
maxillary process
hindlimb
mandibular process
paddle-shaped forelimb

a. 35 ± 1 day (10–12 mm)

developing eye
forebrain
nasal pit
tail
developing ear
elbow
handplate

b. 37 ± 1 day (12.5–15.75 mm)

midbrain
pigmented eye
heart prominence
paddle-shaped foot plate
external auditory meatus
external ear
wrist
digital rays

c. 40 ± 1 day (16–21 mm)

notches between digital rays
toe rays
external ear

d. 45 ± 1 day (22–24 mm)

eyelid
webbed fingers
notches between toe rays
ear

e. 49 ± 1 day (28–30 mm)

fingers separated
fan-shaped webbed toes

f. 52 ± 1 day (32–34 mm)

g. Three- to four-month-old fetus

h. Seven- to eight-month-old fetus

_____ 1. What results from the fertilization of an egg by a single sperm?

_____ 2. What is cell division that lacks growth called?

_____ 3. What is the solid cluster of 16 to 32 embryonic cells called?

_____ 4. What is the hollow cavity of a blastula called?

_____ 5. Which part of a human blastocyst becomes the embryo, the inner cell mass or the trophoblast?

_____ 6. During what developmental stage do tissues begin to form?

_____ 7. What does the archenteron become?

_____ 8. What primary germ layer gives rise to the pseudostratified ciliated columnar epithelium that lines the trachea?

_____ 9. What replaces the notochord?

_____ 10. Which extraembryonic membrane has a similar function in chicks and humans?

_____ 11. What extraembryonic membrane is the first site of blood cell formation in humans?

_____ 12. Where is metabolic waste stored in a chick?

_____ 13. During what month of human development is it possible to distinguish the gender of the fetus?

_____ 14. What is the name of the protective cheesy coating that begins to be deposited in the fifth month of human development?

Thought Questions

15. Some invertebrate animals (covered in an upcoming laboratory exercise) have only two primary germ layers, ectoderm and endoderm. What structures are likely to be missing from these animals? Justify your reply.

16. Chorionic villus sampling and amniocentesis are prenatal tests that can be done to check for certain genetic conditions or a chromosomal condition.

   a. What part of the human blastocyst gives rise to the chorion and amnion?

   b. What do these layers have in common?

17. a. If the notochord is removed, will ectoderm in this location become a neural tube? Explain.

   b. If ectoderm above the notochord is removed and replaced with belly ectoderm, will the belly ectoderm become a neural tube? Explain.

20
Patterns of Inheritance

Learning Outcomes

20.1 One-Trait Crosses
- State Mendel's law of segregation, and relate this law to laboratory exercises.   272–276
- Describe and predict the results of a one-trait cross in both tobacco seedlings and corn.   273–276

20.2 Two-Trait Crosses
- State Mendel's law of independent assortment, and relate this law to laboratory exercises.   276–281
- Explain and predict the results of a two-trait cross in corn plants and *Drosophila*.   277–281

20.3 X-Linked Crosses
- Explain and predict the results of an X-linked cross in *Drosophila*.   281–283

Introduction

Gregor Mendel, sometimes called the "father of genetics," formulated the basic laws of genetics examined in this laboratory. He determined that individuals have two alternate forms of a gene (now called **alleles**) for each trait in their body cells. Today, we know that alleles are on the chromosomes. An individual can be **homozygous dominant** (two dominant alleles, *GG*), **homozygous recessive** (two recessive alleles, *gg*), or **heterozygous** (one dominant and one recessive allele, *Gg*). **Genotype** refers to an individual's genes, while **phenotype** refers to an individual's appearance (Fig. 20.1). Homozygous dominant and heterozygous individuals show the dominant phenotype; homozygous recessive individuals show the recessive phenotype.

Figure 20.1   Genotype versus phenotype.
Only with homozygous recessive do you immediately know the genotype.

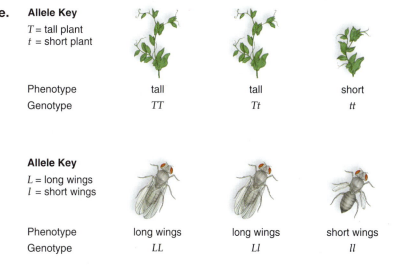

Allele Key
T = tall plant
t = short plant

Phenotype	tall	tall	short
Genotype	*TT*	*Tt*	*tt*

Allele Key
L = long wings
l = short wings

Phenotype	long wings	long wings	short wings
Genotype	*LL*	*Ll*	*ll*

Punnett Squares

Punnett squares, named after the man who first used them, allow you to easily determine the results of a cross between individuals whose genotypes are known. Consider that when fertilization occurs, two gametes, such as a sperm and an egg, join together. Whereas individuals have two alleles for every trait, gametes have only one allele because alleles are on the chromosomes and homologues separate during meiosis. Heterozygous parents with the genotype *Aa* produce two types of gametes: 50% of the gametes contain an *A* and 50% contain an *a*. A Punnett square allows you to vertically line up all possible types of sperm and to horizontally line up all possible types of eggs. Every possible combination of gametes occurs within the squares, and these combinations indicate the genotypes of the offspring. In Figure 20.2, one offspring is *AA* = homozygous dominant, two are *Aa* = heterozygous, and one is *aa* = homozygous recessive. Therefore, three of the offspring will have the dominant phenotype and one will have the recessive phenotype. This is said to be a **phenotypic ratio** of 3:1.

A Punnett square can be used for any cross regardless of the trait(s) and the genotypes of the parents. All you need to do is use the correct letters for the particular trait(s), and make sure you have given the parents the correct genotypes and correct proportion of each type of gamete. Then you can determine the genotypes of the offspring and the resulting phenotypic ratio among the offspring.

Figure 20.2 What are the expected results of a cross?
A Punnett square allows you to determine the expected phenotypic ratio for a cross.

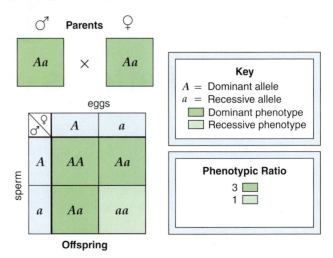

Offspring

20.1 One-Trait Crosses

A single pair of alleles is involved in one-trait crosses. Mendel found that reproduction between two heterozygous individuals (*Aa*), called a **monohybrid cross,** results in both dominant and recessive phenotypes among the offspring. In Figure 20.2, the expected phenotypic ratio among the offspring is 3:1. Three offspring have the dominant phenotype for every one that has the recessive phenotype.

Mendel realized that these results are obtainable only if the alleles of each parent segregate (separate from each other) during meiosis. Therefore, Mendel formulated his first law of inheritance:

Law of Segregation

Each organism contains two alleles for each trait, and the alleles segregate during the formation of gametes. Each gamete (egg or sperm) then contains only one allele for each trait. When fertilization occurs, the new organism has two alleles for each trait, one from each parent.

Inheritance is a game of chance. Just as there is a 50% probability of heads or tails when tossing a coin, there is a 50% probability that a sperm or egg will have an *A* or an *a* when the parent is *Aa*. The chance of an equal number of heads or tails improves as the number of tosses increases. In the same way, the chance of an equal number of gametes with *A* and *a* improves as the number of gametes increases. Therefore, the 3:1 ratio among offspring is more likely the more offspring you count for the same type of cross.

Color of Tobacco Seedlings

In tobacco plants, a dominant allele (*C*) for chlorophyll gives the plants a green color, and a recessive allele (*c*) for chlorophyll causes a plant to appear white. If a tobacco plant is homozygous for the recessive allele (*c*), it cannot manufacture chlorophyll and thus appears white (Fig. 20.3).

Figure 20.3 Monohybrid cross.
These tobacco seedlings are growing on an agar plate. The white plants cannot manufacture chlorophyll.

Experimental Procedure: Color of Tobacco Seedlings

1. Obtain a numbered agar plate on which tobacco seedlings are growing. They are the offspring of a cross between heterozygous parents: the cross *Cc* × *Cc*. Complete the Punnett square to determine the expected phenotypic ratio.

   What is the expected phenotypic ratio? _____

2. Record the plate number and using a stereomicroscope, view the seedlings. Count the number that are green and the number that are white. Record your results in Table 20.1.

3. Repeat this procedure for two additional plates. Total the number that are green and the number that are white.

4. Complete Table 20.1 by recording the class data. Total the number that are green and the number that are white per class.

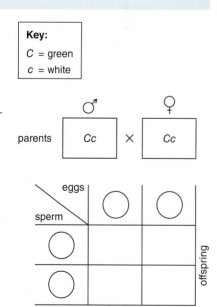

Key:
C = green
c = white

Laboratory 20   Patterns of Inheritance   **273**

Table 20.1 Color of Tobacco Seedlings

| | Number of Offspring | | |
	Green Color	White Color	Phenotypic Ratio
Plate # _____			
Plate # _____			
Plate # _____			
Totals			
Class data			

Conclusions: Color of Tobacco Seedlings

- In the last column of Table 20.1, record the actual phenotypic ratio per observed plate; per total number of green versus white plants you counted; and per the entire class. To determine the actual phenotypic ratio, divide the number of green color seedlings in a plate by the number of white color seedlings in a plate.

  Do your results differ from the expected phenotypic ratio? If so, explain your answer. _____

  _____

  _____

- Mendel found that the more plants he counted, the closer he came to the expected phenotypic ratio.

  Was your class data closer to the expected phenotypic ratio than your individual data? _____

  This is expected, because the more crosses you observe, the more likely it is that all types of sperm and eggs will have a chance to come together.

Color of Corn Kernels

In corn plants, the allele for purple kernel (*P*) is dominant over the allele for yellow kernel (*p*) (Fig. 20.4).

Figure 20.4 One-trait cross.

Two types of kernels are seen on an ear of corn following a cross between *Pp* and *pp*.

© Evelyn Jo Johnson

1. Obtain an ear of corn from the supply table. You will be examining the results of the cross $Pp \times pp$. Complete the Punnett square to determine the expected phenotypic ratio. Note that when one parent has only one possible type of gamete, only one column is needed in the Punnett square.

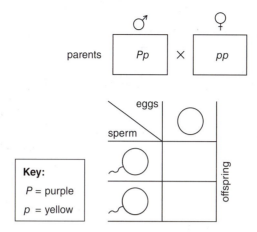

What is the expected phenotypic ratio among the offspring? _____

2. Record the sample number and then count the number of kernels that are purple and the number that are yellow. As before, use two more samples, and record your results in Table 20.2.

Table 20.2   Color of Corn Kernels

	Number of Kernels		
	Purple	**Yellow**	**Phenotypic Ratio**
Sample # _____			
Sample # _____			
Sample # _____			
Totals			
Class data			

Conclusions: Color of Corn Kernels

- In the last column of Table 20.2, record the actual phenotypic ratio per observed sample; per total number of purple versus yellow kernels you counted; and per the entire class. Do your results differ from the expected phenotypic ratio? If so, explain your answer. _____

_____

- Was your class data closer to the expected phenotypic ratio than your individual data? _____ If so, explain your answer. _____

_____

One-Trait Genetics Problems

1. In pea plants, purple flowers (*P*) is dominant and white flowers (*p*) is recessive. What is the genotype of pure-breeding white plants? Pure-breeding means that they produce plants with only one phenotype. _____ If pure-breeding purple plants are crossed with these white plants, what phenotype is expected? _____

2. In pea plants, tall (*T*) is dominant and short (*t*) is recessive. A heterozygous tall plant is crossed with a short plant. What is the expected phenotypic ratio? _____

3. Unexpectedly to the farmer, two tall plants have some short offspring. What is the genotype of the parent plants and the short offspring? parent _____ offspring _____

4. In horses, two trotters are mated to each other and produce only trotters; two pacers are mated to each other and produce only pacers. When one of these trotters is mated to one of the pacers, all the horses are trotters. Create a key and show the cross. key _____ cross _____

5. A brown dog is crossed with two different black dogs. The first cross produces only black dogs and the second cross produces equal numbers of black and brown dogs. What is the genotype of the brown dog? _____ the first black dog? _____ the second black dog? _____

6. In pea plants, green pods (*G*) is dominant and yellow pods (*g*) is recessive. When two pea plants with green pods are crossed, 25% of the offspring have yellow pods. What is the genotype of all plants involved? plants with green pods _____ plants with yellow pods _____

7. A breeder wants to know if a dog is homozygous black or heterozygous black. If the dog is heterozygous, which cross is more likely to produce a brown dog, *Bb* × *bb* or *Bb* × *Bb?* Explain your answer. _____
_____
_____

8. If the cross in problem 6 produces 220 plants, how many offspring have green pods and how many have yellow pods? _____ If the cross in problem 2 produces 220 plants, how many offspring are tall and how many are short? _____

20.2 Two-Trait Crosses

Two-trait crosses involve two pairs of alleles. Mendel found that during a **dihybrid cross,** when two dihybrid individuals (*AaBb*) reproduce, the phenotypic ratio among the offspring is 9:3:3:1, representing four possible phenotypes. He realized that these results could be obtained only if the alleles of the parents segregated independently of one another when the gametes were formed. From this, Mendel formulated his second law of inheritance:

Law of Independent Assortment

Members of an allelic pair segregate (assort) independently of members of another allelic pair. Therefore, all possible combinations of alleles can occur in the gametes.

The FOIL method is a way to determine the gametes. FOIL stands for *First* two alleles from each trait; *Outer* two alleles from each trait; *Inner* two alleles from each trait; *Last* two alleles from each trait. Here is how the FOIL method can help you determine the gametes for the genotype *PpSs:*

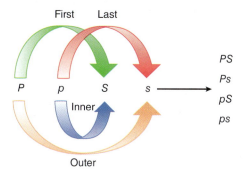

Color and Texture of Corn

In corn plants, the allele for purple kernel (*P*) is dominant over the allele for yellow kernel (*p*), and the allele for smooth kernel (*S*) is dominant over the allele for rough kernel (*s*) (Fig. 20.5).

Figure 20.5 Dihybrid cross.
Four types of kernels are seen on an ear of corn following a dihybrid cross (*PpSs* × *PpSs*): purple smooth, purple rough, yellow smooth, and yellow rough. © Carolina Biological Supply/Phototake

Experimental Procedure: Color and Texture of Corn

1. Obtain an ear of corn from the supply table. You will be examining the results of the cross *PpSs* × *PpSs*.

2. Do the Punnett square on page 278, in order to state the expected phenotypic ratio among the offspring. _____

3. Count the number of kernels of each possible phenotype listed in Table 20.3. Record the sample number and your results in Table 20.3. Use three samples, and total your results for all samples. Also record the class data (i.e., the number of kernels that are the four phenotypes per class).

Key:
P = purple
p = yellow
S = smooth
s = rough

♂   ♀

parents   PpSs   ×   PpSs

eggs

sperm

offspring

Table 20.3   Color and Texture of Corn

	Number of Kernels				
	Purple Smooth	Purple Rough	Yellow Smooth	Yellow Rough	Phenotypic Ratio
Sample # _____					
Sample # _____					
Sample # _____					
Totals					
Class data					

Conclusions: Color and Texture of Corn

- Calculate the actual phenotypic ratios based on the data and record them in Table 20.3. Do the results differ from the expected ratio per individual data? _____ Per class data? _____
  If so, explain your answer. _____
  _____
  _____

Wing Length and Body Color in *Drosophila*

Drosophila are the tiny flies you often see flying around ripe fruit; therefore, they are called fruit flies. If a culture bottle of fruit flies is on display, take a look at it. Because so many flies can be grown in a small culture bottle, fruit flies have contributed substantially to our knowledge of genetics. If you were to examine *Drosophila* flies under the stereomicroscope, they would appear like this:

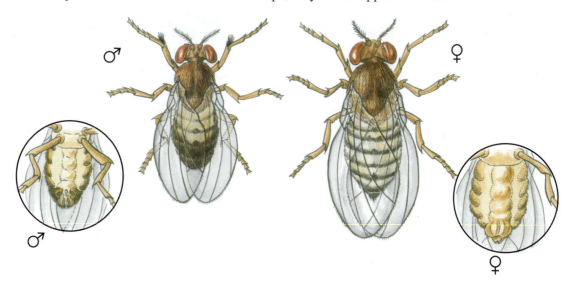

In *Drosophila,* long wings (*L*) are dominant over short (vestigial) wings (*l*), and gray body (*G*) is dominant over black (ebony) body (*g*). Consider the cross *LlGg* × *llgg* and complete this Punnett square:

Key:
L = long wing
l = short (vestigial) wing
G = gray body
g = ebony (black) body

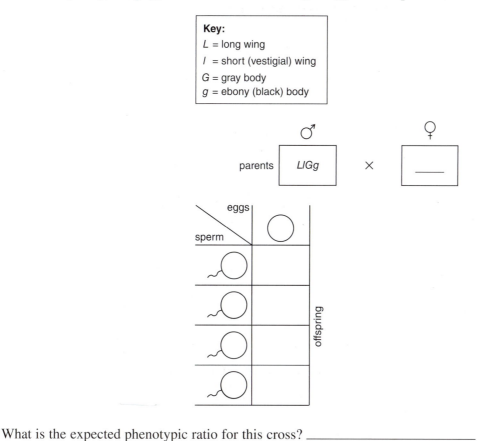

What is the expected phenotypic ratio for this cross? _____

Experimental Procedure: Wing Length and Body Color in Drosophila

If your instructor has frozen flies available, cross out the numbers in Table 20.4 and use the stereomicroscope or a hand lens to count the flies of each type given in Table 20.4. Otherwise simply use the data supplied for you in Table 20.4.

Table 20.4   Wing Length and Body Color in *Drosophila**

	Phenotypes				
	Long Gray	Long Ebony	Short Gray	Short Ebony	Phenotypic Ratio
Number of offspring	28	32	28	30	
Class data	128	120	120	120	

*Wings and body are understood in this table.

Conclusions: Wing Length and Body Color in Drosophila

- Calculate the actual phenotypic ratio based on the data and record in Table 20.4. Do the results differ from the expected ratio per individual data? _____ Per class data? _____ Explain your answer. _____

  _____

Two-Trait Genetics Problems

1. In tomatoes, tall is dominant and short is recessive. Red fruit is dominant and yellow fruit is recessive. Choose a key for height _____ for color of fruit _____ What is the genotype of a plant heterozygous for both traits? _____ What are the possible gametes for this plant? _____

   _____

2. Using words, what are the likely parental genotypes if the results of a two-trait problem are 1:1:1:1 among the offspring? _____ × _____

3. In horses, black (*B*) and a trotting gait (*T*) are dominant, while brown (*b*) and a pacing gait (*t*) are recessive. If a black trotter (homozygous for both traits) is mated to a brown pacer, what phenotypic ratio is expected among the offspring? _____

4. Two black trotters have a brown pacer offspring. What is the genotype of all horses involved?
   black trotter parents _____ brown pacer offspring _____

5. The phenotypic ratio among the offspring for two corn plants producing purple and smooth kernels is 9:3:3:1. (See page 278 of this lab for the key.) What is the genotype of these plants? parental plants _____ the 9 offspring _____ 3 of the offspring _____ the other 3 _____ and the 1 offspring? _____

6. Which matings could produce at least some fruit flies heterozygous in both traits? Write yes or no beside each. (You do not need a key.)

ggLl × *Ggll* _____   *GGLl* × *ggLl* _____   *GGLL* × *ggll* _____

Explain your answer. _____

_____

7. State two new crosses that could not produce fruit flies heterozygous in both traits.

_____ × _____   _____ × _____

8. Chimpanzees are not deaf if they inherit both an allele *E* and an allele *G*. A cross between two deaf chimpanzees produces only chimpanzees that can hear. What are the genotypes of all chimpanzees

involved? parents _____ × _____ offspring _____

20.3   X-Linked Crosses

In animals such as fruit flies, chromosomes differ between the sexes. All but one pair of chromosomes in males and females are the same; these are called **autosomes** because they do not actively determine sex. The pair that is different is called the **sex chromosomes.** In fruit flies and humans, the sex chromosomes in females are XX and those in males are XY.

Some alleles on the X chromosome have nothing to do with gender, and these genes are said to be X-linked. The Y chromosome does not carry these genes and indeed carries very few genes. Males with a normal chromosome inheritance are never heterozygous for X-linked alleles, and if they inherit a recessive X-linked allele it will be expressed.

Red/White Eye Color in *Drosophila*

In fruit flies, red eyes (X^R) are dominant over white eyes (X^r). You will be examining the results of the cross $X^R Y \times X^R X^r$. Complete this Punnett square and state the expected phenotypic ratio for this cross.

females _____   males _____

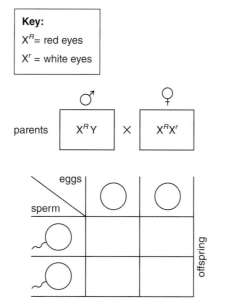

Experimental Procedure: Red/White Eye Color in Drosophila

If your instructor has frozen flies available, cross out the numbers in Table 20.5 and use the stereomicroscope or a hand lens to count the flies of each type given in Table 20.5. Use the art on page 279 to tell males from females, and record male and female data separately. If frozen flies are not available, simply use the data supplied for you in Table 20.5.

Table 20.5 Red/White Eye Color in *Drosophila*			
	Number of Offspring		
Your Data:	**Red Eyes**	**White Eyes**	**Phenotypic Ratio**
Males	16	17	
Females	63	0	
Class Data:			
Males	45	48	
Females	215	0	

Conclusions: Red/White Eye Color in Drosophila

- Calculate the phenotypic ratios based on the data for males and females separately and record them in Table 20.5. Do the results differ from the expected ratio per individual data? _____ Per class data? _____ If so, explain your answer. _____

  _____

  _____

- Using the Punnett square provided, calculate the expected phenotypic results for the cross $X^R Y \times X^r X^r$. What is the expected phenotypic ratio among the offspring? males _____ females _____

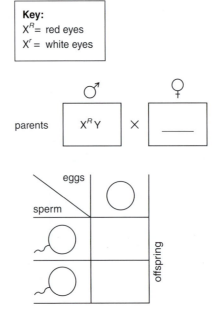

Key:
X^R = red eyes
X^r = white eyes

X-Linked Genetics Problems

1. State the genotypes and gametes for each of these fruit flies:

	genotype	gamete(s)
white-eyed male	_____	_____
white-eyed female	_____	_____
red-eyed male	_____	_____
homozygous red-eyed female	_____	_____
heterozygous red-eyed female	_____	_____

2. What are the phenotypic ratios if a white-eyed female is crossed with a red-eyed male?

   males _____ females _____

3. Regardless of any type of cross, do white-eyed males inherit the allele for white eyes from their father or their mother? _____ Explain your answer. _____

4. In sheep, horns are sex linked; H = horns and h = no horns. Using symbols, what cross do you recommend if a farmer wants to produce hornless males? _____ × _____

5. In *Drosophila,* bar eye is sex linked; B = bar eye and b = no bar eye. What are the genotypes of the parents and the phenotypic ratios for these crosses?

   bar-eyed male × non–bar-eyed female _____, _____

   bar-eyed male × heterozygous female _____, _____

   non–bar-eyed male × heterozygous female _____, _____

6. A female fruit fly has white eyes. What is the genotype of the father? _____ What could be the genotype of the mother? _____ or _____

7. In a cross between fruit flies, all the males have white eyes and the females are 1:1. What is the genotype of the parents? female parent _____ male parent _____

8. In a cross between fruit flies, a white-eyed male and a red-eyed female produce no offspring that have white eyes. What is the genotype of the parents? male parent _____ female parent _____

9. Make up a sex-linked genetic cross using words and parental genotypes. Create a key and show a Punnett square and the phenotypic ratios.

_____ 1. What are the alternate forms of a gene called?

_____ 2. What word is used to describe the possession of one dominant allele and one recessive allele?

_____ 3. What word is used to refer to an individual's appearance?

_____ 4. Which of Mendel's laws says alleles separate during gamete formation so that each gamete contains only one allele for each trait?

_____ 5. What is used to determine the results of a cross between individuals with known genotypes?

_____ 6. If the phenotypic ratio among the offspring is 3:1, what is the genotype of the parents?

_____ 7. How would a tobacco seeding with the genotype *cc* for chlorophyll be described?

_____ 8. If the allele for a purple corn kernel (*P*) is dominant over the allele for a yellow corn kernel (*p*), what is the genotype of the yellow kernel?

_____ 9. What is the cross between two individuals with *GgRr* genotypes called?

_____ 10. How many of the offspring of the cross between two *GgRr* individuals will show the recessive phenotype for both traits?

_____ 11. Which of Mendel's laws says the separation of one pair of alleles into gametes does not impact the separation of a second pair of alleles into gametes?

_____ 12. How many different kinds of gametes can be produced by a parent with a *Ggrr* genotype?

_____ 13. What is the name given to the chromosomes that do not actively determine the gender of an organism?

_____ 14. Can a female fruit fly that is homozygous dominant for red eyes have any offspring (male or female) with white eyes?

Thought Questions

15. If an organism displays the dominant trait phenotype (e.g., green color in tobacco plants), it is uncertain whether the organism's genotype is homozygous dominant (*CC*) or heterozygous (*Cc*). What kind of cross (a known genotype) could be done to determine the genotype of the unknown organism? Explain your reply by describing the expected offsprings' phenotypes.

16. If you count 52 purple rough corn kernels and 50 yellow smooth corn kernels, how many purple smooth corn kernels should you expect to count? How many yellow rough corn kernels should you expect to count? Explain your reply.

17. If a trait is an X-linked recessive trait, why is it unlikely that many females with the recessive trait (e.g., white eyes in fruit flies) phenotype will exist?

21

Human Genetics

Learning Outcomes

21.1 Determining the Genotype
- Determine the genotypes by observation of individuals and their relatives.   285–287

21.2 Determining Inheritance
- Do genetic problems involving autosomal dominant, autosomal recessive, and X-linked recessive alleles.   288–290
- Do genetic problems involving multiple allele inheritance and use blood type to help determine paternity.   292–293

21.3 Genetic Counseling
- Analyze a karyotype to determine if a person's chromosomal inheritance is as expected or whether a chromosome anomaly has occurred.   293–294
- Analyze a pedigree to determine if the pattern of inheritance is autosomal dominant, autosomal recessive, or X-linked recessive.   295–296
- Construct a pedigree to determine the chances of inheriting a particular phenotype when provided with generational information.   297

Introduction

In this laboratory, you will discover that the same principles of genetics apply to humans as they do to plants and fruit flies. A gene has two alternate forms, called **alleles,** for any trait, such as hairline, finger length, and so on. One possible allele, designated by a capital letter, is **dominant** over the **recessive** allele, designated by a lowercase letter. An individual can be **homozygous dominant** (two dominant alleles, *EE*), **homozygous recessive** (two recessive alleles, *ee*) or **heterozygous** (one dominant and one recessive allele, *Ee*). **Genotype** refers to an individual's alleles, and **phenotype** refers to an individual's appearance (Fig. 21.1). Homozygous dominant and also heterozygous individuals show the dominant phenotype; homozygous recessive individuals show the recessive phenotype.

Figure 21.1   Genotype versus phenotype.
Unattached earlobes (*E*) are dominant over attached earlobes (*e*). **a.** Homozygous dominant individuals have unattached earlobes. **b.** Homozygous recessive individuals have attached earlobes. **c.** Heterozygous individuals have unattached earlobes.

a. Unattached earlobe   **b.** Attached earlobe   **c.** Unattached earlobe

21.1   Determining the Genotype

Humans inherit 46 chromosomes that occur in 23 pairs. Twenty-two of these pairs are called autosomes and one pair is the sex chromosomes. Autosomal traits are determined by alleles on the autosomal chromosomes.

Autosomal Dominant and Recessive Traits

Figure 21.2 shows a few human traits.

1. What is the homozygous dominant genotype for type of hairline? _____ What is the phenotype? _____

2. What is the homozygous recessive genotype for finger length? _____ What is the phenotype? _____

3. Why does the heterozygous individual *Ff* have freckles? _____

Figure 21.2 Commonly inherited traits in humans.

The alleles indicate which traits are dominant and which are recessive.   (a) © Superstock; (b) © Dynamic Graphics/PictureQuest RF; (c) © Bob Coyle/McGraw-Hill Education; (d) © Bob Coyle/McGraw-Hill Education; (e) © Bob Coyle/McGraw-Hill Education; (f) © Bob Coyle/McGraw-Hill Education; (g) © Corbis RF; (h) © Creatas/PunchStock RF

a. Widow's peak: *WW* or *Ww*   **b.** Straight hairline: *ww*   **e.** Short fingers: *SS* or *Ss*   **f.** Long fingers: *ss*

c. Unattached earlobes: *EE* or *Ee*   **d.** Attached earlobes: *ee*   **g.** Freckles: *FF* or *Ff*   **h.** No freckles: *ff*

These genetic problems use the alleles from Figure 21.2 and Table 21.1.

4. Maria and the members of her immediate family have attached earlobes. What is Maria's genotype? _____ Her maternal grandfather has unattached earlobes. Deduce the genotype of her maternal grandfather. _____ Explain your answer. _____

5. Moses does not have a bent little finger, but his parents do. Deduce the genotype of his parents. _____ of Moses. _____ Explain your answer. _____

6. Manny is adopted. He has hair on the back of his hand. Could both of his parents have had hair on the back of the hand? _____ Could both of his parents have had no hair on the back of the hand? _____ Explain your answer. _____

1. For this Experimental Procedure, you will need a lab partner to help you determine your phenotype for the traits listed in the first column of Table 21.1.

2. Determine your probable genotype. If you have the recessive phenotype, you know your genotype. If you have the dominant phenotype, you may be able to decide whether you are homozygous dominant or heterozygous by recalling the phenotype of your parents, siblings, or children. Circle your probable genotype in the second column of Table 21.1.

3. Your instructor will tally the class's phenotypes for each trait so that you can complete the third column of Table 21.1.

4. Complete Table 21.1 by calculating the percentage of the class with each trait. Are dominant phenotypes always the most common in a population? _____ Explain your answer. _____

Table 21.1 Autosomal Human Traits

Trait: d = Dominant r = Recessive	Probable Genotypes	Number in Class	Percentage of Class with Trait
Hairline:			
Widow's peak (d)	*WW* or *Ww*	_____	_____
Straight hairline (r)	*ww*	_____	_____
Earlobes:			
Unattached (d)	*UU* or *Uu*	_____	_____
Attached (r)	*uu*	_____	_____
Skin pigmentation:			
Freckles (d)	*FF* or *Ff*	_____	_____
No freckles (r)	*ff*	_____	_____
Hair on back of hand:			
Present (d)	*HH* or *Hh*	_____	_____
Absent (r)	*hh*	_____	_____
Thumb hyperextension—"hitchhiker's thumb":			
Last segment cannot be bent backward (d).	*TT* or *Tt*	_____	_____
Last segment can be bent back to 60° (r).	*tt*	_____	_____
Bent little finger:			
Little finger bends toward ring finger (d).	*LL* or *Ll*	_____	_____
Straight little finger (r)	*ll*	_____	_____
Interlacing of fingers:			
Left thumb over right (d)	*II* or *Ii*	_____	_____
Right thumb over left (r)	*ii*	_____	_____

21.2 Determining Inheritance

Recall that a Punnett square is a means to determine the genetic inheritance of offspring if the genotypes of both parents are known. In a **Punnett square,** all possible types of sperm are lined up vertically, and all possible types of eggs are lined up horizontally, or vice versa, so that every possible combination of gametes occurs within the square. Figure 21.3 shows how to construct a Punnett square when autosomal alleles are involved.

Figure 21.3   Punnett square.
In a Punnett square, all possible sperm are displayed vertically and all possible eggs are displayed horizontally, or vice versa. The genotypes of the offspring (in this case, also the phenotypes) are in the squares.

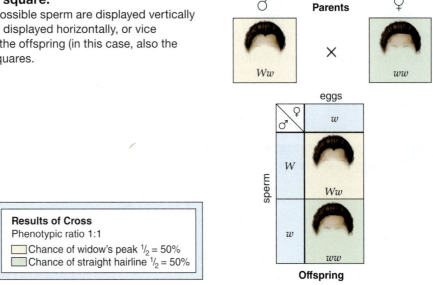

Results of Cross
Phenotypic ratio 1:1
☐ Chance of widow's peak ½ = 50%
☐ Chance of straight hairline ½ = 50%

Inheritance of Genetic Disorders

Figure 21.4 can be used to learn the chances that a particular phenotype will occur.

In Figure 21.4a,

$\frac{1}{4}$ of the offspring have the recessive phenotype = _____ % chance

$\frac{3}{4}$ of the offspring have the dominant phenotype = _____ % chance

In Figure 21.4b,

$\frac{1}{2}$ of the offspring have the recessive or the dominant phenotype = _____ % chance

In all the following genetic problems, use letters to fill in the parentheses with the genotype of the parents.

1. **a.** With reference to Figure 21.4a, if a genetic disorder is recessive and both parents are

    heterozygous (_____), what are the chances that an offspring will have the disorder? _____

    b. With reference to Figure 21.4a, if a genetic disorder is dominant and both parents are heterozygous

    (_____), what are the chances that an offspring will have the disorder? _____

2. **a.** With reference to Figure 21.4b, if the parents are heterozygous (_____) by homozygous recessive

    (_____), and the genetic disorder is recessive, what are the chances that the offspring will have the

    disorder? _____

    b. With reference to Figure 21.4b, if the parents are heterozygous (_____) by homozygous recessive

    (_____), and the genetic disorder is dominant, what are the chances that an offspring will have the

    disorder? _____

Figure 21.5 Two common patterns of X-linked inheritance in humans.

a. The sons of a carrier mother have a 50% chance of being color blind. **b.** A color-blind father has carrier daughters.

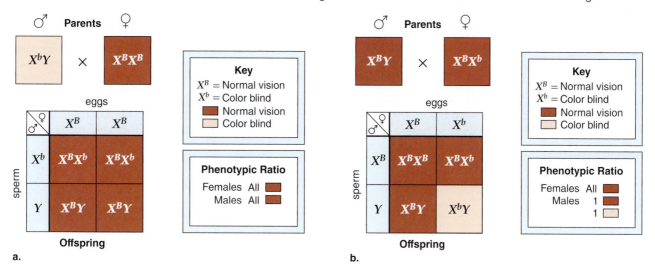

a.

b.

X-Linked Genetics Problems

For **color blindness,** there are two possible X-linked alleles involved. One affects the green-sensitive cones, whereas the other affects the red-sensitive cones. About 6% of men in the United States are color blind due to a mutation involving green perception, and about 2% are color blind due to a mutation involving red perception.

1. A woman with normal color vision (_____), whose father was color blind (_____), marries a man

   with normal color vision (_____). What genotypes could occur among their offspring? _____

   _____

   What genotypes could occur if it was the normal-visioned man's father who was color blind? _____

   _____

2. Antonio's father is color blind (_____) but his mother is not color blind (_____).

   Is Antonio necessarily color blind? _____ Explain your answer. _____

   Could he be color blind? _____ Explain your answer. _____

   _____

   Hemophilia is called the bleeder's disease because the affected person's blood is unable to clot. Although hemophiliacs do bleed externally after an injury, they also suffer from internal bleeding, particularly around joints. Hemorrhages can be checked with transfusions of fresh blood (or plasma) or concentrates of the clotting protein. The most common type of hemophilia is hemophilia A, due to absence or minimal presence of a particular clotting factor called factor VIII.

3. Make up a cross involving hemophilia that could be answered by a Punnett square, as in Figure 21.5a or b.

   _____

   _____

   What is the answer to your genetics problem? _____

Multiple Alleles

When a trait is controlled by **multiple alleles,** the gene has more than two possible alleles. But each person has only two of the possible alleles. For example, ABO blood type is determined by multiple alleles: I^A, I^B, i. Red blood cells have surface molecules called antigens that indicate they belong to the person. The I^A allele causes red blood cells to carry an A antigen, the I^B allele causes red blood cells to carry a B antigen, and the i allele causes the red blood cells to have neither of these antigens. I^A and I^B are dominant to i. Remembering that each person can have any two of the possible alleles, these are possible genotypes and phenotypes for blood types.

> ⚠️ **Protective clothing** Wear protective laboratory clothing, latex gloves, and goggles. If the chemicals touch the skin, eyes, or mouth, wash immediately. If inhaled, seek fresh air.

Genotypes	Antigens on Red Cells	Blood Types
$I^A I^A$, $I^A i$	A	A
$I^B I^B$, $I^B i$	B	B
$I^A I^B$	A and B	AB
ii	none	O

Blood type also indicates whether the person is Rh positve or Rh negative. If the genotype is *DD* or *Dd,* the person is Rh positive, and if the genotype is *dd,* the person is Rh negative. It is customary to simply attach a + or – superscript to the ABO blood type, as in A⁻.

Blood Typing Problems

1. A man with type A blood reproduces with a woman who has type B blood. Their child has blood type O.

   Using I^A, I^B, and i, give the genotype of all persons involved. man _____ woman _____

   child _____

2. If a child has type AB blood and the father has type B blood, what could the genotype of the mother be?

   _____ or _____

3. If both mother and father have type AB blood, they cannot be the parents of a child who has what blood

   type? _____

4. What blood types are possible among the children if the parents are $I^A i \times I^B i$? (*Hint:* Do a Punnett square using the possible gametes for each parent.)

Experimental Procedure: Using Blood Type to Help Determine Paternity

In this Experimental Procedure a mother, Wanda, is seeking support for her child, Sophia. We will use blood typing to decide which of three men could possibly be the father.

1. Obtain three testing plates, each of which contains three depressions; vials of blood from possible fathers 1, 2, and 3 respectively; vials of anti-A serum, anti-B serum, and anti-Rh serum. (All of these are synthetic.)
2. Using a wax pencil, number the plates so you know which plate is for possible father #1, #2, or #3. Look carefully at a plate and notice the wells are designated as A, B, or Rh.
3. Being sure to close the cap to each vial in turn, do the following using plate #1:

   Add a drop of father #1 blood to all three wells—close the cap.
   Add a drop of anti-A (blue) to the well designated A—close the cap.
   Add a drop of anti-B (yellow) to the well designated B—close the cap.
   Add a drop of anti-Rh (clear) to the well designated Rh—close the cap.

Table 21.2	Blood Types of Involved Persons				
	Mother*	Child*		Father?	
	Wanda	Sophia	#1	#2	#3
Blood type					

*Record the blood types of the Mother and Child that your instructor supplies.

4. Stir the contents of each well with a mixing stick of the correct color. After a few minutes, examine the wells for agglutination, i.e., granular appearances that indicate the blood type. (It may take a few minutes for the reaction to occur.) If a person had AB^+ blood, which wells would show agglutination? _____

_____

5. Repeat steps 3 and 4 for plates #2 and #3.
6. Record the blood type results for each of the men in Table 21. 2.

Conclusion

Which man is Sophia's father based on the results of your blood typing experiment? Explain your reply based on what you've learned about the inheritance of alleles for blood type and Rh. _____

_____

_____

_____

21.3 Genetic Counseling

Potential parents are becoming aware that many illnesses are caused by abnormal chromosomal inheritance or by gene mutations. Therefore they are seeking genetic counseling, which is available in many major hospitals. The counselor helps the couple understand the mode of inheritance for a condition of concern so that the couple can make an informed decision about how to proceed.

Determining Chromosomal Inheritance

If a genetic counselor suspects that a condition is due to a chromosome anomaly, he or she may suggest that the parents' chromosomes be examined. It is possible to view the chromosomes of an individual because cells can be microscopically examined and photographed just before cell division occurs. A computer is then used to arrange the chromosomes by pairs. The resulting pattern of chromosomes is called a **karyotype.**

A trisomy occurs when the individual has three chromosomes instead of two chromosomes at one karyotype location. **Trisomy 21** (Down syndrome) is the most common autosomal trisomy in humans. Survival to adulthood is common. Characteristic facial features include an eyelid fold, a flat face, and a large, fissured tongue. Some degree of intellectual impairment is common as is early-onset Alzheimer disease. Sterility due to sexual underdevelopment may be present.

Observation: Sex Chromosome Anomalies

A female with **Turner syndrome** (XO) has only one sex chromosome, an X chromosome; the O signifies the absence of the second sex chromosome. Because the ovaries never become functional, these females do not undergo puberty or menstruation, and their breasts do not develop. Generally, females with Turner syndrome have a short build, folds of skin on the back of the neck, difficulty recognizing various spatial patterns, and normal intelligence. With hormone supplements, they can lead fairly normal lives.

When an egg having two X chromosomes is fertilized by an X-bearing sperm, an individual with **poly-X syndrome** results. The body cells have three X chromosomes and therefore 47 chromosomes. Although they tend to have learning disabilities, poly-X females have no apparent physical anomalies, and many are fertile and have children with a normal chromosome count.

When an egg having two X chromosomes is fertilized by a Y-bearing sperm, a male with **Klinefelter syndrome** results. This individual is male in general appearance, but the testes are underdeveloped, and the breasts may be enlarged. The limbs of XXY males tend to be longer than average, muscular development is poor, body hair is sparse, and many XXY males have learning disabilities.

Jacob syndrome occurs in males who are usually taller than average, suffer from persistent acne, and tend to have speech and reading problems. At one time, it was suggested that XYY males were likely to be criminally aggressive, but the incidence of such behavior has been shown to be no greater than that among normal XY males.

Label each karyotype in Figure 21.6 as one of the syndromes just discussed. Explain your answers on the lines provided.

Figure 21.6   Sex chromosome anomalies.
(a) © CNRI/SPL/Science Source; (b) © CNRI/SPL/Science Source; (c) © CNRI/SPL/Science Source; (d) © CNRI/SPL/Science Source

1. _____   2. _____   3. _____   4. _____

_____   _____   _____   _____

_____   _____   _____   _____

Determining the Pedigree

A pedigree shows the inheritance of a genetic disorder within a family and can help determine the inheritance pattern and whether any particular individual has an allele for that disorder. Then a Punnett square can be done to determine the chances of a couple producing an affected child.

The symbols used to indicate normal and affected males and females, reproductive partners, and siblings in a pedigree are shown in Figure 21.7.

For example, suppose you wanted to determine the inheritance pattern for straight hairline and you knew which members of a generational family had the trait (Fig. 21.8a).

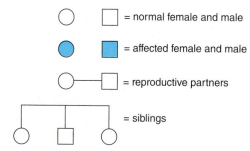

Figure 21.7 Pedigree symbols.

Figure 21.8 Autosomal pedigrees
a. Child with recessive phenotype can have parents without the recessive phenotype. **b.** Child with the dominant phenotype has parent(s) with the dominant phenotype; heterozygous parents can also have a child without the dominant phenotype.

a. Straight hairline is recessive.

b. Widow's peak is dominant.

A pedigree allows you to determine that straight hairline is autosomal recessive because two parents without this phenotype have a child with the phenotype. This can happen only if the parents are heterozygous and straight hairline is recessive. Similarly, a pedigree allows you to determine that widow's peak is autosomal dominant (Fig. 21.8b): a child with this phenotype has at least one parent with the dominant phenotype, but again, heterozygous parents can produce a child without widow's peak. *Give each person in Figure 21.8a and* b *a genotype.*

Not shown is an X-linked recessive pedigree. An X-linked recessive phenotype occurs mainly in males, and it skips a generation because a female who inherits a recessive allele for the condition from her father may have a son with the condition.

Pedigree Analysis

For each of the following pedigrees, decide whether a trait is inherited as an autosomal dominant, autosomal recessive, or X-linked recessive. Then decide the genotype of particular individuals in the pedigree.

1. Study the following pedigree:

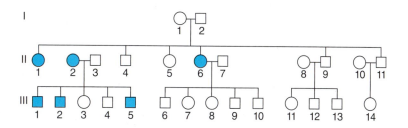

a. Notice that neither of the original parents is affected, but several children are affected. This could happen only if the trait were _____.

b. What is the genotype of the following individuals? Use *A* for the dominant allele and *a* for the recessive allele and explain your answer.

Generation I, individual 1: _____

Generation II, individual 1: _____

Generation III, individual 8: _____

2. Study the following pedigree:

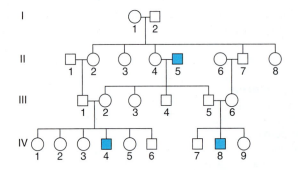

a. Notice that only males are affected. This could happen only if the trait were _____.

b. What is the genotype of the following individuals? Explain your answer.

Generation I, individual 1: _____

Generation II, individual 8: _____

Generation III, individual 1: _____

Construction of a Pedigree

You are a genetic counselor who has been given the following information from which you will construct a pedigree.

1. Your data: <u>Henry</u> has a double row of eyelashes, which is a dominant trait. Both his <u>maternal grandfather</u> and his <u>mother</u> have double eyelashes. <u>Their spouses</u> are normal. Henry is married to <u>Isabella</u> and their <u>first child, Polly,</u> has normal eyelashes. The couple wants to know the chances of any child having a double row of eyelashes.

2. What is your key for this trait?

   Key: _____ normal eyelashes   _____ double row of eyelashes

3. *Construct two pedigrees with symbols (circles and squares) only for the underlined persons in #1.* The pedigrees start with the maternal grandfather and grandmother and end with Polly.

   Pedigree 1   **Pedigree 2**

4. Pedigree 1: Try out a pattern of autosomal dominant inheritance by assigning appropriate genotypes for an autosomal dominant pattern of inheritance to each person in this pedigree. Pedigree 2: Try out a pattern of X-linked dominant inheritance by assigning appropriate genotypes for this pattern of inheritance to each

   person in your pedigree. Which pattern is correct? _____

5. Use correct genotypes to show a cross between Henry and Isabella and from experience with crosses, state the expected phenotypic ratio among the offspring:

   Cross   Henry   Isabella   **Phenotypic ratio:**

   _____ × _____   _____

6. What are the percentage chances that Henry and Isabella will have a child with double eyelashes? _____
   "Chance has no memory," and each child has the same chance for double eyelashes. Explain why.

   _____

   _____

_____   1. What phrase describes an individual with two dominant alleles?

_____   2. What word refers to the alleles possessed by an individual?

_____   3. How many autosomes do humans have?

_____   4. An individual will have a widow's peak if the genotype is *WW* or *Ww*. Is widow's peak an autosomal dominant or autosomal recessive trait?

_____   5. What technique can be used to determine the genetic inheritance of offspring when the parents have known genotypes?

_____   6. Cystic fibrosis is an autosomal recessive disease. What word refers to an individual who has one recessive allele for cystic fibrosis?

_____   7. What is the genotype of individuals who will not develop Huntington disease (autosomal dominant)?

_____   8. From whom does a male child inherit color blindness—his mother or his father?

_____   9. What is the genotype of a carrier for hemophilia?

_____   10. How many dominant alleles does someone with AB blood type have?

_____   11. Can a child with AB blood type have a parent with O blood?

_____   12. What is the presence of three chromosomes (instead of two) called, and what human chromosome is most commonly affected in this fashion?

_____   13. What is created when chromosomes are paired by size and shape?

_____   14. What technique is used to determine the inheritance pattern of specific traits within a family?

Thought Questions

15. The blood types of parents and their two children are determined. The father has type A blood. The mother has type O blood. One child has type A blood and the other has type O blood. What are the genotypes of all people involved? Explain your reply.

16. What are the chances that a color-blind man will have a color-blind grandson (the child of the color-blind man's normal vision daughter and a normal vision male)? What are the genotypes of all people involved?

17. Recall what you learned about the fate of homologous chromosomes during meiosis I and the sister chromatids during meiosis II, and explain how trisomies like trisomy 21 or poly-X syndrome could occur.

22

DNA Biology and Technology

Learning Outcomes

22.1 DNA Structure and Replication
- Explain how the structure of DNA facilitates replication. 300–301
- Explain how DNA replication is semiconservative. 301–302

22.2 RNA Structure
- List the ways in which RNA structure differs from DNA structure. 303–304

22.3 DNA and Protein Synthesis
- Describe how DNA is able to store information. 304
- State the function of transcription and translation during protein synthesis. 305–307

22.4 Isolation of DNA and Biotechnology
- Describe the procedure for isolating DNA. 308
- Describe the process of DNA gel electrophoresis. 308–309

22.5 Detecting Genetic Disorders
- Understand the relationship between an abnormal DNA base sequence and a genetic disorder. 309–310
- Suggest two ways to detect a genetic disorder. 310–311

Introduction

This laboratory pertains to molecular genetics and biotechnology. Molecular genetics is the study of the structure and function of **DNA (deoxyribonucleic acid)**, the genetic material. Biotechnology is the manipulation of DNA for the benefit of humans and other organisms.

First we will study the structure of DNA and see how that structure facilitates DNA replication in the nucleus of cells. DNA replicates prior to mitosis; following mitosis, each daughter nucleus has a complete copy of the genetic material. DNA replication also precedes the production of gametes.

Then we will study the structure of **RNA (ribonucleic acid)** and how it differs from that of DNA, before examining how DNA, with the help of RNA, specifies protein synthesis. The linear construction of DNA, in which nucleotide follows nucleotide, is paralleled by the linear construction of the primary structure of protein, in which amino acid follows amino acid. Essentially, we will see that the sequence of nucleotides in DNA codes for the sequence of amino acids in a protein. We will also review the role of three types of RNA in protein synthesis. DNA's code is passed to messenger RNA (mRNA), which moves to the ribosomes containing ribosomal RNA (rRNA). Transfer RNA (tRNA) brings the amino acids to the ribosomes, and they become linked by peptide bonds in the order directed by mRNA. In this way, a particular polypeptide forms.

We now understand that a mutated gene has an altered DNA base sequence, and altered sequences can cause genetic disorders. You will have an opportunity to carry out a laboratory procedure that detects whether an individual is normal, has sickle-cell disease, or is a carrier.

22.1 DNA Structure and Replication

The structure of DNA lends itself to **replication,** the process that makes a copy of a DNA molecule. DNA replication is a necessary part of chromosome duplication, which precedes mitosis and meiosis. Therefore, DNA replication is needed for growth and repair and for sexual reproduction.

DNA Structure

DNA is a polymer of nucleotide monomers (Fig. 22.1). Each nucleotide is composed of three molecules: deoxyribose (a 5-carbon sugar), a phosphate, and a nitrogen-containing base.

Figure 22.1   Overview of DNA structure.
Diagram of DNA double helix shows that the molecule resembles a twisted ladder. Sugar–phosphate backbones make up the sides of the ladder, and hydrogen-bonded bases make up the rungs of the ladder. Complementary base pairing dictates that A is bonded to T and G is bonded to C and vice versa. *Label the boxed nucleotide pair as directed in the next Observation.*

a. One pair of bases

b. Ladder structure

1. A boxed nucleotide pair is shown in Figure 22.1. If you are working with a kit, draw a representation of one of your nucleotides here. *Label phosphate, base pair, and deoxyribose in your drawing and in spaces 1–3 in Figure 22.1a.*

2. Notice the four types of bases: cytosine (C), thymine (T), adenine (A), and guanine (G). What is the color of each of the four types of bases in Figure 22.1? In your kit? Complete Table 22.1 by writing in the colors of the bases.

Table 22.1 Base Colors		
	In Figure 22.1	**In Your Kit**
Cytosine		
Thymine		
Adenine		
Guanine		

3. Using Figure 22.1 as a guide, join several nucleotides together. Observe the entire DNA molecule.

   What types of molecules make up the backbone (uprights of ladder) of DNA (Fig. 22.1b)? _____

   and _____ In the backbone, the phosphate of one nucleotide is bonded to a sugar of the next nucleotide.

4. Using Figure 22.1 as a guide, join the bases together with hydrogen bonds. *Label a hydrogen bond in Figure 22.1.* Dashes are used to represent hydrogen bonds in Figure 22.1b because hydrogen bonds are

   (strong or weak) _____.

5. Notice in Figure 22.1b and in your model that the base A is always paired with the base _____,

   and the base C is always paired with the base _____. This is called complementary base pairing.

6. In Figure 22.1b, what molecules make up the rungs of the ladder? _____

7. Each half of the DNA molecule is a DNA strand. Why is DNA also called a double helix (Fig. 22.1b)?

   _____

DNA Replication

During replication, the DNA molecule is duplicated so that there are two identical DNA molecules. We will see that complementary base pairing makes replication possible.

Observation: DNA Replication

1. Before replication begins, DNA is unzipped. Using Figure 22.2*a* as a guide, break apart your two DNA strands. What bonds are broken in order to unzip the DNA strands? _____

2. Using Figure 22.2*b* as a guide, attach new complementary nucleotides to each strand using complementary base pairing.

3. Show that you understand complementary base pairing by completing Table 22.2.
   You now have two DNA molecules (Fig. 22.2*c*). Are your molecules identical?

   _____

4. Because of complementary base pairing, each new double helix is

   composed of an _____ strand

   and a _____ strand. *Write old or new in 1–10, Figure 22.2*a, b, *and* c. *Conservative* means to save something from the past. Why is DNA replication called semiconservative?

   _____

   _____

   _____

Figure 22.2   DNA replication.

Use of the ladder configuration better illustrates how replication takes place. **a.** The parental DNA molecule. **b.** The "old" strands of the parental DNA molecule have separated. New complementary nucleotides available in the cell are pairing with those of each old strand. **c.** Replication is complete.

1. _____   2. _____

a.

3. _____   ← 4. _____   6. _____

5. _____   →

b.

7. _____   ← 8. _____   10. _____

9. _____   →

c.

5. Genetic material has to be inherited from cell to cell and organism to organism. Consider that because of DNA replication, a chromosome is composed of two chromatids and each chromatid is a DNA double helix. The chromatids separate during cell division so that each daughter cell receives a copy of each chromosome. In your own words, how does replication provide a means for passing DNA from cell to cell and organism to organism? _____

_____

Table 22.2 DNA Replication																									
Old strand	G	G	G	T	T	C	C	A	T	T	A	A	A	T	T	C	C	A	G	A	A	A	T	C	A	T	A
New strand																											

22.2 RNA Structure

Like DNA, RNA is a polymer of nucleotides (Fig. 22.3). In an RNA nucleotide, the sugar ribose is attached to a phosphate molecule and to a nitrogen-containing base, C, U, A, or G. In RNA, the base uracil replaces thymine as one of the pyrimidine bases. RNA is single stranded, whereas DNA is double stranded.

Figure 22.3   Overview of RNA structure.
RNA is a single strand of nucleotides. *Label the boxed nucleotide as directed in the next Observation.*

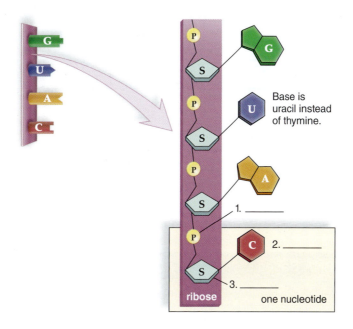

1. Describe the backbone of an RNA molecule. _____

2. Where are the bases located in an RNA molecule? _____

Observation: RNA Structure

1. If you are using a kit, draw a nucleotide for the construction of mRNA. *Label the ribose (the sugar in RNA), the phosphate, and the base in your drawing and in 1–3, Figure 22.3.*

2. Complete Table 22.3 by writing in the colors of the bases for Figure 22.3 and for your kit.

Table 22.3 Base Colors		
	In Figure 22.3	**In Your Kit**
Cytosine		
Uracil		
Adenine		
Guanine		

3. The base uracil substitutes for the base thymine in RNA. Complete Table 22.4 to show the several other ways RNA differs from DNA.

Table 22.4 DNA Structure Compared with RNA Structure

	DNA	RNA
Sugar	Deoxyribose	
Bases	Adenine, guanine, thymine, cytosine	
Strands	Double stranded with base pairing	
Helix	Yes	

Complementary Base Pairing

Complementary base pairing occurs between DNA and RNA. The RNA base uracil pairs with the DNA base adenine; the other bases pair as shown previously. Complete Table 22.5 to show the complementary DNA bases for the RNA bases.

Table 22.5 DNA and RNA Bases

RNA Bases	C	U	A	G
DNA Bases				

22.3 DNA and Protein Synthesis

Protein synthesis requires the processes of transcription and translation. During **transcription,** which takes place in the nucleus, an RNA molecule called **messenger RNA (mRNA)** is made complementary to one of the DNA strands. This mRNA leaves the nucleus and goes to the ribosomes in the cytoplasm. Ribosomes are composed of **ribosomal RNA (rRNA)** and proteins in two subunits. They provide a location for protein synthesis to occur.

During **translation,** RNA molecules called **transfer RNA (tRNA)** bring amino acids to the ribosome, and the amino acids join in the order prescribed by mRNA. In this way, the sequence of amino acids in a new polypeptide was originally specified by DNA. This is the information that DNA, the genetic material, stores.

Explain the role DNA, mRNA, and tRNA have in protein synthesis here:

DNA _____

mRNA _____

tRNA _____

Transcription

During transcription, complementary RNA is made from a DNA template (Fig. 22.4). A portion of DNA unwinds and unzips at the point of attachment of the enzyme **RNA polymerase.** A strand of mRNA is produced when complementary nucleotides join in the order dictated by the sequence of bases in DNA. Transcription occurs in the nucleus, and the mRNA passes out of the nucleus to enter the cytoplasm.

    Label Figure 22.4. For number 1, note the name of the enzyme that carries out mRNA synthesis. For number 2, note the name of this polynucleotide molecule.

Observation: Transcription

1. If you are using a kit, unzip your DNA model so that only one strand remains. This strand is called the **template strand*** because it will be used as a template to construct an mRNA molecule. The **gene strand** is preferred terminology for the complementary DNA strand you discarded because as you can verify in Figure 22.4, it has the same sequence of bases as the mRNA molecule, except that uracil substitutes for thymine.

2. Using Figure 22.4 as a guide, construct a messenger RNA (mRNA) molecule by first lining up RNA nucleotides complementary to the template strand of your DNA molecule. Join the nucleotides together to form mRNA.

3. A portion of DNA has the sequence of bases shown in Table 22.6. *Complete Table 22.6 to show the sequence of bases in mRNA.*

4. If you are using a kit, unzip mRNA transcript from the DNA. Locate the end of the strand

that will move to the _____ in the cytoplasm.

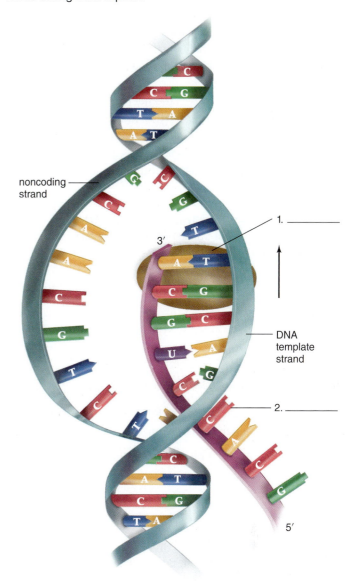

Figure 22.4 Messenger RNA (mRNA).
Messenger RNA, which is complementary to a section of DNA, forms during transcription.

noncoding strand

1. _____

DNA template strand

2. _____

Table 22.6	Transcription
DNA	T A C A C G A G C A A C T A A C A T
mRNA	

Sometimes the template strand is called the sense strand and the gene strand is called the antisense strand.

Translation

During translation, a polypeptide is made. DNA specifies the sequence of amino acids in a polypeptide because every three bases code for an amino acid. Therefore, DNA is said to have a **triplet code.** The bases in mRNA are complementary to the bases in DNA. Every three bases in mRNA are called a **codon.** One codon of mRNA represents one amino acid. Thus, the sequence of DNA bases serves as the blueprint for the sequence of amino acids assembled to make a protein. The correct sequence of amino acids in a polypeptide is the message that mRNA carries.

Messenger RNA leaves the nucleus and proceeds to the ribosomes, where protein synthesis occurs. As previously mentioned, transfer RNA (tRNA) molecules transfer amino acids to the ribosomes. Each RNA has one particular tRNA amino acid at one end and a specific **anticodon** at the other end (Fig. 22.5). *Label Figure 22.5,* where the amino acid is represented as a colored ball, the tRNA is green, and the anticodon is the sequence of three bases. (The anticodon is complementary to the mRNA codon.)

Figure 22.5 Transfer RNA (tRNA).
Each type of transfer RNA with a specific anticodon carries a particular amino acid to the ribosomes.

1. _____
2. _____
3. _____

Observation: Translation

1. Figure 22.6 shows seven tRNA–amino acid complexes. Every amino acid has a name; in the figure, only the first three letters of the name are inside the ball.

2. If you are using a kit, arrange your tRNA–amino acid complexes in the order consistent with Table 22.7.

   Complete Table 22.7. Why are the codons and anticodons in groups of three? _____
   _____

Figure 22.6 Transfer RNA diversity.
Each type of tRNA carries only one particular amino acid, designated here by the first three letters of its name.

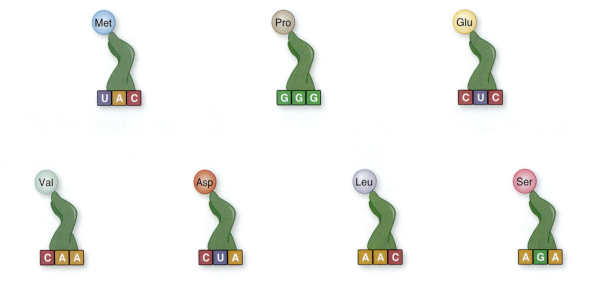

Table 22.7   Translation

mRNA codons		AUG	CCC	GAG	GUU	GAU	UUG	UCU
tRNA anticodons								
Amino acid*								

*Use three letters only. See Table 22.8 for the full names of these amino acids.

Table 22.8   Names of Amino Acids

Abbreviation	Name
Met	methionine
Pro	proline
Asp	aspartate
Val	valine
Glu	glutamic acid
Leu	leucine
Ser	serine

3. Figure 22.7 shows the manner in which the polypeptide grows. A ribosome has three binding sites. From left to right, they are the A (amino acid) site, the P (peptide) site, and the E (exit) site. A tRNA leaves from the E site after it has passed its amino acid or peptide to the newly arrived tRNA–amino acid complex. Then the ribosome moves forward, making room for the next tRNA–amino acid. This sequence of events occurs over and over until the entire polypeptide is borne by the last tRNA to come to the ribosome. Then a release factor releases the polypeptide chain from the ribosome. *In Figure 22.7, label the ribosome, the mRNA, and the peptide.*

Figure 22.7   Protein synthesis.

1. A ribosome has a site for two tRNA–amino acid complexes. **2.** Before a tRNA leaves, an RNA passes its attached peptide to a newly arrived tRNA–amino acid complex. **3.** The ribosome moves forward, and the next tRNA–amino acid complex arrives.

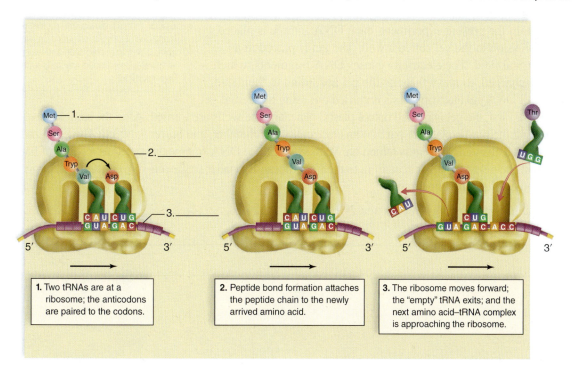

1. Two tRNAs are at a ribosome; the anticodons are paired to the codons.

2. Peptide bond formation attaches the peptide chain to the newly arrived amino acid.

3. The ribosome moves forward; the "empty" tRNA exits; and the next amino acid–tRNA complex is approaching the ribosome.

22.4 Isolation of DNA and Biotechnology

In the following Experimental Procedure, you will isolate DNA from the cells of a fruit or vegetable. It will only be necessary to expose the cells to an agent (dishwasher detergent) that emulsifies membrane in order to "free" the DNA from its enclosures (plasma membrane and nuclear envelope). When transferred to a tube, the presence of NaCl allows DNA to precipitate as a sodium salt. The precipitate forms at the interface between ethanol and the salt solution, and then it floats to the top of the tube where it may be collected.

Experimental Procedure: Isolation of DNA

1. You will need a slice of fruit or vegetable (i.e., tomato, onion, or apple), a mortar and pestle, and a large clean glass test tube on ice.
2. Crush and grind a slice of fruit or vegetable in a mortar and pestle. Remove the pestle and set aside. Add enough 0.9% NaCl solution to achieve a "soupy" consistency.
3. Add two drops of dishwasher detergent (such as Blue Dawn) to the mixture in the mortar. Swirl until the color of the detergent disappears. Wait 3 to 5 minutes. The solution becomes clear and viscous as the DNA escapes its enclosures.
4. Check the viscosity of the solution in the mortar, and if it is too viscous for you to pipet, add a few more ml of NaCl solution. Use a transfer pipet to move 2 to 3 ml of the DNA from the mortar to the glass tube. Try to pick up clear zones of the solution WITHOUT CELL DEBRIS.
5. Slowly add 5 to 7 ml of ice-cold ethanol down the side of the tube on ice. As you do so, DNA will precipitate *between* the ethanol and the NaCl solutions. Then it will float to the top, where it can be picked up by a fresh transfer pipet. Any "white dust" you see consists of RNA molecules.
6. Use a transfer pipet to place the DNA in a small, clean test tube. Pipet away any extra water/ethanol and air-dry the DNA for a few minutes. Dissolve the DNA in 3 to 4 ml distilled H_2O.
7. Add five drops of phenol red, a pH indicator. The resulting dark pink color confirms the presence of nucleic acid (i.e., the DNA).

Experimental Procedure: Gel Electrophoresis

During gel electrophoresis, charged DNA molecules migrate across a span of gel (gelatinous slab) because they are placed in a powerful electrical field. In the present experiment, each DNA sample is placed in a small depression in the gel called a well. The gel is placed in a powerful electrical field. The electricity causes DNA fragments, which are negatively charged, to move through the gel according to their size.

> ⚠ For experimental procedures, wear safety goggles, gloves, and protective clothing. If any chemical spills on your skin, wash immediately with mild soap and water; flood eyes with water only. Report any spills immediately to your instructor.

Almost all DNA gel electrophoresis is carried out using horizontal gel slabs (Fig. 22.8). First, the gel is poured onto a plastic plate, and the wells are formed. After the samples are added to the wells, the gel and the plastic plate are put into an electrophoresis chamber, and buffer is added. The DNA samples begin to migrate after the electrical current is turned on. With staining, the DNA fragments appear as a series of bands spread from one end of the gel to the other according to their size because smaller fragments move faster than larger fragments.

Answer the following questions:

a. What is biotechnology? (See page 299.) _____

_____

_____

Figure 22.8   Equipment and procedure for gel electrophoresis.

a. Agarose solution poured into casting tray

b. Comb that forms wells for samples

c. Wells that can be loaded with samples

d. Electrophoresis chamber and power supply

b. Speculate how the ability to isolate DNA and run gel electrophoresis of DNA relates to biotechnology.

_____

_____

_____

c. Name a biotechnology product someone you know is now using or taking as a medicine.

_____

22.5   Detecting Genetic Disorders

The base sequence of DNA in all the chromosomes is an organism's genome. The Human Genome Project was a monumental effort to determine the normal order of all the 3.6 billion nucleotide bases in the human genome. Someday it will be possible to sequence anyone's genome within a relatively short time; this will enable us to determine what particular base sequence alterations signify that he or she has a disorder or will have one in the future. In this laboratory, you will study the alteration in base sequence that causes a person to have sickle-cell disease.

In persons with sickle-cell disease, the red blood cells aren't biconcave disks like normal red blood cells—they are sickle shaped. Sickle-shaped cells can't pass along narrow capillary passageways. They clog the vessels and break down, causing the person to suffer from poor circulation, anemia, and poor resistance to infection. Internal hemorrhaging leads to further complications, such as jaundice, episodic pain in the abdomen and joints, and damage to internal organs.

Sickle-shaped red blood cells are caused by an abnormal hemoglobin (Hb^S). Individuals with the Hb^AHb^A genotype are normal; those with the Hb^SHb^S genotype have sickle-cell disease, and those with the Hb^AHb^S have sickle-cell trait. Persons with sickle-cell trait do not usually have sickle-shaped cells unless they experience dehydration or mild oxygen deprivation.

Genetic Sequence for Sickle-Cell Disease

Examine Figure 22.9a and b, which show the DNA base sequence, the mRNA codons, and the amino acid sequence for a portion of the gene for Hb^A and the same portion for Hb^S.

1. In what one base does Hb^A differ from Hb^S? Hb^A _____ Hb^S _____

2. What are the codons that contain this base? Hb^A _____ Hb^S _____

3. What is the amino acid difference? Hb^A _____ Hb^S _____

This amino acid difference causes the polypeptide chain in sickle-cell hemoglobin to pile up as firm rods that push against the plasma membrane and deform the red blood cell into a sickle shape:

glutamic acid
(polar *R* group)

valine
(nonpolar *R* group)

Figure 22.9 Sickle-cell disease.

a. When red blood cells are normal, the base sequence (in one location) for Hb^A alleles is CTC. **b.** In sickle-cell disease at these locations, it is CAC.

(a) © Bill Longcore/Science Source; (b) © Bill Longcore/Science Source

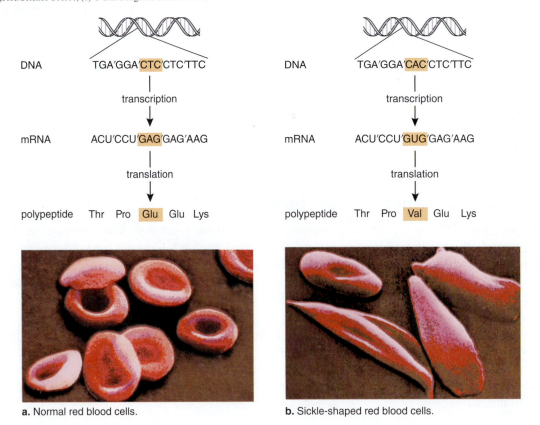

DNA	TGA'GGA'CTC'CTC'TTC

transcription

| mRNA | ACU'CCU'GAG'GAG'AAG |

translation

| polypeptide | Thr Pro Glu Glu Lys |

a. Normal red blood cells.

DNA	TGA'GGA'CAC'CTC'TTC

transcription

| mRNA | ACU'CCU'GUG'GAG'AAG |

translation

| polypeptide | Thr Pro Val Glu Lys |

b. Sickle-shaped red blood cells.

Detection of Sickle-Cell Disease by Gel Electrophoresis

Three samples of hemoglobin have been subjected to protein gel electrophoresis. Protein gel electrophoresis (Fig. 22.10) is carried out in the same manner as DNA gel electrophoresis (see Fig. 22.8) except the gel has a different composition.

1. Sickle-cell hemoglobin (Hb^S) migrates more slowly toward the positive pole than normal hemoglobin (Hb^A) because the amino acid valine has no polar R groups, whereas the amino acid glutamic acid does have a polar R group.

2. In Figure 22.10, which lane contains only Hb^S, signifying that the individual is Hb^SHb^S?

   _____

3. Which lane contains only Hb^A, signifying that the individual is Hb^AHb^A? _____

4. Which lane contains both Hb^S and Hb^A, signifying that the individual is Hb^AHb^S? _____

Figure 22.10 Gel electrophoresis of hemoglobins.

Lane 1   Lane 2   Lane 3

Detection by Genomic Sequencing

You are a genetic counselor. A young couple seeks your advice because sickle-cell disease occurs among the family members of each. You order DNA base sequencing to be done. The results come back that at one of the loci for normal hemoglobin, each has the abnormal sequence CAC instead of CTC. The other locus is normal. What are the chances that this couple will have a child with sickle-cell disease? _____

Conclusion: Detecting Genetic Disorders

• What two methods of detecting sickle-cell disease were described in this section? _____

  _____

• Which method is more direct and probably requires more expensive equipment to do? _____

  _____

• Which method probably preceded the other method as a means to detect sickle-cell disease? _____

  _____

_____ 1. What biomolecule is composed of deoxyribose sugars, phosphates, and the nitrogen-containing bases adenine, guanine, cytosine, and thymine?

_____ 2. What holds the nitrogen bases together in DNA?

_____ 3. What process makes a copy of DNA so that daughter cells that result from mitosis and cytokinesis each have a copy of DNA?

_____ 4. Which five-carbon sugar is characteristic of RNA?

_____ 5. What nitrogen base is associated with RNA but not with DNA?

_____ 6. What kind of RNA is made when transcription occurs?

_____ 7. Which strand of DNA is used to make mRNA?

_____ 8. Where in a cell does protein synthesis occur?

_____ 9. What molecule transports amino acids to the ribosomes?

_____ 10. How many nitrogen bases make up a codon, and what does each codon represent?

_____ 11. What connects one amino acid to another in the polypeptide chain?

_____ 12. Which size DNA fragments, smaller or larger, move more quickly through a gel during electrophoresis?

_____ 13. How are many genetic disorders now identified?

_____ 14. What protein is affected when an individual has sickle-cell disease?

Thought Questions

15. Students sometimes get the impression that DNA must be replicated prior to protein synthesis. Explain why DNA is only replicated prior to the start of mitosis and meiosis and what the result would be if DNA was replicated each time a new protein needed to be made.

16. If the portion of DNA to be transcribed has 66 bases, how many amino acids will the resulting protein have? Explain your reply.

17. **a.** What concerns should high school and college athletic coaches have about the potential for their players to have sickle-cell trait?

    b. How could players be identified as having sickle-cell trait or not?

23

Evidences of Evolution

Learning Outcomes

23.1 Evidence from the Fossil Record
- Use the geologic timescale to trace the evolution of life in broad outline.   314–316
- Describe several types of fossils and explain how fossils help establish the sequence in the evolution of life.   316–319
- Explain how scientists use fossils to establish that organisms are related by common descent.   320

23.2 Evidence from Comparative Anatomy
- Explain how comparative anatomy provides evidence that humans are related to other groups of vertebrates and also other groups of primates.   320–322
- Compare the human skeleton with the chimpanzee skeleton, and explain the differences on the basis of their different ways of life.   322–325
- Compare hominid skulls and hypothesize a possible evolutionary sequence.   325–326

23.3 Molecular Evidence
- Explain how molecular evidence aids the study of how humans are related to all other groups of organisms on Earth.   327
- Explain how molecular evidence also helps show the degree to which humans are related to other vertebrates and primates.   328–329

Introduction

Evolution is the process by which organisms are related by **common descent**: All organisms can trace their ancestry to the first cells. The process of evolution is amazingly simple. A group of organisms changes over time because the members of the group most suited to the natural environment have more offspring than the others in the group. So, for example, among bacteria, those which can withstand an antibiotic leave more offspring, and with time the entire group of bacteria becomes resistant to the antibiotic. Reproduction and therefore evolution have been going on since the first cells appeared on Earth. By studying (1) the fossil record, (2) anatomical and embryological comparative anatomy, and (3) molecular evidence, science is able to show that all organisms are related to one another.

Students digging for fossils.
© Mel Evans/AP/Corbis

23.1 Evidence from the Fossil Record

The geologic timescale, which was developed by both geologists and paleontologists, depicts the history of life based on the fossil record (Table 23.1). A **fossil** is any evidence of the existence of an organism in ancient times as opposed to modern times. Paleontologists specialize in removing fossils from the Earth's crust (see page 313). In this section, we will study the geologic timescale and then examine some fossils.

Geologic Timescale

Divisions of the Timescale

Notice that the timescale divides the history of Earth into eras, then periods, and then epochs. The four eras span the greatest amounts of time, and the epochs are the shortest time frames. Notice that only the periods of the Cenozoic era are divided into epochs, meaning that more attention is given to the evolution of primates and flowering plants than to the earlier evolving organisms. List the four eras in the timescale starting with

Precambrian time: _____

1. Using the geologic timescale, you can trace the history of life by beginning with Precambrian time at the bottom of the timescale. The timescale indicates that the first cells (the prokaryotes) arose some 3,500 million years ago (MYA). The prokaryotes evolved before any other group. Why do you read the timescale

    starting at the bottom? _____

2. The Precambrian time was very long, lasting from the time the Earth first formed until 542 MYA. The fossil record during the Precambrian time is meager, but the fossil record from the Cambrian period onward is rich (for reasons still being determined). This helps explain why the timescale usually does not show any periods until the Cambrian period of the Paleozoic era. You can also use the timescale to check when certain groups evolved and/or flourished.
    Example: During the Ordovician period, the nonvascular plants appear on land, and the first jawless and jawed fishes appear in the seas.

    During the _____ era and the _____ period, the first flowering plants appear. How many

    million years ago was this? _____

3. On the timescale, note the Carboniferous period. During this period great swamp forests covered the land. These are also called coal-forming forests because, with time, they became the coal we burn today. How do you know that the plants in this forest were not flowering trees as most of our trees

    are today? _____

    What type of animal was diversifying at this time? _____

4. You should associate the Cenozoic era with the evolution of humans. Among mammals, humans are

    primates. During what period and epoch did primates appear? _____

    Among primates, humans are hominids. During what period and epoch did hominids appear? _____

    _____ The scientific name for humans is *Homo sapiens*.

    What period and epoch is the age of *Homo sapiens?* _____

Dating within the Timescale

The timescale provides both relative dates and absolute dates. When you say, for example, "Flowering plants evolved during the Jurassic period," you are using relative time, because flowering plants evolved earlier or later than groups in other periods. If you use the dates that are given in millions of years (MYA), you are using absolute time. Absolute dates are usually obtained by measuring the amount of a radioactive isotope in the rocks surrounding the fossils. Why wouldn't you expect to find human fossils and dinosaur fossils together in rocks

dated similarly? _____

Table 23.1 The Geologic Timescale: Major Divisions of Geological Time and Some of the Major Evolutionary Events That Occurred

Era	Period	Epoch	Millions of Years Ago	Plant Life	Animal Life
Cenozoic*	Quaternary	Holocene	0.01–present	Human influence on plant life	Age of *Homo sapiens*
			Significant Mammalian Extinction		
		Pleistocene	1.8–0.01	Herbaceous plants spread and diversify.	Presence of Ice Age mammals Modern humans appear.
		Pliocene	5.3–1.8	Herbaceous angiosperms flourish.	First hominids appear.
		Miocene	23–5.3	Grasslands spread as forests contract.	Apelike mammals and grazing mammals flourish; insects flourish.
	Tertiary	Oligocene	33.9–23	Many modern families of flowering plants evolve.	Browsing mammals and monkeylike primates appear.
		Eocene	55.8–33.9	Subtropical forests with heavy rainfall thrive.	All modern orders of mammals are represented.
		Paleocene	65.5–55.8	Flowering plants continue to diversify.	Primitive primates, herbivores, carnivores, and insectivores appear.
Mesozoic			**Mass Extinction: Dinosaurs and Most Reptiles**		
	Cretaceous		145.5–65.5	Flowering plants diversify; conifers persist.	Placental mammals appear; modern insect groups appear.
	Jurassic		199.6–145.5	Flowering plants appear.	Dinosaurs flourish; birds appear.
			Mass Extinction		
	Triassic		251–199.6	Forests of conifers and cycads dominate.	First mammals appear; first dinosaurs appear; corals and molluscs dominate seas.
Paleozoic			**Mass Extinction**		
	Permian		299–251	Gymnosperms diversify.	Reptiles diversify; amphibians decline.
	Carboniferous		359.2–299	Age of great coal-forming forests: Ferns, club mosses, and horsetails flourish.	Amphibians diversify; first reptiles appear; first great radiation of insects.
			Mass Extinction		
	Devonian		416–359.2	First seed plants appear. Seedless vascular plants diversify.	Jawed fishes diversify and dominate the seas; first insects and first amphibians appear.
	Silurian		443.7–416	Seedless vascular plants appear.	First jawed fishes appear.
			Mass Extinction		
	Ordovician		488.3–443.7	Nonvascular land plants appear. Marine algae flourish.	Invertebrates spread and diversify; jawless fishes (first vertebrates) appear.
	Cambrian		542–488.3	First plants appear on land. Marine algae flourish.	All invertebrate phyla present; first chordates appear.
Precambrian Time			630	Oldest soft-bodied invertebrate fossils	
			1,000–700	Protists evolve and diversify.	
			2,100	Oldest eukaryotic fossils	
			2,700	O_2 accumulates in atmosphere.	
			3,500	Oldest known fossils (prokaryotes)	
			4,570	Earth forms.	

*Many authorities divide the Cenozoic era into the Paleogene period (contains the Paleocene, Eocene, and Oligocene epochs) and the Neogene period (contains the Miocene, Pliocene, Pleistocene, and Holocene epochs).

Limitations of the Timescale

Because the timescale tells when various groups evolved and flourished, it might seem that evolution has been a series of events leading only from the first cells to humans. This is not the case; for example, prokaryotes (bacteria and archaea) never declined and are still the most abundant and successful organisms on Earth. Even today, they constitute up to 90% of the total weight of organisms.

Then, too, the timescale lists mass extinctions, but it doesn't tell when specific groups became extinct. **Extinction** is the total disappearance of a species or a higher group; **mass extinction** occurs when a large number of species disappear in a few million years or less. For lack of space, the geologic timescale can't depict in detail what happened to the members of every group mentioned. Figure 23.1 does show how mass extinction affected a few groups of animals. Which of the animals shown in Figure 23.1 suffered the most during the

P-T extinction (Permian-Triassic extinction)? _____

The **K-T extinction** occurred between the Cretaceous and the Tertiary periods. Which animals shown in

Figure 23.1 became extinct during the K-T extinction? _____

Figure 23.1 shows only periods and no eras. *Fill in the eras on the lines provided in the figure.*

Figure 23.1 Mass extinctions.

Five significant mass extinctions and their effects on the abundance of certain forms of marine and terrestrial life. The width of the horizontal bars indicates the varying number of each life-form considered.

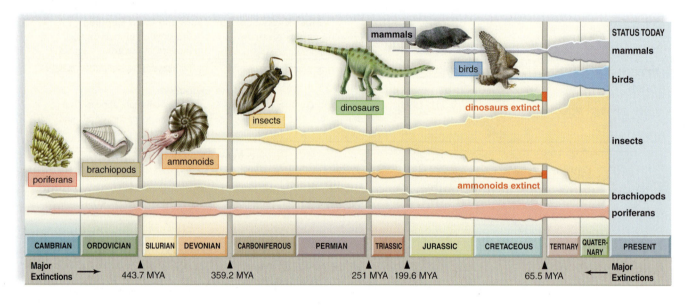

Era _____   _____   _____

Fossils

The fossil record depends heavily on anatomical data to show the evolutionary changes that have occurred as one

group gives rise to another group. Why would that be? _____

_____

Invertebrate Fossils

The **invertebrates** are animals without a backbone (Fig. 23.2). Even the chordates, which is the phylum that contains the **vertebrates** (animals with a backbone), contain a couple of insignificant groups of invertebrates.

In order to familiarize yourself with possible invertebrate fossils, a few of the most common fossilized invertebrate groups are depicted in Figure 23.2.

Figure 23.2 Invertebrate fossils.

(ammonite) © Carl Pendle/Getty Images; (echinoderms) © DEA/G. Nimatallah/Getty Images; (mayfly) © Quan Long/iStock/Getty Images Plus RF; (horseshoe crab): © koi88/Alamy RF; (snails) © Ed Reschke/Getty Images; (trilobites): © Danita Delimont/Getty Images

Fossilized ammonite, a mollusc

Fossilized brittle stars, echinoderms

Fossilized mayfly, an arthropod

Fossilized horseshoe crab, an arthropod

Fossilized snails, mollusks

Fossilized trilobites, arthropods

Observation: Invertebrate Fossils

1. Obtain a box of selected fossils. If the fossils are embedded in rocks, examine the rock until you have found the fossil. Fossils are embedded in rocks because the sediment that originally surrounded them hardened over time. Most fossils consist of hard parts such as shells, bones, or teeth because these parts are not consumed or decomposed over time. One possible reason the Cambrian might be rich in fossils is that

   organisms now had _____ whereas before they did not.

2. The kit you are using, or your instructor, will identify which of the fossils are invertebrate animals. List the names of these fossils in Table 23.2 and give a description of the fossil. Use the Geologic Timescale (Table 23.1) to tell the era and period they were most abundant; list the fossils from the latest (*top*) to earliest (*bottom*).

Table 23.2 Invertebrate Fossils		
Type of Fossil	**Era, Period**	**Description**

Laboratory 23 Evidences of Evolution   **317**

Observation: Vertebrate Fossils

The various groups of vertebrates are shown in Figure 23.3. Today it is generally agreed that birds are reptiles rather than being a separate group. That means that the major groups of vertebrates are (1) various types of fishes (jawless, jawed, cartilaginous, and bony fishes), (2) amphibians such as frogs and salamanders, (3) reptiles such as lizards, crocodiles, and birds, and (4) mammals. There are many types of mammals, from whales to mice to humans. The fossil record can rely on skeletal differences, such as limb structure, to tell if an animal is a mammal.

In order to familiarize yourself with possible vertebrate fossils, a few of the most common fossilized vertebrate groups are depicted in Figure 23.3. Which of the fossils available to you are vertebrates? _____

_____

Use Table 23.1 to associate each fossil with the particular era and period when this type of animal was most abundant. Fill in Table 23.3 according to sequence of the time frames from the latest (*top*) to earliest (*bottom*).

Fossilized bird, a reptile

Fossilized bony fish

Fossilized deerlike mammal

Fossilized frog, an amphibian

Fossilized duckbill dinosaur, a reptile

Fossilized turtle, a reptile

Figure 23.3   Vertebrate fossils.

(bird) © WaterFrame/Alamy; (deerlike mammal): © Gary Ombler/Getty Images; (duckbill dinosaur) © Kevin Schafer/Getty Images; (fish) © National Geographic RF/Getty RF; (frog) © DEA/G. Cigolini/Getty Images; (turtle) © Kevin Schafer/Getty Images

Table 23.3 Vertebrate Fossils		
Type of Fossil	**Era, Period**	**Description**

Observation: Plant Fossils

See Figure 23.4, which shows the evolution of plants including the bryophytes, ferns and their allies, gymnosperms, and angiosperms. The fossil record for plants is not as good as that for invertebrates and vertebrates because plants have no hard parts that are easily fossilized. In order to familiarize you with possible plant fossils, a few of the most common fossilized plant groups are depicted in Figure 23.4.

Fossilized fern

Fossilized flower

Fossilized maple leaf

Fossilized gingko leaf

Fossilized sassafras leaf

Fossilized early seed plant leaves

Figure 23.4   Plant fossils.

(fern) © Ekaterina Andreeva/123RF; (flower) © Barbara Strnadova/Science Source; (maple leaf) © Biophoto Associates/Science Source; (gingko leaf) © Corbin17/Alamy; (sassafras leaf) © Jonathan Blair/Getty Images; (seed plant leaves) © Sinclair Stammers/Science Source

Plants that have no hard parts become fossils when their impressions are filled in by minerals. Use Table 23.1 to associate each fossilized plant in your kit with a particular era and period. Assume trees are flowering plants and associate them with the era and period when flowering plants were most abundant. Fill in Table 23.4 with a description of each fossil in sequence from the latest (*top*) to earliest (*bottom*).

Table 23.4 Plant Fossils		
Type of Fossil	**Era, Period**	**Description**

Summary of Evidence from the Fossil Record

In this section, you studied the geologic timescale and various fossils represented in the record. The geologic timescale gives powerful evidence of evolution because

1. Fossils are _____.

2. Fossils can be arranged _____.

3. Younger fossils and not older fossils are more like _____.

4. In short, the fossil record shows that _____.

23.2 Evidence from Comparative Anatomy

In the study of evolutionary relationships, parts of organisms are said to be **homologous** if they exhibit similar basic structures and embryonic origins. If parts of organisms are similar in function only, they are said to be **analogous.** Only homologous structures indicate an evolutionary relationship and are used to classify organisms.

Comparison of Adult Vertebrate Forelimbs

The limbs of vertebrates are homologous structures (Fig. 23.5). The similarity of homologous structures is explainable by descent from a common ancestor.

humerus		carpals	
radius		metacarpals	
ulna		phalanges	

Figure 23.5 Vertebrate forelimbs.
Because all vertebrates evolved from a common ancestor, their forelimbs share homologous structures.

(bird) © McGraw-Hill Education; (human) © J.W. Ramsey/McGraw-Hill Education; (lizard) © Mauricio Handler/Getty Images; (bat) © Jack Milchanowski/Getty Images; (cat) © Marc Henrie/Getty Images

1. Find the forelimb bones of the ancestral vertebrate in Figure 23.5. The basic components are the humerus (h), ulna (u), radius (r), carpals (c), metacarpals (m), and phalanges (p) in the five digits.
2. *Label the corresponding forelimb bones of the lizard, the bird, the bat, the cat, and the human.*
3. Fill in Table 23.5 to indicate which bones in each specimen appear to most resemble the ancestral condition and which most differ from the ancestral condition.
4. Adaptation to a way of life can explain the modifications that have occurred. Relate the change in bone structure to mode of locomotion in two examples.

Example 1: _____

Example 2: _____

Table 23.5 Comparison of Vertebrate Forelimbs		
Animal	**Bones That Resemble Common Ancestor**	**Bones That Differ from Common Ancestor**
Lizard		
Bird		
Bat		
Cat		
Human		

Conclusion: Vertebrate Forelimbs

- Vertebrates are descended from a _____, but they are adapted to _____.

Comparison of Vertebrate Embryos

The anatomy shared by vertebrates extends to their embryological development. During early developmental stages, all animal embryos resemble each other closely, but as development proceeds the different types of vertebrates take on their own shape and form. In Figure 23.6, the reptile and bird embryo resemble each other more than either resembles a fish. What does that tell you about their evolutionary relationship? _____

_____

_____

In this observation, you will see that as embryos all vertebrates have a postanal tail, a dorsal spinal cord, pharyngeal pouches, and various organs. In aquatic animals, pharyngeal pouches become functional gills (Fig. 23.6). In humans, the first pair of pouches becomes the cavity of the middle ear and auditory tube, the second pair becomes the tonsils, and the third and fourth pairs become the thymus and parathyroid glands.

Figure 23.6 Vertebrate embryos.

Fish — eye, brain, heart, yolk stalk, fin bud, pharyngeal pouches, dorsal spinal cord, tail

Reptile — ear, heart, pharyngeal pouches, eye, brain, yolk stalk, dorsal spinal cord, limb buds, tail

Bird — brain, ear, eye, pharyngeal pouches, heart, yolk stalk, dorsal spinal cord, limb buds, tail

Observation: Vertebrate Embryos

1. Obtain prepared slides of vertebrate embryos at comparable stages of development. Observe each of the embryos using a stereomicroscope.
2. List five similarities of the embryos:

   a. _____

   b. _____

   c. _____

   d. _____

   e. _____

Conclusion: Vertebrate Embryos

Vertebrate embryos resemble one another because _____

_____.

Comparison of Chimpanzee and Human Skeletons

Chimpanzees and humans are closely related, as is apparent from an examination of their skeletons. However, they are adapted to different ways of life. Chimpanzees are adapted to living in trees and are herbivores—they eat mainly plants. Humans are adapted to walking on the ground and are omnivores—they eat both plants and meat.

Observation: Chimpanzee and Human Skeletons

Posture

Chimpanzees are arboreal and climb in trees. While on the ground, they tend to knuckle-walk, with their hands bent. Humans are terrestrial and walk erect. In Table 23.6, compare:

1. **Head and torso:** Where are the head and trunk with relation to the hips and legs—thrust forward over the

   hips and legs or balanced over the hips and legs (Fig. 23.7)? _____

2. **Spine:** Which animal has a long and curved lumbar region, and which has a short and stiff lumbar region? How

   does this contribute to an erect posture in humans? _____

   _____

3. Pelvis: Chimpanzees sway when they walk because lifting one leg throws them off balance. Which animal has a narrow and long pelvis, and which has a broad and short pelvis? Record your observations in Table 23.6.

Table 23.6 Comparison of Chimpanzee and Human Postures		
Skeletal Part	Chimpanzee	Human
1. Head and torso		
2. Spine		
3. Pelvis		
4. Femur		
5. Knee joint		
6. Foot:		
Opposable toe _____		
Arch _____		

4. Femur: In humans, the femur better supports the trunk. In which animal is the femur angled between articulations with the pelvic girdle and the knee? In which animal is the femur straight with no angle? Record your observations in Table 23.6.

5. Knee joint: In humans, the knee joint is modified to support the body's weight. In which animal is the femur larger at the bottom and the tibia larger at the top? Record your observations in Table 23.6.

6. Foot: In humans, the foot is adapted for walking long distances and running with less chance of injury.

In which animal is the big toe opposable? _____ How does an opposable toe assist chimpanzees?

_____

Which foot has an arch? _____
How does an arch assist humans?

_____

Record your observations in Table 23.6.

7. How does the difference in the position of the foramen magnum, a large opening in the base of the skull for the spinal cord, correlate with the posture and stance of the two organisms?

_____

Figure 23.7 Human and chimpanzee skeletons.

spine

spine

pelvis

pelvis

femur

femur

knee joint

knee joint

foot

foot

Skull Features

Humans are omnivorous. A diet rich in meat does not require strong grinding teeth or well-developed facial muscles. Chimpanzees are herbivores, and a vegetarian diet requires strong teeth and strong facial muscles that attach to bony projections. Compare the skulls of the chimpanzee and the human in Figure 23.8 and answer the following questions:

1. **Supraorbital ridge:** For which skull is the supraorbital ridge (the region of frontal bone just above the eye socket) thicker? Record your observations in Table 23.7.
2. **Sagittal crest:** Which skull has a sagittal crest, a projection for muscle attachments that runs along the top of the skull? Record your observation in Table 23.7.
3. **Frontal bone:** Compare the slope of the frontal bones of the chimpanzee and human skulls. How are they different? Record your observations in Table 23.7.
4. **Teeth:** Examine the teeth of the adult chimpanzee and adult human skulls. Are the incisors (two front teeth) vertical or angled? Do the canines overlap the other teeth? Are the molars larger or moderate in size? Record your observations in Table 23.7.
5. **Chin:** What is the position of the mouth and chin in relation to the profile for each skull? Record your observations in Table 23.7.

Figure 23.8 Chimpanzee and human skulls.

a. Adult chimpanzee

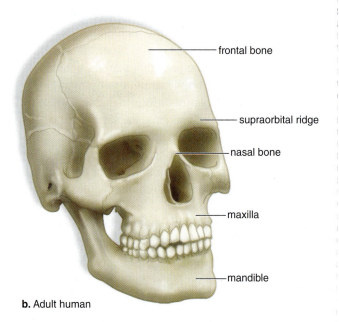

b. Adult human

Table 23.7	Skull Features of Chimpanzees and Humans	
Feature	**Chimpanzee**	**Human**
1. Supraorbital ridge		
2. Sagittal crest		
3. Slope of frontal bone		
4. Teeth		
5. Chin		

Conclusion: Chimpanzee and Human Skeletons

- Do your observations show that the skeletal differences between chimpanzees and humans can be related to posture? _____ Explain your answer. _____

- Do your observations show that diet can be related to the skull features of chimpanzees and humans? _____ Explain your answer. _____

Comparison of Hominid Skulls

The designation *hominid* includes humans and primates that are humanlike. Paleontologists have uncovered several fossils dated from 7.5 MYA to 30,000 years BP (before present), when humans called Cro-Magnons arose that are virtually identical to modern humans (*Homo sapiens*). (Fig. 23.9).

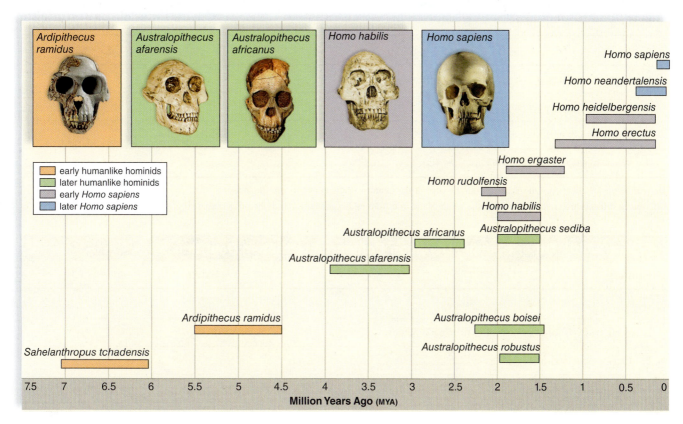

Figure 23.9   Human evolution.

(A. ramidus) © Richard T. Nowitz/Science Source; *(A. afarensis)* © Scott Camazine/Alamy; *(A. africanus)* © Philippe Plailly/Science Source; *(H. habilis)* © Kike Calvo VWPics/Superstock; *(H. sapiens)* © Kenneth Garrett/Getty Images

Observation of Hominid Skulls

Several of the skulls noted in Figure 23.9 are on display. Use Tables 23.8, 23.9, and 23.10 to record data pertaining to the cranium (or braincase), the face, and the teeth. Compare the early skulls on display with a modern human skull. For example, is the forehead like or more flat than the human skull? List at least three skulls you examined for each feature.

Table 23.8   Other Hominid Craniums Compared to Human Cranium

Feature	Skulls		
	1.	2.	3.
a. Frontal bone (like or more flat?)			
b. Supraorbital ridge (divided or continuous?)			
c. Sagittal crest (present?)			
d. Mastoid process (flat or projecting?)			

Table 23.9   Other Hominid Faces Compared to Human Face

Feature	Skulls		
	1.	2.	3.
a. Nasal bones (raised or flat?)			
b. Nasal opening (larger?)			
c. Chin (projecting forward?)*			
d. Width of face (wider?)**			

*If your instructor directs you to, measure from the edge of the foramen magnum to between the incisors.

**If your instructor directs you to, measure the width of the face from mid-zygomatic arch to the other mid-arch.

Figure 23.10   Other Hominid Dentition Compared to Human Dentition

Feature	Skulls		
	1.	2.	3.
a. Teeth rows (parallel or diverging from each other?)			
b. Incisors (vertical or angled?)			
c. Canine teeth (overlapping other teeth?)			
d. Molars (more massive?)			

Conclusions

- Do your data appear to be consistent with the evolutionary sequence of the hominids in Figure 23.9? Explain your answer. _____

- Report here any data you collected that would indicate a particular hominid was closer in time to humans than indicated in Figure 23.9. _____

  _____

- Report here any data you collected that would indicate a particular hominid was more distant in time from humans than indicated in Figure 23.9. _____

  _____

23.3 Molecular Evidence

Molecular data substantiate the comparative and developmental data that biologists have accumulated over the years. The activities of *Hox* genes differ, and this can account, for example, for why vertebrates have a dorsally placed nerve cord while it is ventrally placed in invertebrates. Sequencing DNA data show which organisms are closely related. Molecular data among primates is of extreme interest because it can help determine which of the primates we are most closely related to. Chromosomal and genetic data allow us to conclude that we are more closely related to chimpanzees than to other types of apes.

In this section, we note that scientists can compare the amino acid sequence in proteins to determine the degree to which any two groups of organisms are related. The sequence of amino acids in cytochrome *c*, a carrier of electrons in the electron transport chain found in mitochondria, has been determined per a variety of organisms. On the basis of the number of amino acid *differences* reported in Figure 23.10, it is concluded that the evolutionary relationship between humans and these organisms decreases in the order stated: monkeys, pigs, ducks, turtles, fishes, moths, and yeast. This conclusion agrees with the sequence of dates these organisms are found in the fossil record. Why can comparing amino acid data lead to the same conclusions as comparing DNA data? _____

_____

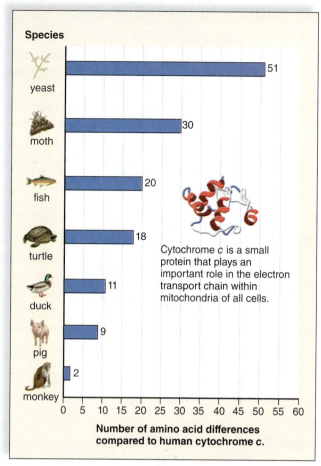

Cytochrome *c* is a small protein that plays an important role in the electron transport chain within mitochondria of all cells.

Figure 23.10 Amino acid differences in cytochrome *c*.
The few differences found in cytochrome *c* between monkeys and humans show that of these organisms, humans are most closely related to monkeys.

Protein Similarities

The immune system makes **antibodies** (proteins) that react with foreign proteins, termed **antigens.** Antigen-antibody reactions are specific. An antibody will react only with a particular antigen. In today's procedure, it is assumed rabbit antibodies to human antigens are in rabbit serum (Fig. 23.11). When these antibodies are allowed to react against the antigens of other animals, the stronger the antibody-antigen reaction (determined by the amount of precipitate), the more closely related the animal is to humans.

Figure 23.11 Antigen-antibody reaction.
When antibodies react to antigens, a precipitate appears.

rabbit antibody
human antigen

Antigen-Antibody Complex

Experimental Procedure: Protein Similarities

1. Obtain a chemplate (a clear glass tray with wells), one bottle of synthetic human blood serum, one bottle of synthetic rabbit blood serum, and five bottles (I–V) of blood serum test solution.

2. Put two drops of synthetic rabbit blood serum in each of the six wells in the chemplate. *Label the wells 1–6.* See yellow circles in Figure 23.12.

3. Add two drops of synthetic human blood serum to each well. See red circles in Figure 23.12. Stir with the plastic stirring rod that was attached to the chemplate. The rabbit serum now contains antibodies against human antigens.

4. Rinse the stirrer. (The large cavity of the chemplate may be filled with water to facilitate rinsing.)

5. Add four drops of blood serum test solution III (contains human antigens) to well 6. Describe what

   you see. _____

   _____

   This well will serve as the basis by which to compare all the other samples of test blood serum.

6. Now add four drops of blood serum test solution I to well 1. Stir and observe. Rinse the stirrer. Do the same for each of the remaining blood serum test solutions (II–V)—adding II to well 2, III to well 3, and so on. Be sure to rinse the stirrer after each use.

7. At the end of 10 and 20 minutes, record the amount of precipitate in each of the six wells in Figure 23.12. Well 6 is recorded as having ++++ amount of precipitate after both 10 and 20 minutes. Compare the other wells with this well (+ = trace amount; 0 = none). Holding the plate slightly above your head at arm's length and looking at the underside toward an overhead light source will allow you to more clearly determine the amount of precipitate.

Figure 23.12 Protein similarity.
The greater the amount of precipitate, the more closely related an animal is to humans.

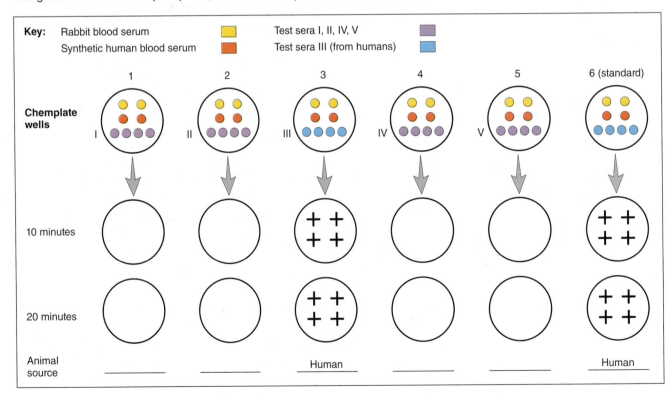

Conclusions: Protein Similarities

- The last row in Figure 23.12 tells you that the test serum in well 3 is from a human. How do your test results confirm this? _____

- Aside from humans, the test sera (supposedly) came from a pig, a monkey, an orangutan, and a chimpanzee. Which is most closely related to humans—the pig or the chimpanzee? _____

- Judging by the amount of precipitate, complete the last row in Figure 23.12 by indicating which serum you believe came from which animal. On what do you base your conclusions? _____

  _____

- In section 23.2, a comparison of bones in vertebrate forelimbs showed that vertebrates share a common ancestor. Molecular evidence shows us that of the vertebrates studied, _____

  and _____ are most closely related. In evolution, closely related organisms share a recent common ancestor. Humans share a more recent common ancestor with chimpanzees than they do with pigs.

Laboratory Review 23

_____ 1. What has occurred when a group of bacteria becomes resistant to a specific antibiotic?

_____ 2. In addition to comparative anatomy and molecular evidence, what shows that all organisms are related to one another?

_____ 3. What era of geologic time are we in currently?

_____ 4. During what period of geologic time did the coal-forming forests flourish?

_____ 5. During what era of geologic time did the first cells evolve?

_____ 6. How many significant mass extinctions have occurred over time?

_____ 7. What do most fossils consist of because these parts are not consumed or decomposed over time?

_____ 8. For which group of organisms—invertebrates, vertebrates, or plants—is the fossil record not as good as that for the other two?

_____ 9. What term describes parts of organisms that exhibit similar basic structures and embryonic origins?

_____ 10. In addition to a postanal tail and a dorsal nerve cord, what structures are observed in vertebrate embryos?

_____ 11. Do the skulls of chimpanzees or humans have a sagittal crest?

_____ 12. What hominid genus flourished 2.8 MYA?

_____ 13. The amino acid sequence of what electron carrier protein has been determined for a variety of organisms?

_____ 14. Are two species closely or distantly related if the antibodies to the antigens of one species form very little precipitate when mixed with the antigens of the other species?

15. Analogous structures perform similar functions but do not arise from common ancestry. What explains the existence of analogous structures?

16. Why are there differences in the axial skeletons and skulls of chimpanzees and humans?

17. People with sickle-cell trait (heterozygotes $Hb^A Hb^S$) are immune to malaria, but people who are $Hb^S Hb^S$ will have sickle-cell disease.

    a. If two people with sickle-cell trait have offspring, what are the chances their offspring will have sickle-cell disease?

    b. Explain where, in the world, the Hb^S allele is mostly likely to be prevalent and why the allele persists in spite of the danger associated with having two Hb^S alleles.

24

Microbiology

Learning Outcomes

24.1 Bacteria
- Relate the structure of a bacterium to its ability to cause disease.   332–333
- Recognize colonies of bacteria on agar plates.   333–334
- Identify the three commonly recognized shapes of bacteria.   335
- Recognize *Gloeocapsa, Oscillatoria,* and *Anabaena* as cyanobacteria.   335–337

24.2 Protists
- Identify and give examples of green, brown, and red algae.   337–341
- Describe the structure and importance of diatoms and dinoflagellates.   341–342
- Describe various types of locomotion used by protozoans.   342–345

24.3 Fungi
- Describe the general characteristics of fungi.   346
- Compare sexual reproduction of black bread mold with that of a mushroom.   347–349
- Give evidence that fungi are agents of human disease.   349

Introduction

The history of life began with the evolution of the prokaryotic cell. Although the prokaryotic cell contains genetic material, it is not located in a nucleus, and the cell also lacks any other type of membranous organelle. At one time prokaryotes were believed to be a unified group, but based on molecular data, they are now divided into two major groups—**domain Bacteria** and **domain Archaea.** Eukaryotes in **domain Eukarya** have a membrane-bound nucleus and membranous organelles. Eukaryotes are more closely related to archaea than bacteria. Prokaryotes resemble each other structurally but are metabolically diverse. Eukaryotes, on the other hand, are structurally diverse and exist as protists, fungi, plants, and animals (Fig. 24.1).

Figure 24.1   The world of organisms.
Prokaryotes are represented in this illustration by the bacteria. The protists, fungi, plants, and animals are all eukaryotes.
(bacteria) © Dr. Dennis Kunkel/Phototake; (paramecium) © Michael Abbey/Science Source; (morel) © Corbis RF; (sunflower) © Photodisc Green/Getty RF; (woodpecker) © Flickr Open/Getty RF

Salmonella,
a bacterium

Paramecium,
a protist

Morel, a fungus

Sunflower, a plant

Great spotted woodpecker, an animal

24.1 Bacteria

In this laboratory you will first relate the general structure of a bacterium to its ability to cause disease. The specific shape, growth habit, and staining characteristics of bacteria are often used to identify them. Therefore, you will observe a variety of bacteria using the microscope. Aside from their medical importance, bacteria are essential in ecosystems because, along with fungi, they are decomposers that break down dead organic remains and thereby return inorganic nutrients to plants.

Pathogenic Bacteria

Pathogenic bacteria are infectious agents that cause disease. Infectious bacteria are able to invade and multiply within a host. Some also produce a toxin. Antibiotic therapy is often an effective treatment against a bacterial infection.

We will explore how it is possible to relate the structure of a bacterium to its ability to be invasive and avoid destruction by the immune system. We will also consider what characteristics allow bacteria to be resistant to antibiotics and to pass the necessary genes on to other bacteria.

Observation: Structure of a Bacterium

1. Study the generalized structure of a bacterium in Figure 24.2 and, if available, examine a model of a bacterium.
2. Identify the following:

   capsule, a gel-like coating outside the cell wall. Capsules often allow bacteria to stick to surfaces such as teeth. They also prevent phagocytic white blood cells from taking them up and destroying them.

   fimbriae, hairlike bristles that allow adhesion to surfaces. This can be how a bacterium clings to and gains access to the body prior to an infection.

   conjugation pilus, an elongated, hollow appendage used to transfer DNA to other cells. Genes that allow bacteria to be resistant to antibiotics can be passed in this manner.

   flagellum, a rotating filament that pushes the cell forward.

   cell wall, a structure that provides support and shapes the cell. Antibiotics that prevent the formation of a cell wall are most effective against Gram-positive bacteria because they have more peptidoglycan in the cell wall than Gram-negative bacteria.

   plasma membrane, a sheet that surrounds the cytoplasm and regulates the entrance and exit of molecules. Resistance to antibiotics can be due to plasma membrane alterations that do not allow the drug to bind to the membrane or cross the membrane, or to a plasma membrane that increases the elimination of the drug from the bacteria.

   ribosomes, site of protein synthesis. Some bacteria possess antibiotic-inactivating enzymes that make them resistant to antibiotics.

   nucleoid, the location of the bacterial chromosome.

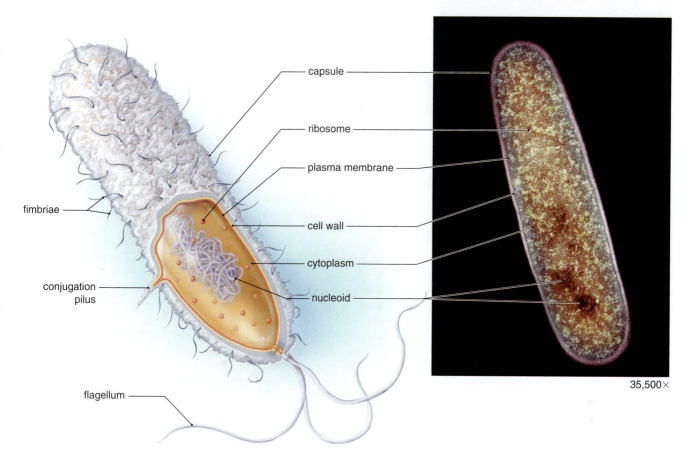

Figure 24.2   Generalized structure of a bacterium.
© Dr. Kari Lounatmaa/Science Source

Also, some bacteria contain plasmids, small rings of DNA that replicate independently of the chromosomes and can be passed to other bacteria. Genes that allow bacteria to be resistant to antibiotics are often located in a plasmid.

Conclusions: Structure of a Bacterium

- Which portions of a bacterial cell aid the ability of a bacterium to cause infections? _____

_____

- Which portions of a bacterial cell aid the ability of a bacterium to be resistant to antibiotics? _____

_____

Colony Morphology

On a nutrient material called agar, bacteria grow as colonies. A **colony** contains cells descended from one original cell. Sometimes, it is possible to identify the type of bacterium by the appearance of the colony (Fig. 24.3).

Figure 24.3   Colony morphology.
Colonies of bacteria on agar plates. Note the variation in features like size and color that helps researchers identify the type of bacterium.   (a) © Kathy Park Talaro; (b) © Kathy Park Talaro

a.

b.

Observation: Colony Morphology

1. View agar plates that have been inoculated with bacteria and then incubated. Notice the "colonies" of bacteria growing on the plates.
2. Compare the colonies' color, surface, and margin, and note your observations in Table 24.1. It is not necessary to identify the type of bacteria.

Table 24.1 Agar Plates	
Plate Number	**Description of Colonies**

3. If available, obtain a sterile agar plate, and inoculate the plate with your thumbprint, or use a sterile swab and inoculate the plate with material from around your teeth or inside your nose. Put your name on the plate, and place it where directed by your instructor. Remember to view the plate next laboratory period. Describe your plate:

4. If available, obtain a sterile agar plate, and expose it briefly (at most for 10 minutes) anywhere you choose, such as in the library, your room, or your car. No matter where the plate is exposed, it subsequently will show bacterial colonies. Describe your plate:

Shape of Bacterial Cell

Most bacteria are found in three basic shapes: **spirillum** (spiral or helical), **bacillus** (rod), and **coccus** (round or spherical) (Fig. 24.4). Bacilli may form long filaments, and cocci may form clusters or chains. Some bacteria form endospores. An **endospore** contains a copy of the genetic material encased by heavy protective spore coats. Spores survive unfavorable conditions and germinate to form vegetative cells when conditions improve.

Observation: Shape of Bacterial Cell

1. View the microscope slides of bacteria on display. What magnification is required to view bacteria?

   _____

2. Using Figure 24.4 as a guide, identify the three different shapes of bacteria.

3. Do any of the slides on display show bacterial cells with endospores? _____

   What is an endospore, and why does it have survival value? _____

   _____

Figure 24.4 Diversity of bacteria.
a. Spirillum, a spiral-shaped bacterium. **b.** Bacilli, rod-shaped bacteria. **c.** Cocci, round bacteria.

(a) © Ed Reschke/Getty Images; (b) © Eye of Science/Science Source; (c) © SciMAT/Science Source

| **a.** Spirillum: LM 25× | **b.** Bacilli: SEM 13,300× | **c.** Cocci: SEM 6,250× |
| *Spirillum volutans* | *Escherichia coli* | *Streptococcus thermophilus* |

Cyanobacteria

Cyanobacteria were formerly called blue-green algae because their general growth habit and appearance through a compound light microscope are similar to green algae. Electron microscopic study of cyanobacteria, however, revealed that they are structurally similar to other bacteria, particularly other photosynthetic bacteria. Although cyanobacteria do not have chloroplasts, they do have thylakoid membranes, where photosynthesis occurs.

Observation: Cyanobacteria

Gloeocapsa

1. Prepare a wet mount of a *Gloeocapsa* culture, if available, or examine a prepared slide, using high power (45×) or oil immersion (if available). The single cells adhere together because each is surrounded by a sticky, gelatinous sheath (Fig. 24.5).

2. What is the estimated size of a single cell? _____

a. Micrograph at low magnification. 150×   **b.** Micrograph at high magnification. 250×

one organism   gelatinous sheath

Figure 24.5 *Gloeocapsa.*

(a) © Steven R. Spilatro, Marietta College, Marietta, OH; (b) © Michael Abbey/Science Source

Oscillatoria

1. Prepare a wet mount of an *Oscillatoria* culture, if available, or examine a prepared slide, using high power (45×) or oil immersion (if available). This is a filamentous cyanobacterium with individual cells that resemble a stack of pennies (Fig. 24.6).

2. *Oscillatoria* takes its name from the characteristic oscillations that you may be able to see if your sample is alive. If you have a living culture, are oscillations visible? _____

Figure 24.6 *Oscillatoria.*

© M.I. Walker/Science Source

250×

Anabaena

1. Prepare a wet mount of an *Anabaena* culture, if available, or examine a prepared slide, using high power (45×) or oil immersion (if available). This is also a filamentous cyanobacterium, although its individual cells are barrel-shaped (Fig. 24.7).

2. Note the thin nature of this strand. If you have a living culture, what is its color? _____

160×

Figure 24.7 *Anabaena.*
© R. Knauft/Science Source

24.2 Protists

Protists were the first eukaryotes to evolve. Their diversity and complexity make it difficult to categorize them, but on the basis of molecular data, they are now placed in supergroups. Animals, fungi, and land plants are also included in supergroups (Fig. 24.8). However, the diagram in Figure 24.8 may be helpful.

Supergroup		Members	Distinguishing Features
Archaeplastids	*Volvox*, colonial green alga with embedded daughter colonies	Green algae, red algae, land plants, charophytes	Plastids; unicellular, colonial, and multicellular
Chromalveolates	Assorted fossilized diatoms	Stramenopiles: brown algae, diatoms, golden brown algae, water molds	Most with plastids; unicellular and multicellular
		Alveolates: ciliates, apicomplexans, dinoflagellates	Alveoli support plasma membrane; unicellular
Excavates	*Giardia*, a single-celled flagellated diplomonad	Euglenids, kinetoplastids, parabasalids, diplomonads	Feeding groove; unique flagella; unicellular
Amoebozoans	*Amoeba proteus*, a protozoan	Amoeboids, plasmodial and cellular slime molds	Pseudopods; unicellular
Rhizarians	Radiolarians (assorted), produce a calcium carbonate shell	Foraminiferans, radiolarians	Thin pseudopods; some with tests; unicellular
Opisthokonts	Choanoflagellate (unicellular), animal-like protist	Choanoflagellates, animals, *nucleariids*, fungi	Some with flagella; unicellular and colonial

Figure 24.8 Protist diversity. Supergroups into which protists are classified along with animals, fungi, and land plants. (volvox) © Stephen Durr; (diatoms) © M.I. Walker/ Science Source; (Giardia) © CDC/Dr. Stan Erlandsen and Dr. Dennis Feely; (amoeba) © Melba Photo Agency/PunchStock RF; (radiolarians) © Eye of Science/Science Source; (choanoflagellate) © D. J. Patterson, provided courtesy microscope.mbl.edu

Traditionally, protists were discussed according to their mode of nutrition. This section will continue to do so while also noting the supergroup to which each type of protist belongs. The algae do photosynthesis like that done by land plants. Some green algae share a common ancestor with land plants; both groups possess chlorophylls *a* and *b*, store reserve food as starch, and have cell walls containing cellulose. Heterotrophic protists, commonly called protozoans, must acquire food from an external source by ingesting it as animals do, or by absorbing it as fungi do. Some of the protozoans may be ancestral to animals; the relationship of any of the protists to fungi is currently unclear.

Photosynthetic Protists

Algae is a term that has been used for aquatic organisms that photosynthesize in the same manner as land plants. All photosynthetic protists contain green chlorophyll, but they also may contain other pigments that mask the chlorophyll color, and this accounts for their common names—the green algae, red algae, brown algae, and golden-brown algae.

Observation: Green Algae

If available, view a video showing the many forms of green algae. Notice that green algae can be single cells, filaments, colonies, or multicellular sheets. You will examine a filamentous form (*Spirogyra*) and a colonial form (*Volvox*). A **colony** is a loose association of cells.

1. Make a wet mount of live *Spirogyra* or observe a prepared slide. *Spirogyra* is a filamentous alga, lives in fresh water, and often is seen as a green scum on the surface of ponds and lakes. The most prominent feature of the cells is the spiral, ribbonlike chloroplast (Fig. 24.9). How do you think *Spirogyra* got its

   name? _____

   _____

Spirogyra's chloroplast contains a number of circular bodies, the **pyrenoids,** centers of starch polymerization. The nucleus is in the center of the cell, anchored by cytoplasmic strands.

2. Your slide may show **conjugation,** a sexual means of reproduction illustrated in Figure 24.9*b*. If it does not, obtain a slide that does show this process. Conjugation tubes form between two adjacent filaments, and the contents of one set of cells enter the other set. As the nuclei fuse, a zygote is formed. The zygote overwinters, and in the spring, meiosis and, subsequently, germination occur. The resulting adult protist is therefore haploid.

Volvox

1. Using a depression slide, prepare a wet mount of live *Volvox,* or study a prepared slide. *Volvox* is a green algal colony. It is motile (capable of locomotion) because the thousands of cells that make up the colony have flagella. These cells are connected by delicate cytoplasmic extensions (Fig. 24.10).

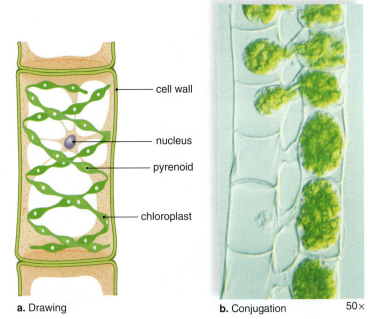

a. Drawing   **b.** Conjugation   50×

Figure 24.9 *Spirogyra.*
a. *Spirogyra* is a filamentous green alga, in which each cell has a ribbonlike chloroplast. **b.** During conjugation, the cell contents of one filament enter the cells of another filament. Zygote formation follows.
(b) © M.I. Walker/Science Source

Figure 24.10 *Volvox.*
Volvox is a colonial green alga. The adult *Volvox* colony often contains daughter colonies, asexually produced by special cells.

© Manfred Kage/Science Source; (inset) © Manfred Kage/Science Source

17×

daughter colony

50×

vegetative cells

Volvox is capable of both asexual and sexual reproduction. Certain cells of the adult colony can divide to produce **daughter colonies** (Fig. 24.10) that reside for a time within the parental colony. A daughter colony escapes the parental colony by releasing an enzyme that dissolves away a portion of the matrix of the parental colony. During sexual reproduction, some colonies of *Volvox* have cells that produce sperm, and others have cells that produce eggs. The resulting zygote undergoes meiosis, and the adult *Volvox* is haploid.

2. In Table 24.2, list the genus names of each of the green algae specimens available, and give a brief description.

Table 24.2 Green Algae Diversity	
Specimen	**Description**

Observation: Brown Algae and Red Algae

Brown algae and red algae are commonly called *seaweed,* along with the multicellular green algae.

1. If available, study preserved specimens of brown algae (Fig. 24.11), which have brown pigments that mask chlorophyll's green color. These algae are large and have specialized parts. *Laminaria* algae are called kelps. ***Fucus*** is called rockweed because it is seen attached to rocks at the seashore when the tide is out (Fig. 24.11). If available, view a preserved specimen. Note the dichotomously branched body plan, so called

Figure 24.11 Brown algae.

Laminaria and *Fucus* are seaweeds known as kelps. They live along rocky coasts of the north temperate zone. The other brown algae featured, *Macrocystis* and *Nereocystis,* form spectacular underwater "forests" at sea.

(rockweed) © D.P Wilson/Eric & David Hosking/Science Source

stipe

holdfast

air bladder

blade

Laminaria

Fucus

Rockweed, *Fucus*

Nereocystis

Macrocystis

because the **stipe** repeatedly divides into two branches (Fig. 24.11). Note also the **holdfast** by which the alga anchors itself to the rock; the **air vesicles,** or bladders, that help hold the thallus erect in the water; and the **receptacles,** or swollen tips. The receptacles are covered by small raised areas, each with a hole in the center. These areas are cavities in which the sex organs are located, with the gametes escaping to the outside through the holes. *Fucus* is unique among algae in that as an adult it is diploid (2n) and always reproduces sexually.

2. If available, study preserved specimens of red algae. Like most brown algae, the red algae (Fig. 24.12) are multicellular, but they occur chiefly in warmer seawater, growing both in shallow waters and as deep as light penetrates. Some forms of red algae are filamentous, but more often they are complexly branched with a feathery, flat, and expanded or ribbonlike appearance. Coralline algae are red algae that have cell walls impregnated with calcium carbonate ($CaCO_3$).

3. In Table 24.3, list the genus names of each of the brown and red algae specimens available, and give a brief description.

Figure 24.12 Red algae.
Generally, red algae are smaller and more delicate than brown algae. **a.** *Sebdenia* has a pronounced filamentous structure. **b.** Calcium carbonate is deposited in the walls of the red alga *Corallina*.   (a) © Steven P. Lynch; (b) © Jeffrey L. Rotman/Corbis

a.

b.

Table 24.3 Brown and Red Algae		
Specimen	Genus	Description
1		
2		
3		
4		

Observation: Diatoms and Dinoflagellates

Diatoms (golden-brown algae) have a yellow-brown pigment that, in addition to chlorophyll, gives them their color.

1. Make a wet mount of live diatoms, or view a prepared slide (Fig. 24.13). Describe what you see: _____

   _____

   _____

   _____

130×

Figure 24.13 Diatoms.
Diatoms, photosynthetic protists.   © Jan Hinsch/Science Source

The cell wall of diatoms is in two sections, with the larger one fitting over the smaller as a lid fits over a box. Since the cell wall is impregnated with silica, diatoms are said to "live in glass houses." The glass cell walls of diatoms do not decompose, so they accumulate in thick layers subsequently mined as diatomaceous earth and used in filters and as a natural insecticide. Diatoms, being photosynthetic and extremely abundant, are important food sources for the small heterotrophs in both marine and freshwater environments.

2. Make a wet mount of live dinoflagellates or view a prepared slide (Fig. 24.14). Describe what you see:

_____

_____

_____

_____

Dinoflagellates are photosynthetic, but they have two flagella; one is free, but the other is located in a transverse groove that encircles the organism. The beating of these flagella causes the organism to spin like a top. The cell wall, when present, is frequently divided into closely joined polygonal plates of cellulose. At times there are so many of these organisms in the ocean that they cause a condition called "red tide." The toxins given off in these red tides cause widespread fish kills and can cause paralysis in humans who eat shellfishes that have fed on the dinoflagellates.

Figure 24.14 Dinoflagellates.
Dinoflagellates such as *Gonyaulax* have cellulose plates.

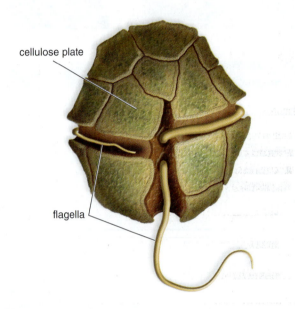

cellulose plate

flagella

Heterotrophic Protists

The term *protozoan* refers to unicellular eukaryotes and is often restricted to heterotrophic organisms that ingest food by forming **food vacuoles.** Other vacuoles, such as **contractile vacuoles** that rid the cell of excess water, are also typical. Usually protozoans have some form of locomotion, and as shown in Figure 24.15, some use **pseudopodia,** some move by **cilia,** and some use flagella.

Plasmodium vivax, which causes malaria, is an apicomplexan. Apicomplexans have a special organelle called an apicoplast that assists in penetrating host cells. *Plasmodium vivax* spends a portion of its life cycle in mosquitoes (sexual phase) and the other part in human hosts (asexual phase). During the asexual phase of their life cycle, all apicomplexans exist as particulate spores, which is why they are also called sporozoans. Malaria is caused when the *Plasmodium* spores multiply in and rupture red blood cells. In general, how do sporozoans

differ from the other protozoans shown in Figure 24.15? _____

_____

_____

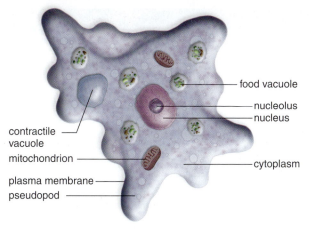

food vacuole
nucleolus
nucleus
contractile vacuole
mitochondrion
plasma membrane
pseudopod
cytoplasm

a. *Amoeba* moves by pseudopods.

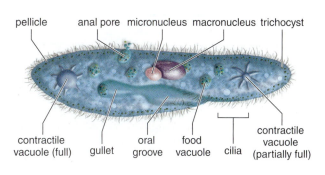

pellicle   anal pore   micronucleus   macronucleus   trichocyst

contractile vacuole (full)   gullet   oral groove   food vacuole   cilia   contractile vacuole (partially full)

b. *Paramecium* moves by cilia.

red blood cell

drawing   flagellum

undulating membrane

3,000×

c. *Trypanosoma*, which lives in bloodstream of host, moves by flagella.

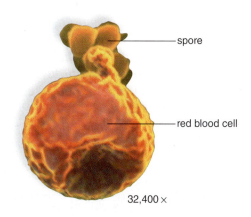

spore

red blood cell

32,400×

d. *Plasmodium* bursts forth as spores from infected red blood cells.

Figure 24.15 Protozoan diversity.

Many protozoans are motile. **a.** *Amoeba* moves by pseudopodia. **b.** *Paramecium* moves by cilia. **c.** *Trypanosoma* move by flagella. A blood infection is the cause of African sleeping sickness. **d.** *Plasmodium*, a cause of malarias, exists as nonmotile spores inside the red blood cells of humans.   (c) © Eye of Science/Science Source; (d) © Dr. Gopal Murti/Science Source

Observation: Heterotrophic Protists

Individual Protozoans

1. Watch a video if available, and note the various forms of protozoans.
2. Prepare wet mounts or examine prepared slides of protozoans as directed by your instructor. You may already have had the opportunity to observe a protozoan such as *Euglena* or *Paramecium* in Laboratory 2. However, your instructor may want you to observe these organisms again.
3. Complete Table 24.4, listing the structures for locomotion in the types of protozoans you have observed.

Table 24.4 Heterotrophic Protists

Name	Structures for Locomotion	Observations

Observation: Euglena

Euglena (Fig. 24.16) typifies the problem of classifying protists. One-third of all *Euglena* genera have chloroplasts; the rest do not. This discrepancy can be explained: the chloroplasts are probably green algae taken up by phagocytosis (engulfing them). A pyrenoid is a region of the chloroplast where a special type of carbohydrate is formed.

    Euglena has a long flagellum that projects out of a vaselike indentation and a much shorter one that does not project out. It moves very quickly, and you will be advised to add Protoslo to your wet mount to slow it down. Like some of the protozoans discussed previously, *Euglena* is bounded by a flexible pellicle made of protein. This means it can also assume all sorts of shapes. *Euglena* lives in fresh water and contains a contractile vacuole that collects water and then contracts, ridding the body of excess water.

    Make a wet mount of *Euglena* by using a drop of a *Euglena* culture and adding a drop of Protoslo (methyl

cellulose solution) onto the slide to slow it down. Describe what you see: _____

_____

_____

Figure 24.16 *Euglena.*
Euglena is a unicellular, flagellated protist.
(b) © M.I. Walker/Alamy

a. Drawing

b. Photomicrograph   150×

Pond water typically contains various examples of protists studied in this laboratory. Your instructor may have an illustrated manual that will help you identify the ones unfamiliar to you.

1. Make a wet mount of pond water by taking a drop from the bottom of a container of pond water. Be sure to select some of the sediment or strands of filamentous algae if present.
2. Scan the slide for organisms: Start at the upper left-hand corner, and move the slide forward and back as you work across the slide from left to right.
3. Experiment by using all available objective lenses, by focusing up and down with the fine-adjustment knob, and by adjusting the light so that it is not too bright.
4. Identify the organisms you see by consulting Figure 24.17, and use any pictorial guides provided by your instructor.

Figure 24.17   Microorganisms found in pond water. Drawings are not actual sizes of the organisms.

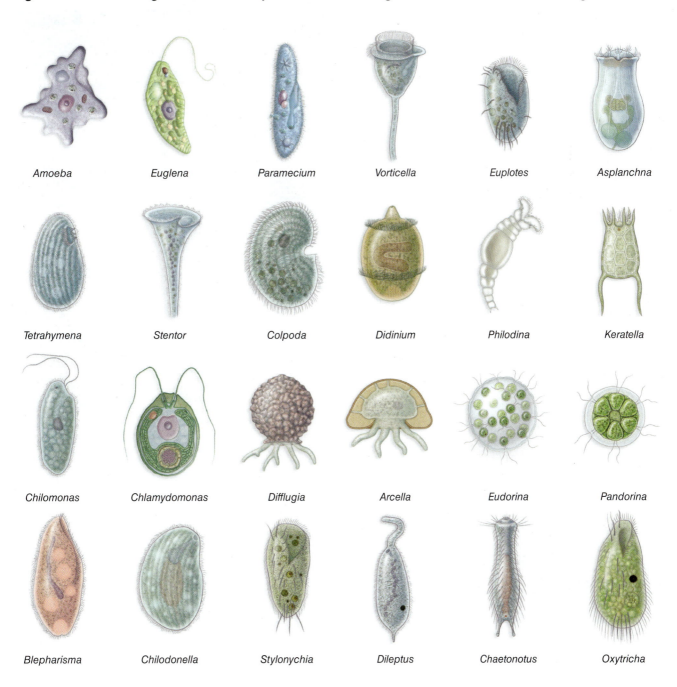

Amoeba	Euglena	Paramecium	Vorticella	Euplotes	Asplanchna
Tetrahymena	Stentor	Colpoda	Didinium	Philodina	Keratella
Chilomonas	Chlamydomonas	Difflugia	Arcella	Eudorina	Pandorina
Blepharisma	Chilodonella	Stylonychia	Dileptus	Chaetonotus	Oxytricha

24.3   Fungi

Fungi (kingdom Fungi) (Fig. 24.18) are multicellular heterotrophs that release digestive juices into the environment and then absorb the resulting nutrients. Many fungi and heterotrophic bacteria are decomposers that break down waste and dead organisms, making inorganic nutrients available to plants. A fungal body, called a **mycelium,** is composed of many strands, called **hyphae** (Fig. 24.19). Sometimes, the nuclei within a hypha are separated by septae.

Fungi produce windblown **spores** (small, haploid bodies with a protective covering) when they reproduce sexually or asexually. The types of sexual spores produced by fungi enable biologists to further classify the various fungi. In this part of the lab exercise, you'll examine some fungi most familiar to students.

Figure 24.18   Diversity of fungi.
a. Scarlet hood, an inedible mushroom. **b.** Spores exploding from a puffball. **c.** Common bread mold.
d. Morel, an edible fungus.

(a) © Biophoto Assoc./Science Source; (b) © RF Company/Alamy RF;
(c) © imageBroker/Alamy; (d) © Carol Wolfe/photographer

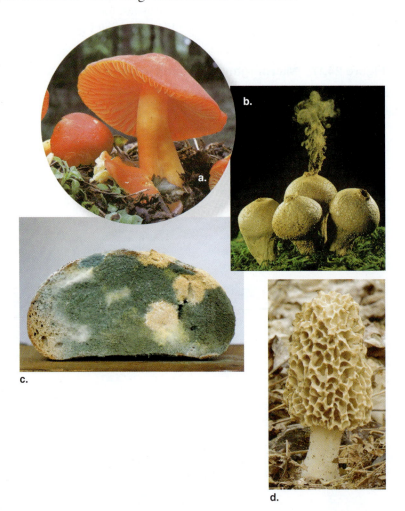

Figure 24.19   Body of a fungus.
a. The body of a fungus is called a mycelium. **b.** A mycelium contains many individual chains of cells, and each chain is called a hypha.   (a) © Chanus/Shutterstock RF

a.   1,000×   b.

Black Bread Mold

In keeping with its name, black bread mold grows on bread and any other type of baked goods. Notice in Figure 24.20 the sporangia at the tips of aerial hyphae that produce spores in the asexual portion of the life cycle. A **zygospore** is diploid (2n). The other structures associated with the asexual and sexual portions of bread mold's life cycle are haploid (n).

Figure 24.20   Black bread mold.
The mycelium of this mold (1) uses sporangia to produce windblown spores. (2) During sexual reproduction, the ends of plus and minus hyphae fuse as (3) fertilization occurs. (4) The resulting zygospore undergoes meiosis to produce spores that germinate on the bread to produce a new mycelium (5).

© Garry DeLong/Getty Images

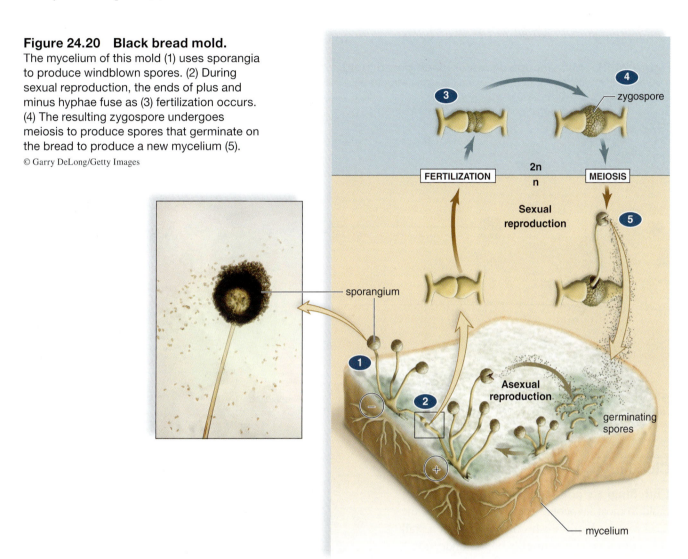

1. If available, examine bread that has become moldy. Do you recognize black bread mold on the bread?
2. Obtain a petri dish that contains living black bread mold. Observe with a stereomicroscope. *Label the mycelium and a sporangium in Figure 24.21a.*
3. View a prepared slide of *Rhizopus,* using both a stereomicroscope and the low-power setting of a light microscope. The absence of cross walls in the hyphae is an identifying feature of zygospore fungi. *Label the mycelium and zygospore in Figure 24.21b.*

Figure 24.21   Microscope slides of black bread mold.
a. Asexual life cycle structures. **b.** Sexual life cycle structures.

(a) © Carolina Biological Supply Company/Phototake; (b) © Carolina Biological Supply Company/Phototake

Club Fungi

Club fungi are just as familiar as black bread mold to most laypeople because they include the mushrooms. A gill mushroom consists of a stalk and a terminal cap with gills on the underside (Fig. 24.22). The cap, called a **basidiocarp,** is a fruiting body that arises following the union of + and – hyphae. The gills bear basidia, club-shaped structures where nuclei fuse, and meiosis occurs during spore production. The spores are called **basidiospores.**

Figure 24.22   Sexual reproduction produces mushrooms.
Fusion of + and – hyphae tips results in hyphae that form the mushroom (a fruiting body). The nuclei fuse in clublike structures attached to the gills of a mushroom, and meiosis produces basidiospores.

1. Obtain an edible mushroom—for example, *Agaricus*—and identify as many of the following structures as possible:
   a. **Stalk:** The upright portion that supports the cap.
   b. **Annulus:** A membrane surrounding the stalk where the immature (button-shaped) mushroom was attached.
   c. **Cap:** The umbrella-shaped basidiocarp of the mushroom.
   d. **Gills:** On the underside of the cap, radiating lamellae on which the basidia are located.
   e. **Basidia:** On the gills, club-shaped structures where basidiospores are produced.
   f. **Basidiospores:** Spores produced by basidia.
2. View a prepared slide of a cross section of *Coprinus*. Using all three microscope objectives, look for the gills, basidia, and basidiospores.
3. Can you see individual hyphae in the gills? _____
4. Are the basidiospores inside or outside of the basidia? _____
5. What type of nuclear division does the zygote undergo to produce the basidiospores? _____

   _____
6. Can you suggest a reason for some of the basidia having fewer than four basidiospores? _____

   _____
7. What happens to the basidiospores after they are released? _____

   _____

Fungi and Human Diseases

Fungi cause a number of human diseases. Oral thrush is a yeast infection of the mouth common in new-borns and AIDS patients (Fig. 24.23a). Ringworm is a group of related diseases caused by the fungus *Tinea*. The fungal colony grows outward, forming a ring of inflammation (Fig. 24.23b). Athlete's foot is a form of *Tinea* that affects the foot, mainly causing itching and peeling of the skin between the toes (Fig. 24.23c).

Figure 24.23 Human fungal diseases.
a. Thrush, or oral candidiasis, is characterized by the formation of white patches on the tongue.
b. Ringworm and **c.** athlete's foot are caused by *Tinea* spp.

(a) © Dr. M. A. Ansary/Science Source; (b) © John Hadfield/SPL/Science Source; (c) © P. Marazzi/ SPL/Science Source

Laboratory Review 24

_____ **1.** Into which domain are protists, fungi, plants, and animals classified?

_____ **2.** What kind of cell lacks a nucleus and other membrane-bound organelles?

_____ **3.** Where are proteins synthesized in bacteria?

_____ **4.** What shape are the bacteria classified as *Bacillus subtilis?*

_____ **5.** Where does photosynthesis occur in cyanobacteria?

_____ **6.** Where does photosynthesis occur in green algae?

_____ **7.** What kind of algae has silica in its cell walls?

_____ **8.** What type of protist spins like a top when its flagella beat?

_____ **9.** What word refers to unicellular eukaryotes that are typically heterotrophic consumers?

_____ **10.** What kind of protozoan moves using pseudopods?

_____ **11.** Which organisms studied in this lab exercise are eukaryotic, multicellular, and heterotrophic?

_____ **12.** What is an important ecological "job" done by many fungi and bacteria that ensures plants have a source of inorganic nutrients?

_____ **13.** When black bread mold sexually reproduces, what kind of spore does it make?

_____ **14.** Where are the spore-producing structures located on a mushroom?

Thought Questions

15. Compare and contrast cyanobacteria and algae with regard to their importance to other organisms and their cellular characteristics.

16. Many people have the misconception that all bacteria are "germs." Briefly make a case that many bacteria do good work; be sure to mention the group of organisms that relies on this good work.

17. When fungi reproduce (asexually or sexually) they produce many, many spores. Speculate about why it is beneficial/essential for them to produce so many spores and why the number of spores is detrimental to many people.

25

Seedless Plants

Learning Outcomes

25.1 The Evolution and Diversity of Land Plants
- Understand the relationship of charophytes to land plants and the main events in the evolution of plants as they became adapted to life on land.   352–353
- Describe the plant life cycle and the concept of the dominant generation.   353
- Contrast the role of meiosis in the plant and animal life cycles.   353

25.2 Nonvascular Plants
- Describe, in general, the nonvascular plants and how they are adapted to reproducing on land.   354
- Describe the life cycle of a moss, including the appearance of its two generations.   354–356
- Describe the liverworts, another group of nonvascular plants.   357

25.3 Seedless Vascular Plants
- Describe, in general, the seedless vascular plants and how they are adapted to reproducing on land.   357
- Describe the lycophytes, whisk ferns, and horsetails.   358–359
- Describe the life cycle of a fern and the appearance of its two generations.   360–363

Introduction

Plants are multicellar photosynthetic eukaryotes; their evolution is marked by adaptations to a land existence. Among aquatic green algae, the **charophytes** are most closely related to the plants that now live on land. The charophytes have several features that would have promoted the evolution of land plants, including retention of and care of the zygote. The most successful land plants are those that protect all phases of reproduction (sperm, egg, zygote, and embryo) from drying out and that have an efficient means of dispersing offspring on land.

In this laboratory, you will have the opportunity to examine the various adaptations of plants to living on land. Although this lab will concentrate on land plant reproduction, you will also see that much of a land plant's body is covered by a **waxy cuticle** that prevents water loss, and that land plants require structural support to oppose the force of gravity and to lift their leaves up toward the sun. **Vascular tissue** offers this support and transports water to, and nutrients from, the leaves.

Also, you will examine the adaptations of nonvascular plants and the seedless vascular plants to a land existence, and you will see that reproduction in these plants requires an outside source of water. In Laboratory 26, we will note that reproduction in seed plants is not dependent on water at all.

25.1 The Evolution and Diversity of Land Plants

Figure 25.1 shows the evolution of land plants. *Circle the evolutionary events that led to the adaptation of plants to a land existence.*

Figure 25.1   Evolutionary relationships among the plants.

Algal Ancestor of Land Plants

In the evolutionary tree, the charophytes are green algae that share a common ancestor with land plants. This common ancestor may have resembled a Charales, such as *Chara,* which can live in warm, shallow ponds that occasionally dry up. Adaptation to such periodic desiccation may have facilitated the ability of certain members of the common ancestor population to invade land, and with time, become the first land plants.

Observation: Chara

Examine a living *Chara* (Fig. 25.2). How does it superficially resemble a land plant? _____
_____

 Chara is a filamentous green alga that consists of a primary branch and many side branches (Fig. 25.2). Each branch has a series of very long cells. Note where one cell ends and the other begins. Measure the length of one cell. _____ Gently pick up and handle *Chara* while you are examining it. What does it feel like? _____ Its cell walls are covered with calcium carbonate deposit.

Figure 25.2 *Chara.*

Chara is an example of a stonewort, the type of green alga believed to be most closely related to the land plants. (left) © Bob Gibbons/Alamy; (right) © Kingsley Stern

Chara, several individuals   One individual

branch

main axis

node

Conclusion: Chara

- What characteristics cause *Chara* to resemble land plants? _____

  _____

- Why are *Chara* called stoneworts? _____

Alternation of Generations

Land plants have a two-generation life cycle called **alternation of generations.**

1. The **sporophyte** (diploid) **generation** produces haploid spores by meiosis. Spores develop into a haploid generation by mitosis. *Label the appropriate arrow "mitosis" in the first part of Figure 25.3.*
2. The **gametophyte** (haploid) **generation** produces **gametes** (eggs and sperm) by mitosis. The gametes then unite to form a diploid zygote. The zygote becomes the sporophyte by mitosis. *Label the appropriate arrow "mitosis" in Figure 17.3a.*

In this life cycle, the two generations are dissimilar, and one dominates the other. The dominant generation is larger and exists for a longer period. Figure 25.3 contrasts the plant life cycle (alternation of generations) with the animal life cycle (**diploid life cycle**).

1. In the plant life cycle, meiosis occurs during the production of _____.
2. In the human life cycle, meiosis occurs during the production of _____.
3. In the plant life cycle, the generation that produces gametes is (n or 2n) _____.
4. In the human life cycle, the individual that produces gametes is (n or 2n) _____.

Figure 25.3   Plant and animal life cycles.

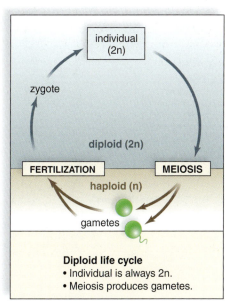

25.2 Nonvascular Plants

The nonvascular plants include the **mosses** (phylum Bryophyta) and **liverworts** (phylum Hepaticophyta). The gametophyte is dominant in all nonvascular plants. The gametophyte produces eggs within archegonia and flagellated sperm in **antheridia.** The sperm swim to the egg, and the embryo develops within **archegonia.** The nonvascular sporophyte grows out of the archegonium. The sporophyte is dependent on the female gametophyte (Fig. 25.4) for nourishment.

3. The zygote:
The zygote and developing sporophyte are retained within the archegonium.

developing sporophyte

Mitosis

zygote

4. The sporophyte:
The mature sporophyte has a foot buried in female gametophyte tissue, a stalk, and an upper capsule (the sporangium), where meiosis occurs and spores are produced.

capsule

Sporangium

calyptra

Sporophyte — stalk

teeth

operculum

5. The spores:
When the calyptra and lid (operculum) of a capsule fall off, the spores are mature. One or two rings of teeth project inward from the margin of the capsule. The teeth close the opening, except when the weather is dry.

diploid (2n)

haploid (n)

FERTILIZATION

MEIOSIS

2. Fertilization:
Flagellated sperm produced in antheridia swim in external water to archegonia, each bearing a single egg.

egg

sperm

Archegonia

Antheridia

foot (n)

Spores

Mitosis

6. Spore dispersal:
Spores are released when they are most likely to be dispersed by air currents.

buds

Protonema

♀

♂

1. The mature gametophytes:
In mosses, the leafy gametophyte shoots bear either antheridia or archegonia, where gametes are produced by mitosis.

♂

♀

7. The immature gametophyte:
A spore germinates into a male or female protonema, the first stage of the male and the female gametophytes.

Gametophytes

rhizoids

Figure 25.4   Moss life cycle.
In mosses, the haploid generation (gametophyte) is dominant.

(sporophyte) © Heather Angel/Natural Visions; (gametophyte) © Steven P. Lynch

1. Put a check mark beside the phrases that describe nonvascular plants:

I   II

_____ No vascular tissue to transport water   _____ Vascular tissue to transport water

_____ Flagellated sperm that swim to egg   _____ Sperm protected from drying out

_____ Dominant gametophyte   _____ Dominant sporophyte

2. Which listing of features (**I** or **II**) would you expect to find in a plant fully adapted to a land environment? _____ Explain. _____

3. In nonvascular plants, windblown spores are dispersal agents, and some species forcefully expel their spores. How are windblown spores an adaptation to reproduction on land? _____

_____

Observation: Moss Gametophyte

Living or Plastomount

Obtain a living moss gametophyte or a plastomount of this generation. Describe its appearance. _____

_____

The leafy green shoots of a moss are said to lack true roots, stems, and leaves because, by definition, roots, stems, and leaves are structures that contain vascular tissue.

Microscope Slide

1. Study a slide of the top of a male moss shoot that contains antheridia, the reproductive structures where sperm are produced (Fig. 25.5). What is the chromosome number (choose 2n or n) of the sperm (see Fig. 25.4)?

   _____ Are the surrounding cells haploid or diploid? _____

2. Study a slide of the top of a female moss shoot that contains archegonia, the reproductive structures where

   eggs are produced (Fig. 25.6). What is the chromosome number of the egg? _____

   Are the surrounding cells haploid or diploid? _____ When sperm swim from the antheridia to the archegonia, a zygote results. The zygote develops into the sporophyte. Is the sporophyte haploid or

   diploid? _____

— sperm

— egg

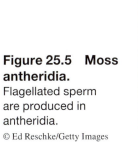

Figure 25.5 Moss antheridia.
Flagellated sperm are produced in antheridia.
© Ed Reschke/Getty Images

Figure 25.6 Moss archegonia.
Eggs are produced in archegonia.
© Robert Knauft/Biology Pix/
Science Source

Living Sporophyte

1. Examine the living sporophyte of a moss in a minimarsh, or obtain a plastomount of a female shoot with the sporophyte attached. Identify the **sporangium,** a capsule where spores are produced and released.
2. *Bracket and label the gametophyte and sporophyte in Figure 25.7*a. *Place an* n *beside the gametophyte and a* 2n *beside the sporophyte.*

Figure 25.7   Moss sporophyte.
a. The moss sporophyte is dependent on the female gametophyte. **b.** The sporophyte produces spores by meiosis.
(b) © Ed Reschke

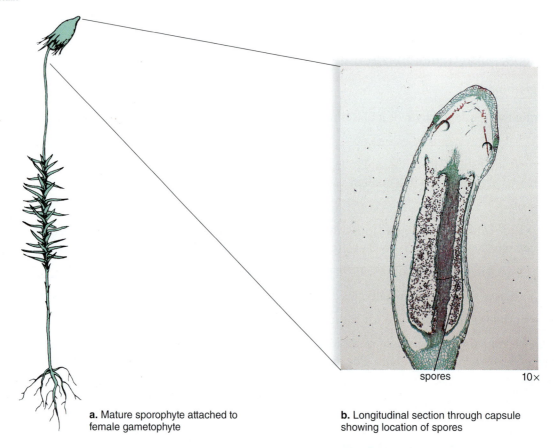

spores   10×

a. Mature sporophyte attached to female gametophyte

b. Longitudinal section through capsule showing location of spores

Microscope Slide

Examine a slide of a longitudinal section through a moss sporophyte (Fig. 25.7*b*). Identify the stalk and the

sporangium, where spores are being produced. By what process are the spores being produced? _____

When spores germinate, what generation begins to develop? _____

Why is it proper to say that spores are dispersal agents? _____

_____

Obtain a living sample of the liverwort *Marchantia* (Fig. 25.8) thallus, body of the plant. Examine it under the stereomicroscope. What generation is this sample? _____ Identify the following:

1. The gametophyte consisting of lobes. Each lobe is about a centimeter or so in length; the upper surface is smooth, and the lower surface bears numerous **rhizoids** (rootlike hairs).
2. **Gemma cups** on the upper surface of the thallus. These contain groups of cells called **gemmae** that can asexually start new plants.
3. Disk-headed stalks bear antheridia, where flagellated sperm are produced.
4. Umbrella-headed stalks bear archegonia, where eggs are produced. Following fertilization, tiny sporophytes arise from the archegonia.

Figure 25.8   Liverwort, *Marchantia*.
a. Gemmae can detach and start new plants. **b.** Antheridia are present in disk-shaped structures. **c.** Archegonia are present in umbrella-shaped structures.   (a) © Ed Reschke; (b) © J.M. Conrarder/Nat'l Audubon Society/Science Source; (c) © Ed Reschke/Oxford Scientific/Getty Images

a. Thallus with gemma cups

b. Male gametophytes bear antheridia

c. Female gametophytes bear archegonia

25.3   Seedless Vascular Plants

Seedless vascular plants include the **lycophytes** (club mosses) and **pteridophytes**—ferns and their allies, the whisk ferns and horsetails. These plants were prevalent and quite large during the swampy forest Carboniferous period. At that time, the coal deposits still used today were formed.

The sporophyte is dominant in the seedless vascular plants. The dominant sporophyte has adaptations for living on land; it has vascular tissue and produces windblown spores. The spores develop into a separate gametophyte generation that is very small (less than 1 cm). The gametophyte lacks vascular tissue and produces flagellated sperm.

1. Place a check mark beside the phrases that describe seedless vascular plants:

   I

   _____ Independent gametophyte

   _____ Flagellated sperm

   II

   _____ Gametophyte dependent on sporophyte, which has vascular tissue

   _____ Sperm protected from drying out

2. Which listing (**I** or **II**) would you expect to find in a plant fully adapted to a land environment? _____

   Explain. _____

   _____

   Are seedless vascular plants fully adapted to living on land? Explain your answer. _____

   _____

Lycophytes (phylum Lycophyta) are commonly called **club mosses.** Lycophytes are representative of the first vascular plants. They have an aerial stem and a horizontal root (rhizome with attached rhizoids) both of which have vascular tissue. The leaves are called **microphylls** because they have only one strand of vascular tissue.

Ground Pines

1. Examine a living or preserved specimen of *Lycopodium* (Fig. 25.9).
2. Note the shape and the size of the microphylls and the branches of the stems.
3. Note the terminal clusters of leaves, called **strobili,** that are club-shaped and bear sporangia.
4. *Label strobili, leaves, stem, and rhizoids in Figure 25.9.*
5. Examine a prepared slide of a *Lycopodium* that shows the sporangia with spores inside. The spore develops into a tiny microscopic gametophyte that remains in the soil.

Figure 25.9 *Lycopodium.*
In the club moss *Lycopodium,* green photosynthetic stems are covered by scalelike leaves, and spore-bearing leaves are clustered in strobili.   © Steve Solum/Photoshot

Observation: Pteridophytes

Molecular (DNA) studies tell us that the whisk ferns, horsetails, and ferns are closely related.

Whisk Ferns

Psilotum is representative of **whisk ferns,** named for their resemblance to whisk brooms.

1. Examine a preserved specimen of *Psilotum,* and note that it has no leaves. The underground stem, called a **rhizome,** gives off upright, aerial stems with a dichotomous branching pattern, where bulbous sporangia are located (Fig. 25.10).

   What generation are you examining? _____

2. *Label a sporangium, the stem, and the rhizome in Figure 25.10b.*

Figure 25.10 Whisk fern, *Psilotum.*

This whisk fern has no roots or leaves—the branches carry on photosynthesis. The sporangia are yellow.

(a) © Carolina Biological Supply Company/Phototake

scale

rhizoid

1.

2.

3.

a.

b.

Horsetails

In **horsetails,** a rhizome produces aerial stems that stand about 1.3 meters tall.

1. Examine *Equisetum,* a horsetail, and note the minute, scalelike leaves (Fig. 25.11).
2. Feel the stem. *Equisetum* contains a large amount of silica in its stem. For this reason, these plants are sometimes called scouring rushes and may be used by campers for scouring pots.
3. Strobili appear at the tips of the stems, or else special buff-colored stems bear the strobili. Sporangia are in the strobili.
4. *Label the strobilus, branches, leaves, and rhizome of the horsetail in Figure 25.11.*

Figure 25.11 Horsetail.

In the horsetail *Equisetum,* whorls of branches or tiny leaves appear in the joints of the stem. The sporangia are borne in strobili.　© Robert P. Carr/Photoshot

1.

2.

3.

4.

Ferns

Ferns are quite diverse and range in size from those that are low growing and resemble mosses to those that are as tall as trees. The rhizome grows horizontally, which allows ferns to spread without sexual reproduction (Fig. 25.12). The gametophyte (called a **prothallus**) is small (about 0.5 cm) and usually heart-shaped. The prothallus contains both archegonia and antheridia. Ferns are largely restricted to moist, shady habitats because

sexual reproduction requires adequate moisture. Why? _____

_____

Figure 25.12   Fern life cycle.
The sporophyte is the frond, and the gametophyte is the heart-shaped prothallus.

(dryopterus) © Matt Meadows/Photolibrary/Getty Images

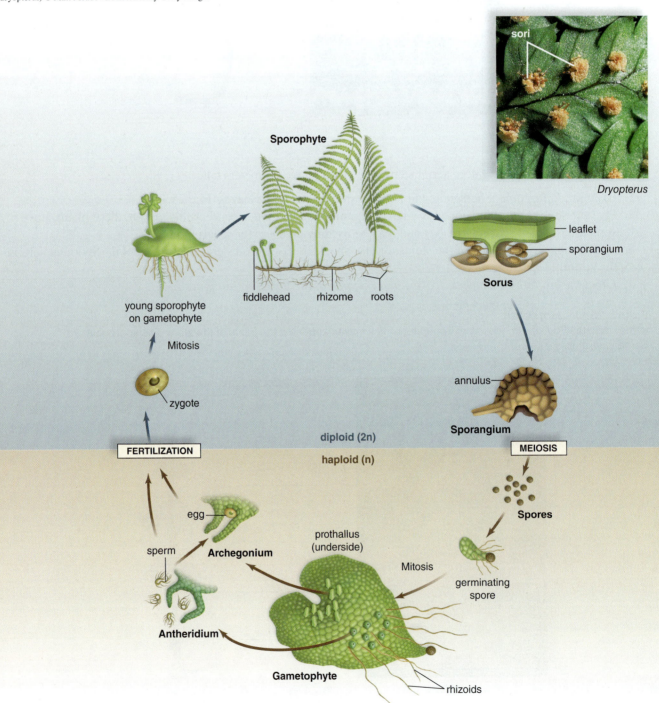

Fern spores are produced by meiosis in sporangia, which in many species occur on the underside of large leaves called **fronds.** These leaves are called **megaphylls** because they are broad leaves with several strands of vascular tissue. The spores are released when special cells of the **annulus** (a line or ring of thickened cells on the outside of the sporangium) dry out, and the sporangium opens. How do ferns move to new locations? _____

_____

Observe the ferns on display, and then complete Table 25.1.

Table 25.1 Fern Diversity	
Type of Fern	**Description of Frond**

Observation: Fern Sporophyte

Study the life cycle of the fern (Fig. 25.12), and find the sporophyte generation. This large, complexly divided leaf is known as a frond. Fronds arise from an underground stem called a rhizome.

Living or Preserved Frond

Examine a living or preserved specimen of a frond, and on the underside, notice a brownish clump called a **sorus** (pl., sori), each a cluster of many sporangia (Fig. 25.13).

What is being produced in the sporangia? _____

Given that this is the generation we call the fern, what generation is dominant in ferns? _____

Figure 25.13   Underside of frond leaflets.
Sori occur on the underside of frond leaflets.   © Carlyn Iverson/McGraw-Hill Education

sorus

1. Examine a prepared slide of a cross section of a frond leaflet. Using Figure 25.14 as a guide, locate the fern leaf above and the sorus below.

Figure 25.14   Micrograph of cross section of a frond leaflet.
Micrograph of the internal anatomy of a sorus depicts many sporangia, where spores are produced.   © Steven P. Lynch

leaflet

sporangium

spore

2. Within the sorus, find the sporangia and spores. Look for an **indusium** (not present in all species), a shelflike structure that protects the sporangia until they are mature. Does this fern have

an indusium? _____

Observation: Fern Gametophyte

Plastomount

1. Examine a plastomount showing the fern life cycle.

2. Notice the prothallus, a small, heart-shaped structure. Can you find this structure in your fern minimarsh

(if available)? _____ The prothallus is the gametophyte generation of the fern. Most people do not

realize that this structure exists as a part of the fern life cycle. What is the function of this structure? _____

_____

Microscope Slide

1. Examine whole-mount slides of fern prothallium-archegonia and fern prothallium-antheridia (Fig. 25.15).

2. If you focus up and down very carefully on an archegonium, you may be able to see an egg inside. What is being produced inside an antheridium? _____ When sperm produced in an antheridium swim to the archegonium in a film of water, what results? _____

   This structure develops into what generation? _____

Conclusions: Ferns

- How are ferns dispersed from one area to another? _____

- Is either generation in the fern dependent for any length of time on the other generation? _____
  Explain. _____

Figure 25.15   Fern prothallus.
The underside of the heart-shaped fern prothallus contains archegonia, where eggs are produced, and antheridia, where flagellated sperm are produced.   (right side, top) © Robert Knauft/Biology Pics/Science Source; (right side, bottom) © Carolina Biological Supply Company/Phototake

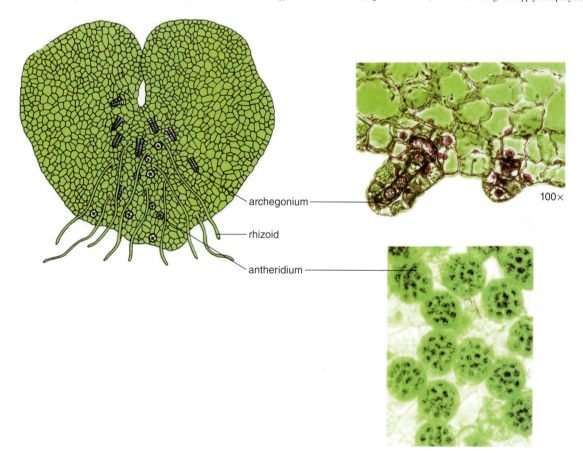

archegonium

rhizoid

antheridium

100×

Laboratory Review 25

_____ 1. What kind of tissue provides support and moves water to, and nutrients from, the leaves?

_____ 2. What generation of the plant life cycle undergoes meiosis, sporophyte or gametophyte?

_____ 3. What occurs, meiosis or mitosis, when spores develop into the gametophytes?

_____ 4. When two gametes fuse, what plant generation is formed?

_____ 5. When a plant identified as moss is observed, what generation of the moss life cycle is it?

_____ 6. What kind of gamete is produced by moss antheridia?

_____ 7. Do the rhizoids that anchor moss in the soil contain vascular tissue?

_____ 8. How are mosses dispersed to new locations?

_____ 9. Do seedless vascular plants have dominant sporophytes or gametophytes?

_____ 10. Which generation in the life cycle of ferns lacks vascular tissue?

_____ 11. Are fern zygotes dispersed to new locations?

_____ 12. Do fern roots and rhizomes contain vascular tissue?

_____ 13. What are the brownish clumps on the underside of a fern frond called?

_____ 14. What structure produces eggs in the life cycle of ferns?

Thought Questions

15. Is there anything comparable to a spore in the life cycle of animals? Justify your reply.

16. Compare and contrast the sporophyte generation of mosses and ferns. Your reply could take into consideration vascular tissue, dominant form, diploid or haploid, and what happens to it as the life cycle advances.

17. During the life cycle of animals, gametes are produced when meiosis occurs. Why must gametes be produced by mitosis during the plant life cycle?

26

Seed Plants

Learning Outcomes

26.1 Life Cycle of Seed Plants
- Compare the life cycles of nonseed and seed plants.   366
- List significant innovations of seed plants.   366

26.2 Gymnosperms
- Name three groups of gymnosperms and give the distinguishing features of each group.   367
- Distinguish between pollination and fertilization in the life cycle of a pine tree.   368–369
- Identify the structure and function of pollen cones and seed cones, and tell where you would find the male and female gametophytes and seeds.   370–371

26.3 Angiosperms
- Identify the parts of a flower and the two major groups of flowering plants.   372–373
- Identify the innovations that make the life cycle of a flowering plant different from that of a pine tree.   373
- Identify the male and female gametophyte and their roles in the life cycle of a flowering plant.   374–376

26.4 Comparison of Gymnosperms and Angiosperms
- Compare the adaptations of flowering plants to those of conifers.   377

Introduction

Among plants, **gymnosperms** and **angiosperms** are seed plants. Review the plant evolutionary tree (see Fig. 25.1) and note that the gymnosperms and angiosperms share a common ancestor, which produced **seeds,** structures that contains the next sporophyte generation. We shall see that the use of seeds to disperse the next generation requires a major overhaul of the alternation-of-generations life cycle. Nonseed plants disperse the gametophyte, the n generation, when spores are carried by the wind to a new location. Seed plants disperse the sporophyte, the 2n generation. Which generation—gametophyte or sporophyte—is better adapted to a land

environment because it contains vascular tissue? _____

Nonseed plants utilize an archegonium to protect the egg. After fertilization, the sporophyte develops within and remains attached to the archegonium. Seed plants protect the entire female gametophyte within an **ovule.** The ovule develops into a sporophyte-containing seed following fertilization. You will see that the formation of the ovule and also the **pollen grains** (contain male gametophytes) are radical innovations in the life cycle of seed plants. In gymnosperms, pollen grains are windblown, but many flowering plants have a mutualistic relationship with animals, particularly flying insects, which disperse pollen within the species. Animals also disperse the seeds of many flowering plants. Relationships with animals help explain why flowering plants are so diverse and widespread today.

ovule   egg   100×

© Ed Reschke

26.1 Life Cycle of Seed Plants

Figure 26.1 shows alternation of generations as it occurs in nonseed plants, and Figure 26.2 shows alternation of generations as it occurs in seed plants. Note which structures are haploid and which are diploid in these diagrams.

1. In which life cycle, nonseed or seed, do you note pollen sacs (microsporangia) and ovules

   (megasporangia)? _____ In which life cycle, nonseed or seed, do you note two types of spores,

   microspores and megaspores? _____ The formation of **heterospores** (unlike spores) is an

   innovation that leads to the production of pollen grains and the formation of ovules in seed plants.
   Label heterospores where appropriate in Figure 26.2. In which life cycle do you note male gametophyte

   (pollen grain) and female gametophyte (embryo sac in ovule)? _____

2. In nonseed plants, flagellated sperm must swim in external water to the egg. In seed plants, **pollination** (the transport of male gametophytes by wind or pollen carrier to the vicinity of female gametophytes) does not require external water. The production of pollen grains is an innovation in seed plants. *Label pollination where appropriate in Figure 26.2.*

3. In which life cycle does a seed appear between the zygote and the sporophyte? _____

   What generation is present in a seed? _____ Formation of a seed from an embryo sac

   in the ovule is an innovation in seeds.

Figure 26.2   Alternation of generations in flowering plants which produce seeds.

Figure 26.1   Alternation of generations in nonseed plants.

26.2 Gymnosperms

The term *gymnosperm* means "naked seed." The ovules and seeds of gymnosperms are exposed on a cone scale or some other comparable structure. Most gymnosperms do produce cones. Three groups of gymnosperms are especially familiar: cycads, ginkgoes, and conifers. Representatives of these plants may be available for you to examine. Compare them with Figure 26.3.

Conifers

The **conifers** (phylum Pinophyta) are by far the largest group of gymnosperms. Pines, hemlocks, and spruces are evergreen conifers because their leaves remain on the tree through all seasons. The cypress tree and larch (tamarack) are examples of conifers that are not evergreen.

Figure 26.3 Gymnosperms.
Three divisions of gymnosperms are well known: **a.** cycads, **b.** ginkgoes, and **c.** conifers.

(a) © Hoberman Collection/Alamy; (a, insert) © PlazaCameraman/Getty RF; (b) © Kingsley Stern; (c) © D. Giannechini/Science Source

a. An African cycad

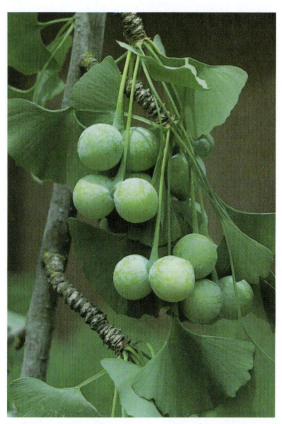

b. Ginkgo, *Ginkgo biloba.* Female maidenhair tree with seeds.

c. Conifer, *Picea.* Spruce with pollen cones and seed cones.

Figure 26.4   Pine life cycle.

The sporophyte is the tree. The male gametophytes are windblown pollen grains shed by pollen cones. The female gametophytes are retained within ovules on seed cones. The ovules develop into windblown seeds.

© Carolina Biological Supply Company/Phototake

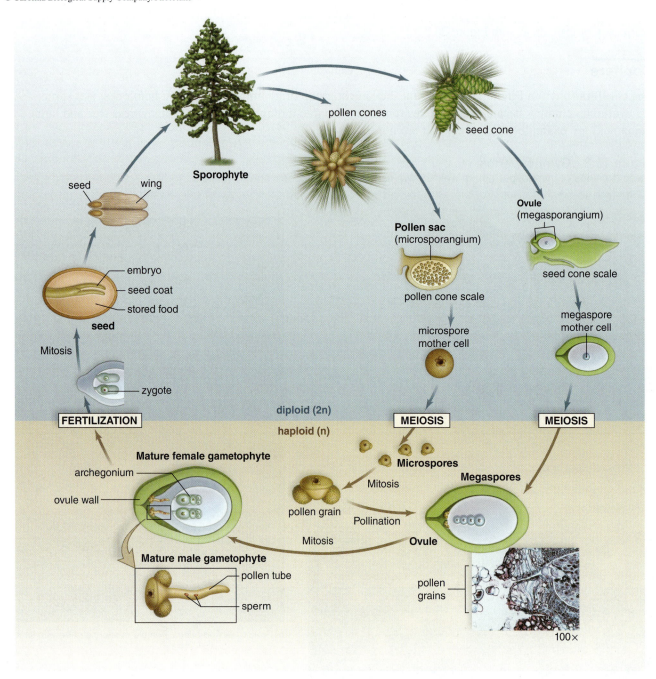

Pine Trees

The pine tree is the dominant sporophyte generation (Fig. 26.4) in the life cycle of a pine. Vascular tissue extends from the roots, through the stem, to the leaves. Pine tree leaves are needlelike, leathery, and covered with a waxy, resinous cuticle. The **stomata,** openings in the leaves for gas exchange, are sunken. The structure of the leaf and the leaf's internal anatomy are adaptive to a drier climate.

In pine trees, the **pollen sacs** and ovules are located in cones. Pollen sacs on pollen cone scales contain cells which undergo meiosis to produce **microspores.** Microspores (n) become sperm-bearing **male gametophytes** (pollen grains). Seed cone scales bear ovules where cells undergo meiosis to produce **megaspores.** A megaspore (n) becomes an egg-bearing **female gametophyte.**

Pollination occurs when pollen grains are windblown to the seed cones. After **fertilization,** the egg becomes a sporophyte (2n) embryo enclosed within the ovule, which develops a seed coat. The seeds are winged and are dispersed by the wind.

1. Which part of the pine life cycle is the sporophyte? _____

2. Which part of the pine life cycle is the male gametophyte? The female gametophyte? _____

_____

_____

3. Where does fertilization occur and what structure becomes a seed? _____

_____

Observation: Pine Leaf

Obtain a cluster of pine leaves (needles). A very short woody stem is at its base. Each type of pine has a typical number of leaves in a cluster (Fig. 26.5). How many leaves are in the cluster you are examining? _____

What is the common name of your specimen? _____

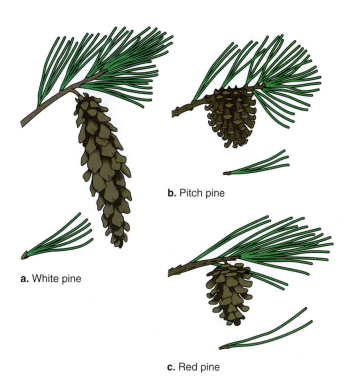

b. Pitch pine

a. White pine

c. Red pine

Figure 26.5 Pine leaves (needles).
a. The needles of white pines are in clusters of five. **b.** The needles of pitch pines are in clusters of three. **c.** The needles of red pines are in clusters of two.

Preserved Cones

1. Compare a pine pollen cone to a pine seed cone. _____

   _____

   _____

   Note the size and texture of the pollen cones relative to the larger, woody, seed cones.

2. Pollen cones. Remove a single scale (sporophyll) from the pollen cone and examine with a stereomicroscope. Note the two pollen sacs on the lower surface of each scale (Fig. 26.6a). What do the pollen sacs produce? _____

   _____

   _____

3. Seed cones.
   a. Seed cones of three distinct ages are present. First-year cones are about 1 cm wide. Second-year cones are about 10 cm long. They are green and the scales are tightly closed together. In third-year cones the scales have opened, revealing the mature seeds at the base of the cone scale.
   b. If available, examine a first-year cone cut lengthwise. The ovules are visible as small, milky-white domes at the base of the cone scale. Ovules will hold the female gametophyte generation, and then they become seeds.

   c. Examine mature seed cones. See if any seeds are present (Fig. 26.6b). Where are they located? _____

   _____ Note the hardness of the seed coat.

   What is the function of the seed coat? _____

   If instructed to do so, use tweezers to pull out a seed and note the wing. What is the wing for? _____

   _____

   Replace the seed in the cone when you are finished.

4. Pine seeds. If available, examine a pine seed (called a pine nut) that has the seed coat removed. These seeds can be used for cooking foods such as pesto. Carefully cut the seed lengthwise and examine. Can you find an embryo inside? _____

Figure 26.6   Pine cones.
a. The scales of pollen cones bear pollen sacs where microspores become pollen grains. **b.** The scales of seed cones bear ovules that develop into winged seeds.

a. Pollen cones

b. Seed cone

Conclusions: Pine Cones

- Are the pine seeds covered by tissue donated by the original sporophyte? _____

  What does gymnosperm mean? _____

  Explain. _____

Microscope Slides

1. Examine a prepared slide of a longitudinal section through a mature pine pollen cone. *Label a pollen sac in Figure 26.7a and a pollen grain in Figure 26.7b.* A pollen grain has a central body and two attached hollow bladders. How do these help in the dispersal of pine pollen? _____

   _____

   One pollen grain cell will divide to become two nonflagellated sperm, one of which fertilizes the egg after pollination. The other cell forms the **pollen tube** through which a sperm travels to the egg.

Figure 26.7   Pine pollen cone.

a. Pollen cones bear pollen sacs (microsporangia) in which microspores develop into pollen grains. **b.** Enlargement of pollen grains.   (a) © Biophoto Associates/Science Source; (b) © Ed Reschke

1. _____

a.

2. _____

b.   100×

2. Examine a prepared slide of a longitudinal section through an immature pine seed cone. Seed cone scales bear ovules. The ovule contains a megaspore mother cell, which undergoes meiosis to produce four megaspores, three of which disintegrate. This megaspore (n) becomes an egg-bearing female gametophyte. *Label the ovule and the megaspore mother cell in Figure 26.8. Also, label the pollen grains that you can see just outside the ovule.*

Figure 26.8   Seed cone.

Seed cones bear ovules, each of which will contain a female gametophyte. Note pollen grains near the entrance.

© Carolina Biological Supply Company/ Phototake

a. _____

c. _____   b. _____   100×

26–7   Laboratory 26   Seed Plants   **371**

26.3 Angiosperms

Flowering plants (phylum Magnoliophyta) are the dominant plants today. They occur as trees, shrubs, vines, and garden plants. At some point in their life cycle, all angiosperms bear flowers as was discussed in Laboratory 10.

Figure 26.9 Generalized flower.
A flower has four main kinds of parts: sepals, petals, stamens, and a carpel. A stamen has an anther and a filament. A carpel has a stigma, a style, and an ovary. An ovary contains ovules.

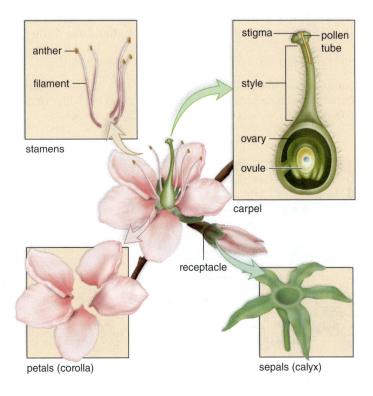

Observation: A Flower

1. With the help of Figure 26.9, identify the following structures on a model of a flower:
   a. **Receptacle:** The portion of a stalk to which the flower parts are attached.
   b. **Sepals:** An outermost whorl of modified leaves, collectively called the **calyx.** Sepals are green in most flowers. They protect a bud before it opens.
   c. **Petals:** Usually colored leaves that collectively constitute the **corolla.**
   d. **Stamen:** A swollen terminal **anther** and the slender supporting **filament.** The anther contains two pollen sacs, where microspores develop into male gametophytes (pollen grains).
   e. **Carpel:** A modified sporophyll consisting of a swollen basal ovary; a long, slender **style** (stalk); and a terminal **stigma** (sticky knob).
   f. **Ovary:** The enlarged part of the carpel that develops into a fruit.
   g. **Ovule:** The structure within the ovary where a megaspore develops into a female gametophyte (embryo sac). The ovule becomes a seed.

2. Carefully inspect a fresh flower. What is the common name of your flower? _____

3. Remove the sepals and petals by breaking them off at the base. How many sepals and petals are there?

   _____

4. Are the stamens taller than the carpel? _____

Flower parts in threes and multiples of three

Table 26.1 Monocots and Eudicots

Monocots	Eudicots
One cotyledon	Two cotyledons
Flower parts in threes or multiples of three	Flower parts in fours or fives or multiples of four or five
Usually herbaceous	Woody or herbaceous
Usually parallel venation	Usually net venation
Scattered bundles in stem	Vascular bundles in a ring
Never woody	Can be woody

Flower parts in fours or fives and their multiples

5. Remove a stamen, and touch the anther to a drop of water on a slide. If nothing comes off in the water, crush the anther a little to release some of its contents. Place a coverslip on the drop, and observe with low- and high-power magnification. What are you observing? _____

6. Remove the carpel by cutting it free just below the base. Make a series of thin cross sections through the ovary. The ovary is hollow, and you can see nearly spherical bodies inside. What are these bodies? _____
_____
_____

7. Flowering plants are divided into two classes, called **monocots** and **eudicots.** Table 26.1 lists significant differences between the two classes of plants. Is your flower a monocot or eudicot? _____

Life Cycle of Flowering Plants

The life cycle of a flowering plant is like that of the pine tree except for these innovations:

- In angiosperms, the often brightly colored flower contains the pollen sacs and ovules. Locate these structures in Figure 26.10, and trace the life cycle of flowering plants from the sporophyte generation (the tree) through the various stages to the sporophyte generation once again.
- Pollination in flowering plants—when pollen is delivered to the stigma of the carpel—is sometimes accomplished by wind but more likely by the assistance of an animal pollinator. The pollinator acquires nutrients (e.g., nectar) from the flower and inadvertently collects pollen, which it takes to the next flower.
- Notice that flowering plants practice **double fertilization.** A mature pollen grain contains two sperm; one fertilizes the egg, and the other joins with the two polar nuclei to form **endosperm** (3n), which serves as food for the developing embryo.
- Also, flowering plants have seeds enclosed in fruits. Fruits protect the seeds and aid in seed dispersal. Sometimes, animals eat the fruits, and after the digestion process, the seeds are deposited far away from the parent plant. The term *angiosperm* means "covered seeds." The seeds of angiosperms are found in fruits, which develop from parts of the flower, particularly the ovary.

Figure 26.10 Flowering plant life cycle.

The parts of the flower involved in reproduction are the anthers of stamens and the ovules in the ovary of a carpel.

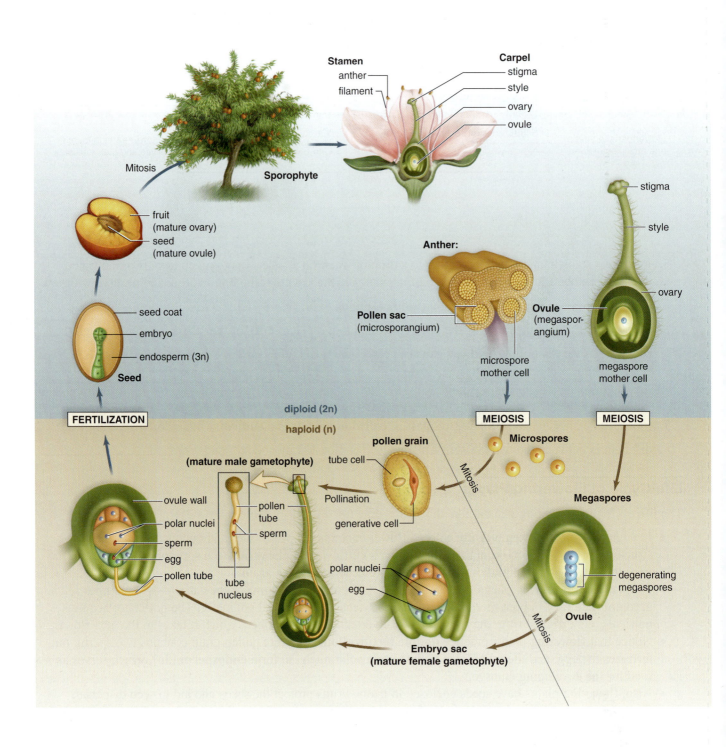

The Male Gametophyte

Notice that in the life cycle of flowering plants (Fig. 26.10), meiosis in pollen sacs produces four microspores, each of which will become a two-celled pollen grain. In flowering plants, **pollination** is the transfer of the pollen grain from the anther to the stigma, where the pollen grain germinates and becomes the mature male gametophyte. How is pollination accomplished in the flowering plant life cycle? _____

_____

Observation: Pollen Grain Slide

1. A nongerminated pollen grain newly released from the anther has two cells. The larger of the two cells is the **tube cell,** and the smaller is the **generative cell.** Examine a prepared slide of pollen grains. Identify the tube cell and the generative cell. Sketch your observation here.

2. During germination, the pollen grain's tube cell gives rise to the pollen tube. As it grows, the pollen tube passes through the stigma and grows through the style and into the ovary. Two sperm cells produced by division of the pollen grain's generative cell migrate through the pollen tube into the embryo sac. Examine a prepared slide of germinated pollen grains with pollen tubes. You should be able to see the tube cell nucleus and two sperm cells. What portion of a germinated pollen grain tells you it is the mature male gametophyte?

_____

*Experimental Procedure: Pollen Grains

Inoculate an agar-coated microscope slide with pollen grains. Invert the slide, and place it onto a pair of wooden supports in a covered petri dish. Leave the inverted slide in the covered petri dish for 1 hour. Then remove the slide, and examine it with the compound microscope. Have any of the pollen grains

germinated? _____

If so, describe. _____

_____

*Experimental Procedure from Richard Carter and Wayne R. Faircloth, *General Laboratory Studies,* Lab 17.2, 1991. Used with permission of Kendall/Hunt Publishing Company.

The Female Gametophyte

In the ovule, a megaspore undergoes three mitotic divisions to produce a seven-celled (eight-nuclei) female gametophyte called an **embryo sac** (Fig. 26.11). One of these cells is an egg cell. The largest cell contains two polar nuclei.

Figure 26.11   Embryo sac in a lily ovule.
An embryo sac is the female gametophyte of flowering plants. It contains seven cells, one of which is the egg. The fate of each cell is noted.   © W. P. Armstrong 2004

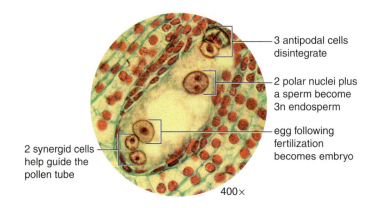

3 antipodal cells disintegrate

2 polar nuclei plus a sperm become 3n endosperm

egg following fertilization becomes embryo

2 synergid cells help guide the pollen tube

400×

Observation: Embryo Sac Slide

Examine the demonstration slide of the mature embryo sac of *Lilium*. Identify the egg labeled in Figure 26.11.

Answer these questions.

1. What portion of the embryo sac tells you that it is the female gametophyte? _____

    _____

2. When the pollen tube delivers sperm to the embryo sac, double fertilization occurs. Due to double

    fertilization, what happens to the egg? _____

    _____

    What happens to the polar nuclei? _____

    _____

3. Following fertilization, the ovules develop into seeds, and the ovary develops into the fruit.

    What are the three parts of a seed (as shown in Figure 26.10)? _____

Label the flower remnants, the fruit, and the seeds in the drawings of a pea pod and apple provided. The flesh of an apple comes from the enlarged receptacle that grows up and around the ovary, while the ovary largely consists of the core.

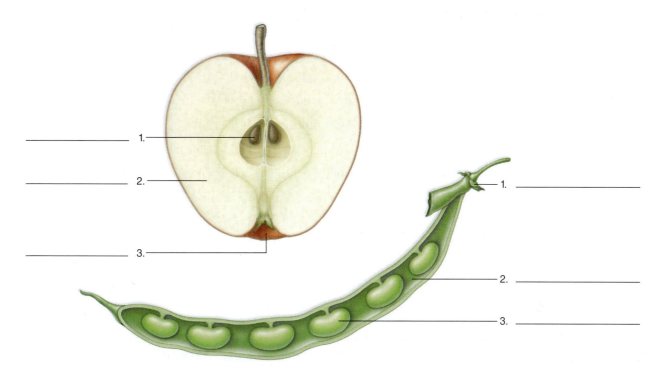

26.4 Comparison of Gymnosperms and Angiosperms

1. Complete Table 26.2 using "yes" or "no" to compare adaptations of gymnosperms and angiosperms.

Table 26.2 Comparison of Gymnosperms and Angiosperms					
	Heterospores	Pollen grains/Ovule	Cones	Flower	Fruit
Gymnosperms					
Angiosperms					

2. What structure in gymnosperms and angiosperms delivers sperm to the vicinity of the egg? _____ Does delivery require external water? _____

3. What structure in gymnosperms and angiosperms becomes a seed? _____

4. The embryo of what generation, sporophyte or gametophyte, is in a seed? _____

5. What innovation in angiosperms led to the seeds being covered by fruit? _____

Laboratory Review 26

_____ 1. Which generation, gametophyte or sporophyte, is dispersed by seed plants?

_____ 2. What protects the entire female gametophyte in seed plants?

_____ 3. When meiosis occurs in a seed plant, what results?

_____ 4. What is the transfer of the seed plants' male gametophyte by wind or a carrier called?

_____ 5. What is the name given to plants that produce "naked seeds"?

_____ 6. How are the seeds of a pine tree dispersed?

_____ 7. In the life cycle of a pine, what is the sporophyte?

_____ 8. In the life cycle of a pine, where are the ovules located?

_____ 9. What flower structure is formed by the anther and filament?

_____ 10. What does mitotic division of the generative cell in a pollen grain produce?

_____ 11. What structure do the sperm move through to reach an angiosperm's female gametophyte?

_____ 12. What is produced when double fertilization occurs in angiosperms?

_____ 13. What is the name of the female gametophyte of flowering plants?

_____ 14. What flower structure forms the covering around the seed of an angiosperm?

Thought Questions

15. **a.** How are the gametophytes of mosses and flowering plants different from one another?

b. How are the gametophytes of mosses and flowering plants similar to one another?

16. Why do pines produce clusters of pollen cones, and why are the pollen cones located at the tips of branches instead of closer to the trunk of the tree?

17. Angiosperm means "covered seed." What covers the seeds of angiosperms, and what role does this covering have in ensuring the success of the embryos in the seeds?

27

Introduction to Invertebrates

Learning Outcomes

Introduction
- Distinguish between invertebrate and vertebrate animals.   379

27.1 Evolution of Animals
- Discuss the evolution of animals in terms of symmetry, number of tissue layers, and pattern of development.   380–381

27.2 Sponges
- Identify a sponge and describe the anatomy of a sponge.   381–383
- Show that the anatomy and behavior of a sponge aids its survival and ability to reproduce.   381–382

27.3 Cnidarians
- Identify various types of cnidarians, and describe the anatomy of *Hydra* in particular.   384–385
- Show that the anatomy and behavior of *Hydra* aid its survival and ability to reproduce.   385–387

27.4 Flatworms
- Identify various flatworms, both free-living and parasitic forms. Contrast their lifestyles.   388–392
- Contrast the structure of a planarian with that of *Hydra*, and show that each is adapted to a particular way of life.   390

27.5 Roundworms
- Identify a roundworm, and contrast the anatomy of a roundworm with that of a planarian.   392–394
- Show that roundworms have successfully exploited various ways of life, including the parasitic way of life.   393–394

27.6 Rotifers
- Identify rotifers and describe their anatomy and behavior.   394–395

Introduction

In our survey of the animal kingdom, we will see that animals are very diverse in structure. Even so, all animals are multicellular and **heterotrophic,** which means their food consists of organic molecules made by other organisms. Consistent with the need to acquire food, animals achieve locomotion by the use of muscle fibers. Animals are always diploid, and during sexual reproduction, the embryo undergoes specific developmental stages.

While we tend to think of animals in terms of **vertebrates** (e.g., dogs, fishes, squirrels), which have a backbone, most animal species are those that lack a backbone, commonly known as **invertebrates.** In this laboratory, we will examine those invertebrates that lack a true body cavity, called a **coelom.** A survey of the rest of the animal kingdom follows in Laboratory 28 and Laboratory 29.

> 🐚 **Planning Ahead** To see hydra and planarians feed, have students observe the animals at the start of lab, add food, and then check frequently until food engulfment occurs.

27.1 Evolution of Animals

Today, molecular data is used to trace the evolutionary history of animals. These data tell us, as shown in the phylogenetic (evolutionary) tree (Fig. 27.1), that all animals share a common ancestor. This common ancestor was most likely a colonial protist consisting of flagellated cells. All but one of the phyla depicted in the tree consist of only invertebrates—the phylum Chordata contains a few invertebrates and also the vertebrates.

Figure 27.1 Evolutionary tree of animals.

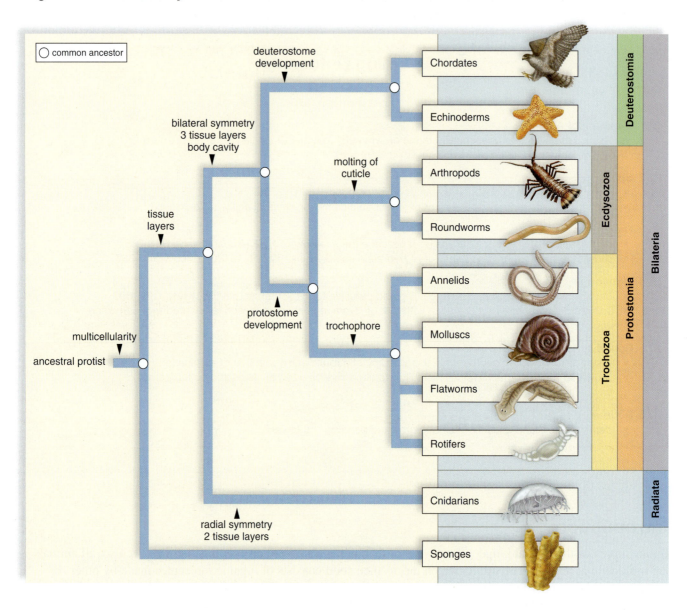

Certain anatomical features of animals are used in the tree. One feature of interest is formation of tissue layers. Sponges have no true tissue layers. Which phyla in the tree have only two tissue layers? _____ The other phyla have three tissue layers.

Another feature of interest is **symmetry. Asymmetry** means the animal has no particular symmetry. Which phyla have **radial symmetry,** in which, as in a wheel, two identical halves are obtained no matter how the animal is longitudinally sliced? _____ The other phyla have **bilateral symmetry,** which means the animals have one plane of symmetry with a definite right half and left half. Animals in phylum Echinodermata are bilateral in the larval form, but they have radial symmetry as adults.

Finally, complex animals are either **protostomes** or **deuterostomes.** Protostomes have a pattern of development in which the first opening of the embryo forms the mouth. The second opening of the embryo forms the mouth in deuterostomes; the first opening becomes the anus. Which pattern of development do the flatworms, rotifers, and roundworms (animals included in this laboratory) have? _____

27.2 Sponges

Sponges (phylum Porifera) live in water, mostly marine, attached to rocks, shells, and other solid objects. An individual sponge is typically shaped like a tube, cup, or barrel. Sponges grow singly or in colonies whose overall appearances vary widely. A single sponge can become a colony by asexual budding.

Anatomy of Sponges

Sponges consist of loosely organized cells and have no well-defined tissues. They are asymmetrical or radially symmetrical and **sessile** (immotile). They can reproduce asexually by budding or fragmentation, but they also reproduce sexually by producing eggs and sperm.

Sponges have a few types of specialized cells. Most notably they have flagellated **collar cells** (**choanocytes**). The movement of their flagella keeps water moving through the pores into the central cavity and out the osculum of a sponge (Fig. 27.2). Collar cells also take in suspended food particles from the water and digest them for the benefit of all the other cells in a sponge.

> ### *Observation: Anatomy of Sponges*

Preserved Sponge

1. Examine a preserved sponge (Fig. 27.2*a*). Note the main excurrent opening (**osculum**) and the multiple incurrent pores. Water is constantly flowing in through the pores and out the osculum.
   Label the arrows in Figure 27.2a to indicate the flow of water. Use the labels *water out* and *water in* through pores.
2. Examine a sponge specimen cut in half. Note the central cavity and the sponge wall. The wall is convoluted in some sponges, and the pores line small canals. Does this particular sponge have pore-lined canals? _____
3. You may be able to see **spicules,** fine projections over the body and especially encircling the osculum. Does this sponge have spicules? _____

Figure 27.2 Sponge anatomy.

a. Movement of water through pores into the central cavity and out the osculum is noted. **b.** Collar cells line the central cavity of a sponge, and the movement of their flagella keeps the water moving through the sponge. **c.** Draw an enlargement of spicules here.

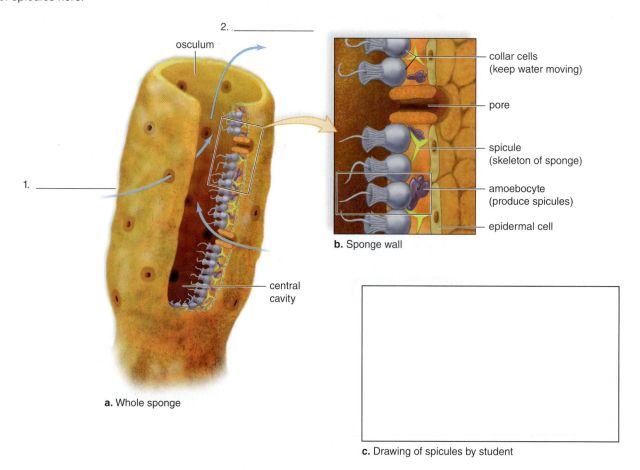

2. _____

osculum

collar cells
(keep water moving)

pore

spicule
(skeleton of sponge)

amoebocyte
(produce spicules)

epidermal cell

b. Sponge wall

1. _____

central
cavity

a. Whole sponge

c. Drawing of spicules by student

Prepared Slides

1. Examine a prepared slide of *Grantia*.

   a. Find the collar cells that line the interior (Fig. 27.2*b*). A sponge is a **sessile filter feeder.** Collar cells phagocytize (engulf) tiny bits of food that come through the pores along with the water flowing through the sponge. They then digest the food in food vacuoles. Explain the expression *sessile filter feeder.* _____
   _____

   b. Do you see any spicules? Do they project from the wall of a sponge? _____

   c. Depending on the sponge, spicules are made of either calcium carbonate, silica (glass), or protein. Calcium carbonate and silica produce hard, sharp spicules. Name two possible advantages of spicules to a sponge. _____
   _____

2. Examine a prepared slide of sponge spicules. What do you see? _____

   Draw a sketch of four spicules, each having a different appearance, in the space provided in Figure 27.2c.

Diversity of Sponges

Sponges are very diverse and come in many shapes and sizes. Some sponges live in fresh water, although most live in the sea and are a prominent part of coral reefs, areas of abundant sea life discussed in the next section. Zoologists have described over 5,000 species of sponges, which are grouped according to the type of spicule (Fig. 27.3).

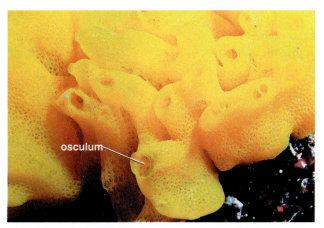

a. Calcareous sponge, *Clathrina canariensis*

b. Bath sponge, *Xestospongia testudinaria*

Figure 27.3 Diversity of sponges.
a. Calcareous (chalk) sponges have spicules of calcium carbonate. **b.** Bath sponges have a skeleton of spongin. **c.** Glass sponges have glassy spicules.

(a) © Amar and Isabelle Guillen, Guillen Photography/Alamy; (b) © Andrew J. Martinez/Science Source; (c) © Kenneth M. Highfill/Science Source

c. Glass sponge, *Euplectella aspergillum*

Conclusions: Anatomy of Sponges

- The anatomy and behavior of a sponge aid its survival and its ability to reproduce. How does a sponge

  a. protect itself from predators? _____

  b. acquire and digest food? _____

  c. reproduce asexually and sexually? _____

27.3   Cnidarians

Cnidarians (phylum Cnidaria) are tubular or bell-shaped animals that live in shallow coastal waters, except for the oceanic jellyfishes. Two basic body forms are seen among cnidarians. The mouth of a **polyp** is directed upward, while the mouth of a jellyfish, or **medusa,** is directed downward. At one time, both body forms may have been a part of the life cycle of all cnidarians. When both are present, the sessile polyp stage produces medusae, and this motile stage produces egg and sperm (Fig. 27.4). Today in some cnidarians, one stage is dominant and the other is reduced; in other species, one form is absent altogether. Regardless, all cnidarians are radially symmetrical. How can radial symmetry be a benefit to an animal? _____

_____

How can a life cycle that involves two forms, called **polymorphism,** be of benefit to an animal, especially if one stage is sessile (stationary)? _____

_____

Figure 27.4   The life cycle of a cnidarian.
Some cnidarians have both a polyp stage and a medusa stage; in others, one stage may be dominant or absent altogether.

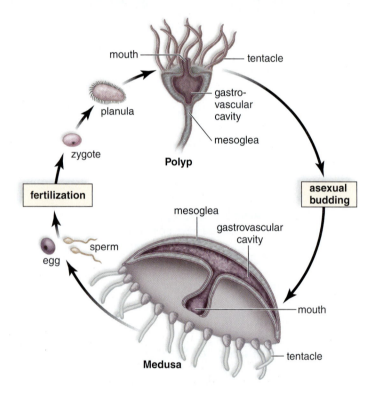

Anatomy of *Hydra*

Figure 27.5 shows the anatomy of a hydra, which will be studied as a typical cnidarian. Hydras exist only as sessile polyps; there is no alternate stage. Note the **tentacles** that surround the **mouth,** the large **gastrovascular cavity,** and the basal disk. A gastrovascular cavity has a single opening that is used as both an entrance for food and an exit for wastes.

Figure 27.5   Anatomy of *Hydra*.
Hydra typifies the anatomy of a cnidarian.
© NHPA/M.I. Walker/Photoshot RF

Observation: Anatomy of Hydra

1. Preserved *Hydra*. With the aid of a hand lens, examine preserved specimens of *Hydra*. Hydras typically reproduce asexually by budding. Do you see any evidence of buds that are developing directly into small

   hydras? _____ The body wall can also produce ovaries and testes that produce eggs and sperm. The testes are generally located near the attachment of the tentacles; the ovaries appear farther down on the trunk, toward the basal disk.

2. Slide of *Hydra*. Examine prepared slides of cross and longitudinal sections of *Hydra*. With the help of Figure 27.5, note the epidermis, the mesoglea (a gelatinous material between the two tissue layers), and the gastrodermis, which lines the gastrovascular cavity. Switch to high power. Do you find any

   cells? _____ Describe them. _____

   _____

Figure 27.6   Hydra feeding.

(both) © Roland Birke/Getty Images

ingested
Daphnia

3. Living specimen. The tentacles of a hydra capture food, which is stuffed into the gastrovascular cavity (Fig. 27.6). Observe a living *Hydra* in a small petri dish for a few minutes. What is the current behavior of

your hydra? _____

_____

Most often a hydra is attached to a hard surface by its basal disk. A hydra can move, however, by turning somersaults:

After a few minutes, tap the edge of the petri dish. What is the reaction of your hydra? _____

_____

_____

4. Mount a living *Hydra* on a depression glass slide with a coverslip and examine a tentacle. Unique to cnidarians are specialized stinging cells, called **cnidocytes,** which give the phylum its name. Each cnidocyte has a fluid-filled capsule called a **nematocyst** (see Fig. 27.5, far right), which contains a long, spirally coiled hollow thread. The threads trap and/or sting prey. Note the cnidocytes as swellings on the tentacles. Add a drop of vinegar (5% acetic acid) and note what happens to the cnidocytes. Did your hydra discard any

nematocysts? _____ Describe. _____

_____

_____

Of what benefit is it to *Hydra* to have cnidocytes? _____

_____

Conclusions: Anatomy of Cnidarians

- The anatomy and behavior of a hydra aid its survival and its ability to reproduce. How does a hydra

  a. acquire and digest food? _____

  _____

  b. protect itself from predators? _____

  _____

  c. reproduce asexually and sexually? _____

  _____

Diversity of Cnidarians

Cnidarians consist of a large number of mainly marine animals (Fig. 27.7). Sea anemones, sometimes called the flowers of the sea, are solitary polyps often found in coral reefs, areas of biological abundance in shallow tropical seas. Stony corals have a calcium carbonate skeleton that contributes greatly to the building of coral reefs. Portuguese man-of-war is a colony of modified polyps and medusae. Jellyfishes are a part of the zooplankton, suspended animals that serve as food for larger animals in the ocean.

Figure 27.7   Cnidarian diversity.

(a) ©; Danita Delimont/Alamy RF; (b) © Ron & Valerie Taylor/Photoshot; (c) © Islands in the Sea 2002, NOAA/OER; (d) © Amos Nachoum/Corbis

a. Sea anemone, *Corynactis*

b. Cup coral, *Tubastrea*

c. Portuguese man-of-war, *Physalia*

d. Jellyfish, *Aurelia*

27.4 Flatworms

Flatworms (phylum Platyhelminthes) are bilaterally symmetrical animals that can be either free-living or parasitic. Free-living flatworms, called planarians, are more complex than the cnidarians. In addition to the germ layers ectoderm and endoderm, mesoderm is also present. Flatworms are usually **hermaphroditic,** which means they possess both male and female sex organs (Fig. 27.8c). Why is it advantageous for an animal to be

hermaphroditic? _____

_____

Planarians practice cross-fertilization when the penis of one is inserted into the genital pore of the other. The fertilized eggs are enclosed in a cocoon and hatch as tiny worms in two or three weeks.

Planarians

Planarians such as *Dugesia* live in lakes, ponds, and streams, where they feed on small living or dead organisms. In planarians, the three germ layers give rise to various organs aside from the reproductive organs (Fig. 27.8). The three-part **gastrovascular cavity** ramifies throughout the body. Excretory organs called **flame cells** (because their cilia reminded early investigators of a flickering flame of a candle) collect fluids from inside the body and send them via a tube to an excretory pore. The nervous system contains a brain and lateral nerve cords connected by transverse nerves, which is why it is called **ladderlike.** *Complete the labels in both 27.8b and d. Label excretory canal, brain, and nerve cord.* Why would you expect an animal that lives in fresh water to have a well-

developed excretory system? _____

_____

Figure 27.8   Planarian anatomy.
a. When a planarian extends the pharynx, food is sucked up into a gastrovascular cavity that branches throughout the body. **b.** The excretory system has flame cells. **c.** The reproductive system has both male (blue) and female (pink) organs. **d.** The nervous system looks like a ladder.

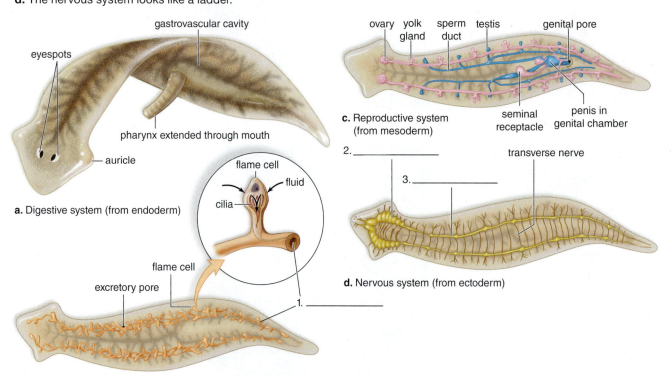

Whole Mount

Examine a whole mount of a planarian that shows the branching gastrovascular cavity (Fig. 27.8*a*). What is the advantage of a gastrovascular cavity that ramifies through the body? _____

Prepared Slide

Examine a cross section of a planarian under the microscope. Can you locate the structures shown in Figure 27.9? Does a planarian have a body cavity?_____ Explain. _____

Figure 27.9   Planarian micrograph.
Cross section of a planarian at the pharynx.
© Carolina Biological Supply Company/Phototake

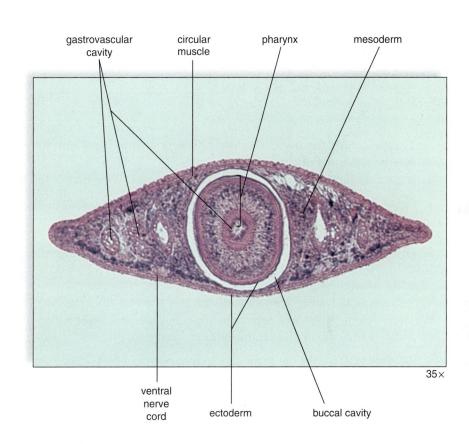

gastrovascular cavity · circular muscle · pharynx · mesoderm

ventral nerve cord · ectoderm · buccal cavity

35×

Living Planarian

1. Examine the behavior of a living planarian (Fig. 27.10) in a petri dish. Describe the behavior of the animal. _____

2. Does the animal move in a definite direction?_____ Why is it advantageous for a predator such as a planarian to have bilateral symmetry and a definite head region? _____

Figure 27.10 Micrograph of planarian feeding and the gastrovascular cavity.
a. The digestive cavity ramifies throughout the body.
b. The pharynx protrudes through the mouth and sucks up food when a planarian feeds.

(a) © NHPA/M. I. Walker/Photoshot RF; (b) © Photography by Marc C. Perkins, Orange Coast College, Costa Mesa, CA; image blending by Heather Bartell, Huntington Beach, CA

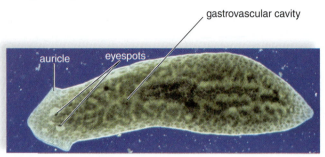

gastrovascular cavity

auricle   eyespots

a. Gastrovascular cavity

10×

b. Planarian feeding

3. Gently touch the animal with a probe. What three types of cells must be present for flatworms to be able to respond to stimuli and move about? _____

_____

The auricles on the side of the head are sense organs. Flatworms have well-developed muscles and a nervous system consisting of a brain and nerves.

4. If a strong light is available, shine it on the animal. What part of the animal would be able to detect light? _____ How does the animal respond to the light? _____

_____

5. Offer the worm some food, such as a small piece of liver or egg yolk, and describe its manner of eating.

_____

Roll the animal away from its food and note the pharynx extending from the body (Fig. 27.10).

6. Transfer your worm to a concave depression slide and cover with a coverslip. Examine with a microscope and note the cilia on the ventral surface. Numerous gland cells secrete a mucous material that assists movement. Describe the mode of locomotion. _____

_____

Planarians Versus Cnidarians

- Planarians, with three germ layers, are more complex than cnidarians. Contrast a hydra with a planarian by stating in Table 27.1 the significant organ differences between them.

- Planarians have no respiratory or circulatory system. As with cnidarians, each individual _____ takes care of its own needs for these two life functions.

Table 27.1	Contrasts Between a Hydra and a Planarian		
	Digestive System	**Excretory System**	**Nervous Organization**
Hydra			
Planarian			

Tapeworms

Tapeworms are parasitic flatworms known as cestodes. They live in the intestines of vertebrate animals, including humans (Fig. 27.11). The worms consist of a **scolex** (head), usually with suckers and hooks, and **proglottids** (segments of the body). Ripe proglottids detach and pass out with the host's feces, scattering fertilized eggs on the ground. If pigs or cattle happen to ingest these, larvae called bladder worms develop and eventually become encysted in muscle, which humans may then eat in poorly cooked or raw meat. A bladder worm that escapes from a cyst develops into a mature tapeworm attached to the intestinal wall.

1. How do humans get infected with the pig tapeworm? _____

2. What is the function of a tapeworm's hooks and suckers? _____

3. Why would you expect a tapeworm to have a reduced digestive system? _____

4. Proglottids mature into "bags of eggs." Given the life cycle of the tapeworm, why might a tapeworm have an expanded reproductive system compared to a planarian? _____

Figure 27.11   Life cycle of the tapeworm *Taenia.*
The secondary host (pig) is the means by which the worm is dispersed to the primary host (humans).

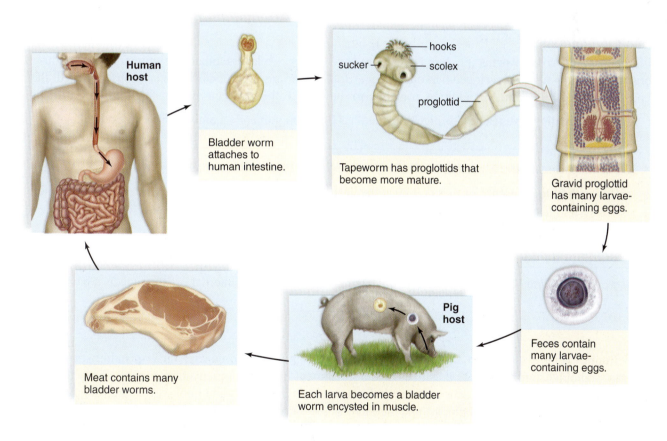

1. Examine a preserved specimen and/or slide of *Taenia pisiformis,* a tapeworm.
2. With the help of Figure 27.12, identify the scolex, with hooks and suckers, and the proglottids.

Figure 27.12 Anatomy of *Taenia*.
The adult worm is modified for its parasitic way of life. It consists of **a.** the scolex and **b.** many proglottids, which become bags of eggs.   (a) © James Webb/Phototake; (b) © Ed Reschke/Getty Images

a. The scolex functions to attach worm to host.   **b.** Proglottids function in reproduction.

27.5 Roundworms

Like planarians, **roundworms** (phylum Nematoda) have three germ layers, bilateral symmetry, and various organs, including a well-developed nervous system. Both planarians and roundworms are nonsegmented; the body has no repeating units. In addition, roundworms have the following features:

1. **Complete digestive tract:** The digestive tract has both a mouth and an anus. What is the advantage

   of this? _____

2. **Pseudocoelom:** A body cavity, which allows space for the organs, is incompletely lined with mesoderm.

   How would the presence of these two features lead to complexity? For example, how would a complete digestive system lead to a greater number of specialized organs such as both a small intestine that assists

digestion and a large intestine that assists elimination? _____

_____

How would a spacious body cavity promote a greater number of diverse internal organs such as a pancreas and a

liver? _____

_____

Roundworms are found in all aquatic habitats and in damp soil. Some even survive in hot springs, deserts, and cider vinegar. They parasitize (take nourishment from) both plants and animals. They are significant crop pests and also cause disease in humans. Both pinworms and hookworms are roundworms that cause intestinal difficulties; trichinosis and elephantiasis are also caused by roundworms.

Ascaris

Ascaris, a large, primarily tropical intestinal parasite, is often studied as an example of this phylum.

Observation: Ascaris

Examine preserved specimens of *Ascaris,* both male and female (Fig. 27.13). In roundworms, the sexes are separate. The male is smaller and has a curved posterior end. Be sure to examine specimens of each sex.

Figure 27.13   Roundworm anatomy.
a. Photograph of male *Ascaris.* **b.** Male reproductive system. **c.** Photograph of female *Ascaris.* **d.** Female reproductive system.

(a) © P&R fotos/age fotostock/Superstock; (c) © P&R fotos/age fotostock/Superstock

a. Male *Ascaris*

c. Female *Ascaris*

b. Male reproductive system

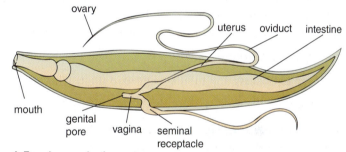

d. Female reproductive system

Trichinella

Trichinella is a parasitic roundworm that causes the disease **trichinosis.** When pigs or humans eat raw or undercooked pork infected with *Trichinella* cysts, juvenile worms are released in the digestive tract, where they penetrate the wall of the small intestine and mature sexually. After male and female worms mate, females produce juvenile worms that migrate and form cysts in various muscles (Fig. 27.14). A human with trichinosis has muscular aches and pains that can lead to death if the respiratory muscles fail.

Figure 27.14   Larva of the roundworm *Trichinella* embedded in a muscle.
A larva coils in a spiral and is surrounded by a sheath derived from a muscle fiber.

© Carolina Biological Supply Company/Phototake

100×

1. Examine preserved, infected muscle or a slide of infected muscle, and locate the *Trichinella* cysts, which contain the juvenile worms.

2. How can trichinosis be prevented in humans? _____

_____

3. How can pig farmers help to stamp out trichinosis so that humans are not threatened by the disease? _____

_____

Filarial Worm

A roundworm called a **filarial worm** infects lymphatic vessels and blocks the flow of lymph. The condition is called **elephantiasis** because when a leg is affected, it becomes massively swollen.

Living Vinegar Eels

Vinegar eels are tiny, free-living nematodes that can live in unpasteurized vinegar.

© Vanessa Vick/The New York Times/Redux

1. Examine live vinegar eels, and observe their active, whiplike swimming movements. This thrashing motion may be a result of nematodes having longitudinal muscles only; they lack circular muscles.

2. Select a few larger vinegar eels for further study, and place them in a small drop of vinegar on a clean microscope slide. If the eels are too active for study, you can slow them by briefly warming them or by adding methyl cellulose.

3. Try to observe the tubular digestive tract, which begins with the mouth and ends with the anus. Also, you may be able to see some of the reproductive organs, particularly in a large female vinegar eel.

Conclusion: Anatomy of Roundworms

• Nematodes are extremely plentiful, in terms of both their variety and their overall number. From your knowledge of adaptive radiation, explain why there might be so many different types of

nematodes. _____

_____

_____

27.6   Rotifers

Rotifers (phylum Rotifera) are common and abundant freshwater animals. They are important constituents of the plankton of lakes, ponds, and streams, and are a significant food source for many species of fish and other animals.

Like roundworms, rotifers have a pseudocoelom and a complete digestive tract with a mouth and anus. The corona is a crown of cilia around the mouth. To some, the movement of the cilia resembles that of a rapid wheel; and in Latin, *rotifer* means "wheel-bearer." The cilia draw water into the mouth, and from there food is ground up by trophi (jaws) before entering the stomach from which nondigested remains pass through the cloaca and anus (Fig. 27.15).

Figure 27.15 Live *Philodina*, a common rotifer.
a. The contents of the digestive system take on a color in a live common rotifer.
b. Line art showing the complex anatomy of a rotifer.
(a) © dpa picture alliance/Alamy

a. Digestive system contents   250×

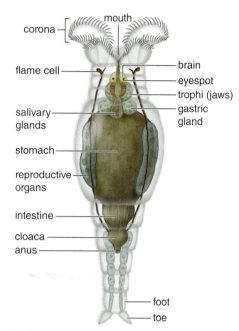

b. Complex anatomy

Observation: Rotifers

Living Rotifers

1. Use a pipet to obtain a living rotifer specimen near a clump of vegetation. Place the liquid and rotifer on a concave depression slide. Do not add a coverslip. Study the animal's behavior and appearance.

2. Describe the rotifer's behavior. _____

   _____

3. Observe the elongated, cylindrical body that can be divided into three general regions: the head, the trunk, and a posterior foot. The rotifers have no true segmentation, but the cuticle covering the body can be divided into a number of superficial segments.

4. Observe the telescoping of the segments when the animal retracts its head.

5. Note also the large, ciliated **corona** at the anterior end. Most rotifers have a conspicuous corona that serves both for locomotion and for feeding. It creates a current that brings smaller microorganisms (e.g., algae, protozoans, bacteria) close enough to be swallowed.

Laboratory Review 27

_____   1. What do invertebrates lack that is present in vertebrates?

_____   2. What is an incompletely lined body cavity called?

_____   3. What type of symmetry is associated with one plane of symmetry and definite right and left halves?

_____   4. What is the name of the main opening in a sponge through which water is expelled?

_____   5. What are the hard and sharp projections of a calcereous sponge called?

_____   6. What is the name given to the body form of a cnidarian that has a mouth directed upward?

_____   7. What kind of symmetry do organisms classified as cnidarians have?

_____   8. What is the fluid-filled capsule in a cnidocyte called?

_____   9. What structure in a planarian digests food and ensures that nutrients get to all the cells?

_____   10. What is the name given to the ciliated cells that perform excretion for a planarian?

_____   11. What structures enable a tapeworm to attach itself to the intestinal wall of its host?

_____   12. What are the anterior and posterior ends of a complete digestive tract called?

_____   13. Which of the phyla covered in this lab exercise have complete digestive tracts?

_____   14. Which phylum of animals covered in this lab exercise relies on a crown of cilia to draw water and food into the mouth?

Thought Questions

15. How does the process of acquiring and digesting food in cnidarians differ from the process in sponges?

16. The anterior end of an animal is often referred to as a head. The anterior end of a tapeworm is referred to as a scolex. Explain why a different word is used for the anterior end of the tapeworm.

17. **a.** Do the animals with a gastrovascular cavity have a body cavity?

b. How do the animals with a pseudocoelom digest their food?

c. What advantage is there to having a body cavity and a separate mouth and anus?

28

Invertebrate Coelomates

Learning Outcomes

Introduction

Animals are distinguished by their body cavity. Acoelomates have no body cavity, and pseudocoeelomates have a body cavity incompletely lined with mesoderm, the last of the germ layers to appear during the evolution of animals. The coelomates have a body cavity completely lined with mesoderm. A coelom offers many advantages:

- The digestive system and body wall can move independently.
- Internal organs can become more complex.
- Coelomic fluid can assist respiration, circulation, and excretion.
- The coelom also serves as a hydrostatic skeleton because muscles can work against a fluid-filled cavity.

All the phyla considered in today's laboratory are coelomates. Molluscs, annelids, and arthropods are protostomes, animals in which the first (*proto*) embryonic opening becomes the mouth (*stoma*). Echinoderms, studied in this lab, and vertebrates studied in Laboratory 29 are deuterostomes. In the deuterostomes, the first opening becomes the anus, and the second (*deutero*) opening becomes the mouth.

28.1 Molluscs

Most **molluscs** (phylum Mollusca) are marine, but there are also some freshwater and terrestrial molluscs (Fig. 28.1). Among molluscs, the grazing marine herbivores, known as **chitons,** have a body flattened dorsoventrally covered by a shell consisting of eight plates (Fig. 28.1*a*). The **bivalves** contain marine and freshwater sessile filter feeders, such as clams and scallops, with a body enclosed by a shell consisting of two valves (Fig. 28.1*b*). The **gastropods** contain marine, freshwater, and terrestrial species. In snails, the shell, if present, is coiled (Fig. 28.1*c*). The **cephalopods** contain marine active predators, such as squids and nautiluses. Tentacles are about the head (Fig. 28.1*d*).

All molluscs have a three-part body consisting of (1) a muscular **foot** specialized for various means of locomotion; (2) **visceral mass** that includes the internal organs; and (3) a **mantle,** a thin tissue that encloses the visceral mass and may secrete a shell. **Cephalization** is the development of a head region. On the lines provided in Figure 28.1, write *cephalization* or *no cephalization* as appropriate for this mollusc.

a. Chiton, *Tonicella*

c. Snail, *Helix* is a gastropod.

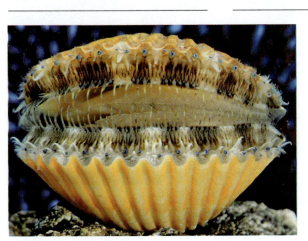

b. Scallop, *Aequipecten* is a bivalve.

d. Nautilus, *Nautilus,* is a cephalopod.

Figure 28.1 Molluscan diversity.
a. You can see the exoskeleton of this chiton but not its dorsally flattened foot. **b.** A scallop doesn't have a foot but it does have strong adductor muscles to close the shell. In this specimen, the edge of the mantle bears tentacles and many blue eyes. **c.** A gastropod, such as a snail, is named for the location of its large foot beneath the visceral mass. **d.** In a cephalopod, such as this nautilus, a funnel (its foot) opens in the area of the tentacles and allows it to move by jet propulsion.
(a) © Randimal/Shutterstock RF; (b) © NHPA/Photoshot; (c) © Rosemary Calvert/Getty Images; (d) © Douglas Faulkner/Science Source

Anatomy of a Clam

Clams are bivalved because they have right and left shells secreted by the mantle. Clams have no head, and they burrow in sand by extending a **muscular foot** between the valves. Clams are **filter feeders** and feed on debris that enters the mantle cavity. In the visceral mass, the blood leaves the heart and enters sinuses (cavities) by way of anterior and posterior aortas. There are many different types of clams. The one examined here is the freshwater clam *Venus.*

External Anatomy

1. Examine the external shell (Fig. 28.2) of a preserved clam (*Venus*). The shell is an **exoskeleton.**
2. Find the posterior and anterior ends. The more pointed end of the **valves** (the halves of the shell) is the posterior end.
3. Determine the clam's dorsal and ventral regions. The valves are hinged together dorsally.
4. What is the function of a heavy shell? _____

Figure 28.2   External view of the clam shell.

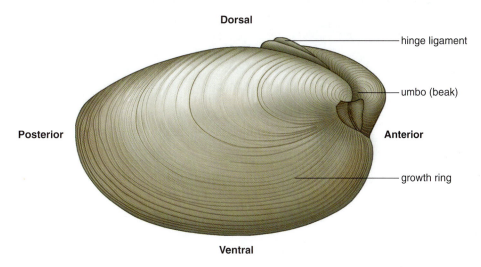

Internal Anatomy

1. Place the clam in the dissecting pan, with the **hinge ligament** and **umbo** (blunt dorsal protrusion) down. Carefully separate the **mantle** from the right valve by inserting a scalpel into the slight opening of the valves. What is a mantle? _____

_____

2. Insert the scalpel between the mantle and the valve you just loosened.
3. The **adductor muscles** hold the valves together. Cut the adductor muscles at the anterior and posterior ends by pressing the scalpel toward the dissecting pan. After these muscles are cut, the valve can be carefully lifted away. What is the advantage of powerful adductor muscles? _____

_____

4. Examine the inside of the valve you removed. Note the concentric lines of growth on the outside, the hinge teeth that interlock with the other valve, the adductor muscle scars, and the mantle line. The inner layer of the shell is mother-of-pearl.
5. Examine the rest of the clam (Fig. 28.3) attached to the other valve. Notice the adductor muscles and the mantle, which lies over the visceral mass and foot.
6. Bring the two halves of the mantle together. Explain the term *mantle cavity.* _____

_____

7. Identify the **incurrent** (more ventral) and **excurrent siphons** at the posterior end (Fig. 28.3).

   Explain how water enters and exits the mantle cavity. _____

_____

Figure 28.3 Anatomy of a bivalve.

The mantle has been removed to reveal the internal organs. **a.** Drawing. **b.** Dissected specimen.

(b) © Ken Taylor/Wildlife Images

a.

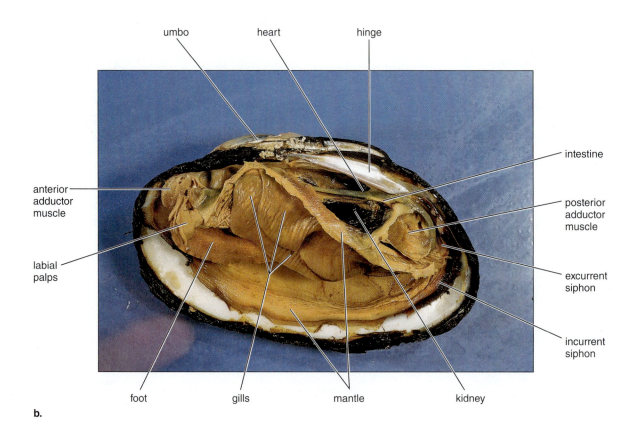

b.

8. Cut away the free-hanging portion of the mantle to expose the **gills.** Does the clam have a respiratory organ? _____ If so, what type of respiratory organ? _____

9. A mucous layer on the gills entraps food particles brought into the mantle cavity, and the cilia on the gills convey these food particles to the mouth. Why is the clam called a filter feeder? _____

_____

_____

10. The nervous system is composed of three pairs of ganglia (located anteriorly, posteriorly, and in the foot), all connected by nerves. The clam does not have a brain. A ganglion contains a limited number of neurons, whereas a brain is a large collection of neurons in a definite head region.

11. Identify the **foot,** a tough, muscular organ for locomotion, and the **visceral mass,** which lies above the foot and is soft and plump. The visceral mass contains the digestive and reproductive organs.

12. Identify the **labial palps** that channel food into the open mouth.

13. Identify the **anus,** which discharges into the excurrent siphon.

14. Find the **intestine** by its dark contents. Trace the intestine forward until it passes into a sac, the clam's only evidence of a coelom.

15. Locate the **pericardial sac (pericardium)** that contains the heart. The intestine passes through the heart. The heart pumps blood into the aortas, which deliver it to blood sinuses (open spaces) in the tissues.

    A clam has an **open circulatory system.** Explain your answer. _____

_____

16. Cut the visceral mass and the foot into exact left and right halves, and examine the cut surfaces. Identify the digestive glands, greenish-brown; the stomach, embedded in the digestive glands; and the intestine, which winds about in the visceral mass. Reproductive organs (gonads) are also present.

Anatomy of a Squid

Squids are cephalopods because they have a well-defined head; the foot became the funnel surrounded by two arms and the many tentacles about the head. The head contains a brain and bears sense organs. The squid moves quickly by jet propulsion of water, which enters the mantle cavity by way of a space that encircles the head. When the cavity is closed off, water exits by means of the funnel. Then the squid moves rapidly in the opposite direction.

The squid seizes fish with its tentacles; the mouth has a pair of powerful, beaklike jaws and a **radula,** a beltlike organ containing rows of teeth. The squid has a **closed circulatory system** composed of vessels and three hearts, one of which pumps blood to all the internal organs, while the other two pump blood to the gills located in the mantle cavity.

Observation: Anatomy of a Squid

1. Examine a preserved squid.
2. Refer to Figure 28.4 for help in identifying the mouth (defined by beaklike jaws and containing a radula) and the tentacles and arms, which encircle the mouth.
3. Locate the head with its sense organs, notably the large, well-developed eye.
4. Find the funnel, where water exits from the mantle cavity, causing the squid to move backward.
5. If the squid has been dissected, note the heart, gills, and blood vessels.

Figure 28.4 Anatomy of a squid.

The squid is an active predator and lacks the external shell of a clam. It captures fish with its tentacles and bites off pieces with its jaws. A strong contraction of the mantle forces water out the funnel, resulting in "jet propulsion." **a.** Drawing. **b.** Dissected specimen. (b) © Ken Taylor/Wildlife Images

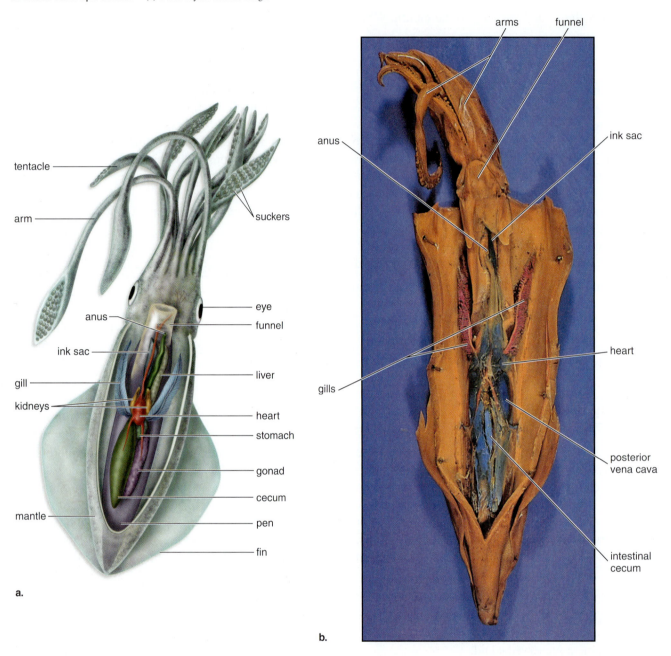

a.

b.

Conclusion: Comparison of Clam to Squid

- Compare clam anatomy to squid anatomy by completing Table 28.1.
- Explain how both clams and squids are adapted to their way of life. _____

_____

_____

_____

Table 28.1 Comparison of Clam to Squid

	Clam	Squid
Feeding mode		
Skeleton		
Circulation		
Cephalization		
Locomotion		

28.2 Annelids

Annelids (phylum Annelida) are the **segmented** worms, so called because the body is divided into a number of segments and has a ringed appearance. The circular and longitudinal muscles work against the fluid-filled coelom to produce changes in width and length (Fig. 28.5). Therefore, annelids are said to have a **hydrostatic skeleton.**

Figure 28.5 Locomotion in the earthworm.
Contraction of first circular muscles and then longitudinal muscles allow the earthworm to move forward.

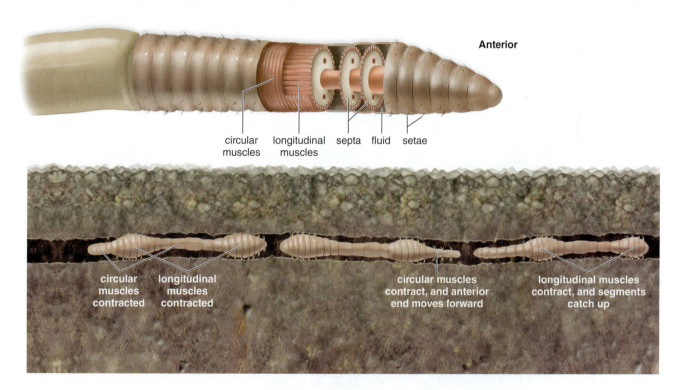

Among annelids (Fig. 28.6), **polychaetes** have many slender bristles called **setae.** The polychaetes, almost all marine, are plentiful from the intertidal zone to the ocean depths. They are quite diverse, ranging from jawed forms that are carnivorous to fanworms that live in tubes and extend feathery filaments when filter feeding. Earthworms are called **oligochaetes** because they have few setae. Earthworms, which have a worldwide distribution in almost any soil, are often present in large numbers, and may reach a length of as much as 3 meters. **Leeches,** annelids without setae, include the medicinal leech, which has been used in the practice of bloodletting for centuries. Most people simply called them bloodsuckers.

Show that the annelids are the segmented worms by *labeling a segment in 28.6a*, b, and d. In which group would you expect the animals to be predators based on the type of head region? _____

Figure 28.6 Annelid diversity.
Aside from **a.** earthworms, which are oligochaetes, there are **b.** clam worms and **c.** fan worms, both of which are polychaetes, and **d.** leeches. Note the obvious segmentation.

(a) © MikeLane45/iStock360/Getty RF; (b) © blickwinkel/Alamy; (c) © Borut Furlan/Getty Images; (d) © Bildagentur/age fotostock

head region

a. Earthworms, *Lumbricus*, mating

head
(sense organs and jaws)

b. Clam worm, *Nereis*

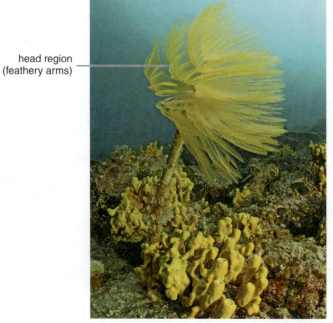

head region
(feathery arms)

c. European fan worm, *Sabella*

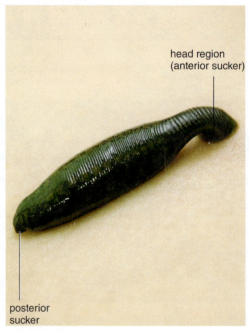

head region
(anterior sucker)

posterior
sucker

d. Leech, *Hirudo*

Anatomy of the Earthworm

Earthworms are segmented in that the body has a series of ringlike segments. Earthworms have no head, and they burrow in the soil by alternately expanding and contracting segments along the length of the body.

Earthworms are scavengers that feed on decaying organic matter in the soil. They have a well-developed coelomic cavity, providing room for a well-developed digestive tract and both sets of reproductive organs. Earthworms are **hermaphroditic.**

Observation: Anatomy of the Earthworm

External Anatomy

1. Examine a live or preserved specimen of an earthworm. Locate the small projection that sticks out over the mouth. Has cephalization occurred? _____ Explain your answer. _____

_____

2. Count the total number of segments, beginning at the anterior end. The sperm duct openings are on segment 15 (somite XV) (Fig. 28.7). The enlarged section around a short length of the body is the **clitellum.** The clitellum secretes mucus that holds the worms together during mating. It also functions as a cocoon, in which fertilized eggs hatch and young worms develop. The anus is located on the worm's terminal segment.

3. Lightly pass your fingers over the earthworm's ventral and lateral sides. Do you feel the setae? _____

_____

Earthworms insert these slender bristles into the soil. Setae, along with circular and longitudinal muscles, enable the worm to locomote. Explain the action. _____

_____

Figure 28.7   External anatomy of an earthworm.
In this drawing, the segments are numbered.

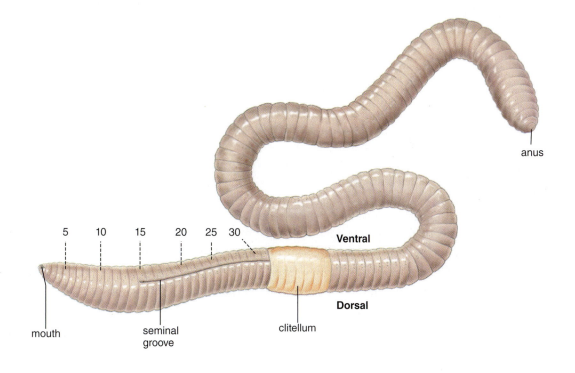

Internal Anatomy

1. If so directed by your instructor, place a preserved earthworm on its ventral side in the dissecting pan and follow these directions; otherwise proceed immediately to "Prepared Slide" on page 408. With a scalpel or razor blade, make a shallow incision slightly to the side of the blackish median dorsal blood vessel (Fig. 28.8a). Start your incision about 10 segments after the clitellum, and proceed anteriorly to the mouth. If you see black ooze, you have accidentally cut the intestine.

2. Identify the thin partitions, the **septa,** between segments.

3. Lay out the body wall, and pin every 10th segment to the wax in your pan. Add water to prevent drying out. While alive, the earthworm body wall is always moist, and this facilitates gas exchange across the body wall. Notice the earthworm has no respiratory organ.

4. An earthworm feeds on **detritus** (organic matter) in the soil. Identify the digestive tract, which begins at the mouth and extends through body segments 1, 2, and 3 (Fig 28.8a and b). It opens into the swollen, muscular, thick-walled **pharynx,** which extends from segment 3 to segment 6. The **esophagus** extends from the pharynx through segment 14 to the **crop.** Next is the **gizzard,** which lies in segments 17 through 28. The intestine extends from the gizzard to the anus. Does the digestive system show specialization

   of parts?_____ Explain your answer. _____

   _____

   _____

5. Identify the earthworm's circulatory system. The blood is always contained within vessels and never runs free. The **dorsal blood vessel** is readily seen along the dorsal side of the digestive tract. A series of aortic arches or "hearts" encircles the esophagus between segments 7 and 11, connecting the dorsal blood vessel with the ventral blood vessel. Does the earthworm have an open or closed circulatory system? _____

   Explain your answer. _____

   _____

6. Locate the earthworm's nervous system. The two-lobed brain is located on the dorsal surface of the pharynx in segment 3. Two nerves, one on each side of the pharynx, connect the brain to a ganglion that lies below the pharynx in segment 4. The **ventral nerve cord** then extends along the floor of the body cavity to the last segment.

7. Find the earthworm's excretory system, which consists of a pair of minute, coiled, white tubules, the **nephridia,** located in every segment except the first three and the last. Each nephridium opens to the outside by means of an excretory pore. Does the excretory system show that the earthworm is segmented? _____ Explain your

   answer. _____

   _____

   _____

8. Identify the earthworm's reproductive system, including **seminal vesicles,** light-colored bodies in segments 9 through 12, which house maturing sperm that have been formed in two pairs of testes within them; **sperm ducts** that pass to openings in segment 15; and **seminal receptacles** (four small, white, special bodies that lie in segments 9 and 10), which store sperm received from another worm. **Ovaries** are located in segment 13 but are too small to be seen.

9. During mating, earthworms are arranged so that the sperm duct openings of one worm are just about, but not quite, opposite the seminal receptacle openings of the other worm. After being released, the sperm pass down a pair of seminal grooves on the ventral surface (see Fig. 28.7) and then cross over at the level of the seminal receptacles of the opposite worm. Once the worms separate, eggs and sperm are released

   into a cocoon secreted by the clitellum. Is the earthworm hermaphroditic? _____ Explain your

   answer. _____

   _____

10. Does the earthworm have a respiratory organ or system (i.e., gills or lungs)? _____ How does the worm exchange gases? _____

_____

Why would you expect an earthworm to lack an exoskeleton? _____

_____

Figure 28.8 Internal anatomy of an earthworm, dorsal view.
a. Drawing shows internal organs and a cross section. The segments are numbered. **b.** Dissected specimen.
(b) © Ken Taylor/Wildlife Images

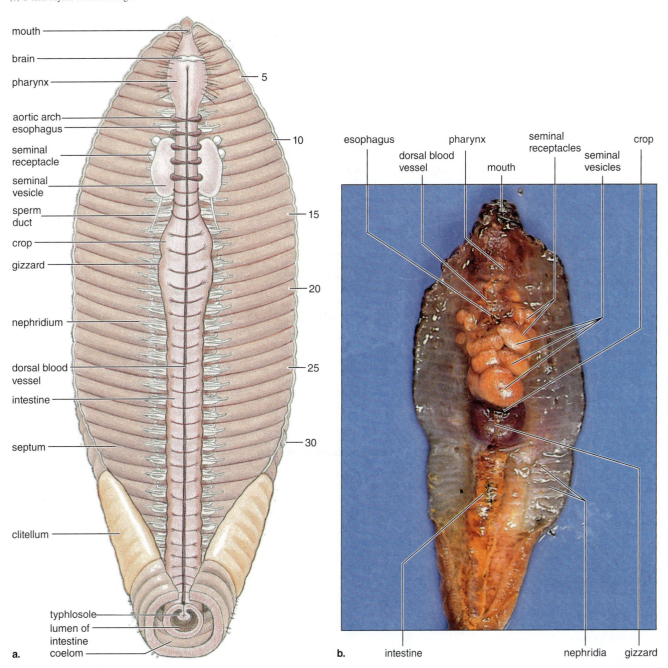

a.

b.

Prepared Slide

1. Obtain a prepared slide of a cross section of an earthworm (Fig. 28.9). Examine the slide under the dissecting microscope and under the light microscope.
2. Identify the following structures.
   a. **Body wall:** A thick outer circle of tissue, consisting of the **cuticle** and the **epidermis.**
   b. **Coelom:** A relatively clear space with scattered fragments of tissue.
   c. **Intestine:** An inner circle with a suspended fold.
   d. **Typhlosole:** A fold that increases the intestine's surface area.
   e. **Ventral nerve cord:** A white, threadlike structure.

3. Does the typhlosole help in nutrient absorption? _____ Explain your answer. _____

Figure 28.9 Cross section of an earthworm.
Cross-section slide as it would appear under the microscope.
© Carolina Biological Supply Company/Phototake

5×

Conclusion: Comparison of Clam to Earthworm

- Complete Table 28.2 to compare the anatomy of a clam to that of an earthworm.

Table 28.2 Comparison of Clam to Earthworm		
	Clam	**Earthworm**
Habitat (where they live)		
Feeding mode		
Skeleton		
Segmentation		
Circulation		
Respiratory organs		
Locomotion		
Reproductive organs		

28.3 Arthropods

Arthropods (phylum Arthropoda) have paired, jointed appendages and a hard exoskeleton that contains chitin. The chitinous exoskeleton consists of hardened plates separated by thin, membranous areas that allow movement of the body segments and appendages.

Figure 28.10*a* features insects and relatives. **Insects** with three pairs of legs, with or without wings, and three distinct body regions comprise 95% of all arthropods. **Millipedes** have two pairs of legs per segment, while **centipedes** have one pair of legs per segment. Figure 28.10*b* features spiders and relatives. Spiders and scorpions have four pairs of legs, no antennae, and a cephalothorax (head and thorax are fused). The horseshoe crab is a living fossil. It has remained unchanged for thousands of years. The **crustaceans** (Fig. 28.10*c*), which include crabs, shrimp, and lobsters, have three to five pairs of legs, and two pairs of antennae. Barnacles are unusual, in that their legs are used to gather food.

For each animal in Figure 28.10, circle the obvious types of appendages.

Figure 28.10 Arthropod diversity.

a. Among arthropods, insects, millipedes, and centipedes are possibly related. **b.** Spiders, scorpions, and horseshoe crabs are related. **c.** Crabs, shrimp, and barnacles, among others, are crustaceans.

(a) (honeybee) © USDA-ARS/Stephen Ausmus; (millipede) © blickwinkel/Alamy; (centipede) © Larry Miller/Science Source; (b) (spider) © Horia Bogdan/Getty RF; (scorpion) © Papilio/Alamy; (horseshoe crab) © U.S. Fish & Wildlife Service/Robert Pos; (c) (crab) © Tom McHugh/Science Source; (shrimp) © imageBroker/Alamy; (barnacles) © L. Newman & A. Flowers/Science Source

a. Insects and relatives

Honeybee, *Apis mellifera*

Millipede, *Rhapidostreptus virgator*

Centipede, *Scolopendra* sp.

b. Spider and relatives

Spider, *Argiope bruennichi*

Scorpion, *Hadrurus arizonensis*

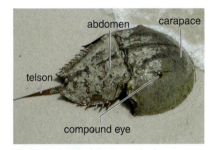

Horseshoe crab, *Limulus polyphemus*

c. Crustaceans

Crab, *Cancer productus*

Shrimp, *Lysmata grabhami*

Barnacles, *Lepas anatifera*

Anatomy of a Crayfish

Crayfish also belong to the group of arthropods called crustaceans. Crayfish are adapted to an aquatic existence. They are known to be scavengers, but they also prey on other invertebrates. The mouth is surrounded by appendages modified for feeding, and there is a well-developed digestive tract. Dorsal, anterior, and posterior arteries carry **hemolymph** (blood plus lymph) to tissue spaces (hemocoel) and sinuses. Therefore, a crayfish has an open circulatory system.

Observation: Anatomy of a Crayfish

External Anatomy

1. In a preserved crayfish, identify the chitinous **exoskeleton.** With the help of Figure 28.11, identify the head, thorax, and abdomen. Together, the head and thorax are called the **cephalothorax;** the

Figure 28.11
Anatomy of a crayfish.
a. Drawing shows external anatomy. **b.** Internal anatomy of a female crayfish.

(b) © Ken Taylor/Wildlife Images

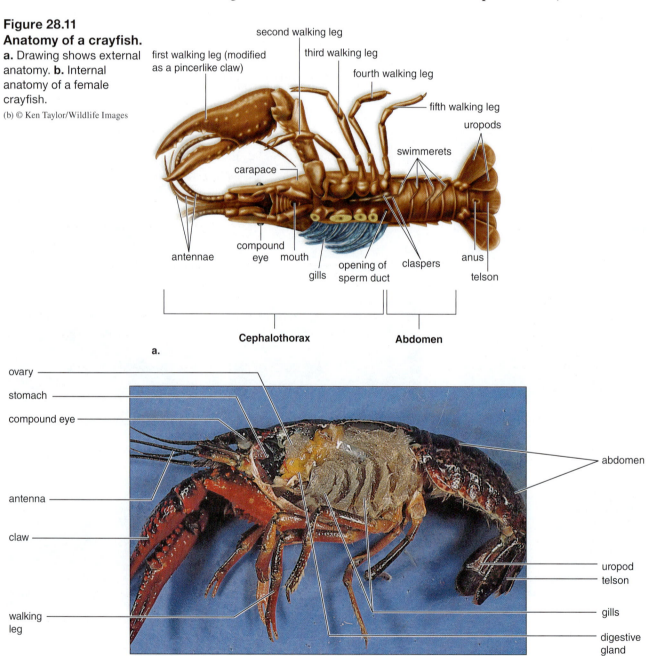

cephalothorax is covered by the **carapace.** Has specialization of segments occurred? _____

Explain your answer. _____

2. Find the **antennae,** which project from the head. At the base of each antenna, locate a small, raised nipple containing an opening for the **green glands,** the organs of excretion. Crayfish excrete a liquid nitrogenous waste.

3. Locate the **compound eyes,** composed of many individual units for sight. Do crayfish demonstrate

cephalization? _____ Explain your answer. _____

4. Identify the six pairs of appendages around the mouth for handling food.

5. Find the five pairs of walking legs attached to the cephalothorax. The most anterior pair is modified as pincerlike claws. In the female, identify the **seminal receptacles,** a swelling located between the bases of the third and fourth pairs of walking legs. Sperm from the male are deposited in the seminal receptacles. In the male, identify the opening of the sperm duct located at the base of the fifth walking leg.

6. Locate the five pairs of **swimmerets** on the abdomen. Crayfish use their swimmerets to swim and move water over their gills. In males, the anterior two pairs are stiffened and folded forward. They are claspers that aid in the transfer of sperm during mating. In females, clusters of eggs and larvae attach to the swimmerets.

7. Find the last abdominal segment, which bears a pair of broad, fan-shaped **uropods** that, together

with a terminal extension of the body, form a tail. Has specialization of appendages occurred? _____

Explain your answer. _____

Internal Anatomy

1. Cut away the lateral surface of the carapace with scissors to expose the **gills** (Fig. 28.11*b*). Observe that

the gills occur in distinct, longitudinal rows. How many rows of gills are there in your specimen? _____

The outer row of gills is attached to the base of certain appendages. Which ones? _____

_____

2. Remove a gill with your scissors by cutting it free near its point of attachment, and place it in a watch glass filled with water. Observe the numerous gill filaments arranged along a central axis.

3. Carefully cut away the dorsal surface of the carapace with scissors and a scalpel. The epidermis that adheres to the exoskeleton secretes the exoskeleton. Remove any epidermis adhering to the internal organs.

4. Identify the diamond-shaped heart lying in the middorsal region. A crayfish has an open circulatory system. Carefully remove the heart.

5. Locate the **gonads** anterior to the heart in both the male and female. The gonads are tubular structures bilaterally arranged in front of the heart and continuing behind it as a single mass. In the male, the testes are highly coiled, white tubes.

6. Find the **mouth;** the short, tubular **esophagus;** and the two-part **stomach,** with the attached **digestive gland,** that precedes the intestine.

7. Identify the **green glands,** two excretory structures just anterior to the stomach, on the ventral segment wall.

8. Remove the thoracic contents previously identified.

9. Identify the **brain** in front of the esophagus. The brain is connected to the ventral nerve cord by a pair of nerves that pass around the esophagus.

10. Remove the animal's entire digestive tract, and float it in water. Observe the various parts, especially the connections of the digestive gland to the stomach.

11. Cut through the stomach, and notice in the anterior region of the stomach wall the heavy, toothlike projections, called the **gastric mill,** which grind up food. Do you see any grinding stones ingested by the

crayfish? _____

If possible, identify what your specimen had been eating. _____

Anatomy of a Grasshopper

The grasshopper is an **insect.** All insects have a head, a thorax, and an abdomen. Their appendages always include (1) three pairs of jointed legs and usually (2) two antennae as sensory organs. Grasshoppers are adapted to live on land. Wings and jumping legs are suitable for locomotion on land; **Malpighian tubules** save water by secreting a solid nitrogenous waste; the **tracheae** are tiny tubules that deliver air directly to the muscles; and the male has a penis with attached claspers to deliver sperm to the seminal receptacles of a female so they do not dry out.

Observation: Anatomy of a Grasshopper

External Anatomy

1. Obtain a preserved grasshopper (*Romalea*), and study its external anatomy with the help of Figure 28.12a. Identify the head, thorax, and abdomen.
2. Use a hand lens or dissecting microscope to examine the grasshopper's special sense organs of the **head.** Identify the **antennae** (a pair of long, jointed feelers), the **compound eyes,** and the three dotlike **simple eyes.** The labial palps, labeled in Figure 28.12a, have sense organs for tasting food.
3. Note the sturdy **mouthparts,** which are used for chewing plant material. A grasshopper's mouthparts are quite different from those of a piercing and sucking insect.
4. Locate the leathery **forewings** and the inner, membranous **hindwings** attached to the **thorax.** Which pair of legs is used for jumping? _____ How many segments does each leg have? _____
5. Is locomotion in the grasshopper adapted to land? _____

   Explain your answer. _____

   _____

6. In the **abdomen,** identify the **tympana** (sing., **tympanum**), one on each side of the first abdominal segment (Fig. 28.12a). The grasshopper detects sound vibrations with these membranes.
7. Locate the **spiracles,** along the sides of the abdominal segments. These openings allow air to enter the tracheae, which constitute the respiratory system.
8. Find the **ovipositors** (Figs. 28.12a and 28.13a), four curved and pointed processes projecting from the abdomen of the female. These are used to dig a hole in which eggs are laid. The male has a **penis** with **claspers** used during copulation (Fig. 28.13b).

Figure 28.12   Female grasshopper.
a. External anatomy. **b.** Internal anatomy.

a. External anatomy

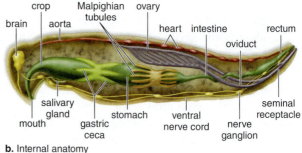

b. Internal anatomy

Figure 28.13 Grasshopper genitalia.

a. Females have an ovipositor, and **b.** males have claspers at the distal end of the penis.

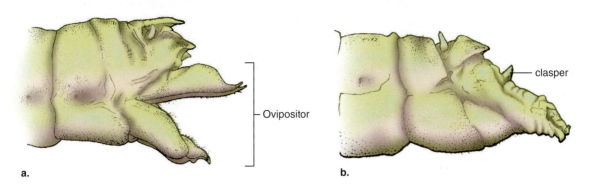

a.

b.

— clasper

— Ovipositor

Internal Anatomy

Observe a longitudinal section of a grasshopper if available on demonstration. Try to locate the structures shown in Figure 28.12*b*. What is the function of Malpighian tubules?_____

Conclusion: Comparison of Crayfish to Grasshopper

Compare the adaptations of a crayfish to those of a grasshopper by completing Table 28.3. Put a star beside each item that indicates an adaptation to life in the water (crayfish) and to life on land (grasshopper). Check with your instructor to see if you identified the maximum number of adaptations.

Table 28.3 Comparison of Crayfish to Grasshopper		
	Crayfish	**Grasshopper**
Locomotion		
Respiration		
Sense organs		
Nervous system		

External reproductive features
    Male
    Female		

Grasshopper Metamorphosis

Metamorphosis means a change, usually a drastic one, in form and shape. Grasshoppers undergo *incomplete metamorphosis,* a gradual change in form rather than a drastic change. The immature stages of the grasshopper are called **nymphs,** and they are recognizable as grasshoppers even though they differ somewhat in shape and form (Fig. 28.14*a*). Some insects undergo what is called *complete metamorphosis,* in which case they have three stages of development: **larvae, pupa,** and **adult** (Fig. 28.14*b*). Metamorphosis occurs during the pupa stage when the animal is enclosed within a hard covering. The animals best known for complete metamorphosis are the butterfly and the moth, whose larval stage is called a caterpillar and whose pupa stage is the cocoon; the adult is the butterfly or moth.

Figure 28.14 Metamorphosis.

a. During incomplete metamorphosis of a grasshopper, a series of **nymphs** leads to a full-grown grasshopper. **b.** During complete metamorphosis of a moth, a series of larvae lead to pupation. The **adult** hatches out of the pupa.

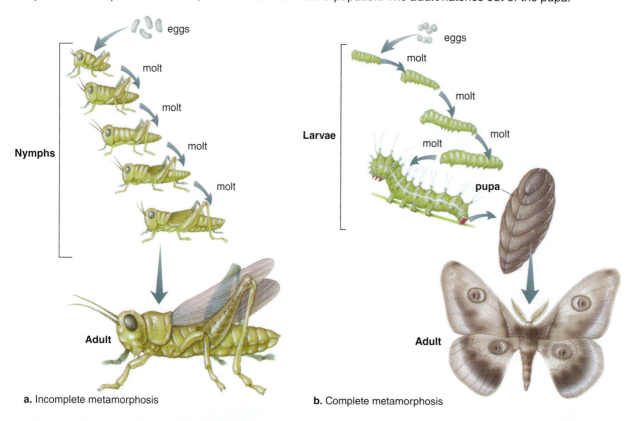

a. Incomplete metamorphosis

b. Complete metamorphosis

Observation: Grasshopper Metamorphosis

1. Use Figure 28.14 to add the grasshopper and the moth to Table 28.4.
2. Examine any specific life cycle displays or plastomounts that illustrate complete and incomplete metamorphosis and add these examples to Table 28.4.

Table 28.4 Insect Metamorphosis	
Specimen	**Complete or Incomplete Metamorphosis**

Conclusions: Insect Metamorphosis

- In insects with incomplete metamorphosis, do the nymphs or the adult have better developed wings? _____ What is the benefit of wings to an insect?_____
_____

- What stage is missing when an insect has incomplete metamorphosis? _____ What happens during this stage? _____

- What form, the larvae or the adult, disperses new individuals in flying insects that exhibit complete metamorphosis? _____

  How is this a benefit? _____

  _____

- In insects that undergo complete metamorphosis, the larvae and the adults utilize different food sources and habitats. Why might this be a benefit? _____

  _____

28.4 Echinoderms

The echinoderms (phylum Echinodermata) are the only invertebrate group that shares deuterostome development with the vertebrates (see Fig. 27.1). Unlike the vertebrates, all **echinoderms** are marine, and they dwell on the seabed, either attached to it, like sea lilies, or creeping slowly over it. The name *echinoderm* means "spiny-skinned," and most members of the group have defensive spines on the outside of their bodies. The spines arise from an **endoskeleton** composed of calcium carbonate plates. The endoskeleton supports the body wall and is covered by living tissue that may be soft (as in sea cucumbers) or hard (as in sea urchins).

Especially note that (1) adult echinoderms have radial symmetry, with generally five points of symmetry arranged around the axis of the mouth but the larvae have bilateral symmetry, and (2) the echinoderms' most unique feature is their **water vascular system.** In those echinoderms in which the arms make contact with the substratum, the **tube feet** associated with the water vascular system are used for locomotion. In other echinoderms, the tube feet are used for gas exchange and food gathering.

Echinoderms belong to one of five groups: sea lilies and feather stars; sea stars; brittle stars; sea urchins and sand dollars; and sea cucumbers (Fig. 28.15). *Where appropriate in Figure 28.15, write "ORS" for obvious radial symmetry or "RSNO" for radial symmetry not obvious on the lines provided.*

Figure 28.15 Echinoderm diversity.

(a) © Borut Furlan/Getty Images; (b) © Jonathan Bird/Getty Images; (c); © Robert Dunne/Science Source; (d) © Neil McDan/Science Source; (e) © Andrew J. Martinez/ Science Source; (f) © Andrey Nekrasov/Alamy

a. Feather star, *Oxymetra erinacea* _____

b. Sea star, *Odontaster validus* _____

c. Brittle star, *Ophiopholis aculeata* _____

d. Sea urchin, *Stronglocentrotus pranciscanus* _____

e. Sand dollar, *Echinarachnius parma* _____

f. Sea cucumber, *Apostichopus japonicus* _____

Anatomy of a Sea Star

Sea stars (starfish) usually have five arms that radiate from a central disk. The mouth is normally oriented downward, and when sea stars feed on clams, they use the suction of their tube feet to force the shells open a crack. Then they evert the cardiac portion of the stomach, which releases digestive juices into the mantle cavity. Partially digested tissues are taken up by the pyloric portion of the stomach; digestion continues in this portion of the stomach and in the digestive glands found in the arms.

Observation: Anatomy of a Sea Star

External Anatomy

1. Place a preserved sea star in a dissecting pan so that the aboral side is uppermost.
2. With the help of Figure 28.16, identify the **central disk** and five arms. What type of symmetry does an adult sea star have? _____

Figure 28.16   Anatomy of a sea star.
a. Diagram and **(b)** image of dissected sea star. Both show the aboral side. **c.** Image of cut arm. **d.** Canals and tube feet of water vascular system. Seen from aboral side.   (b) © BiologyImaging.com; (c) © BiologyImaging.com

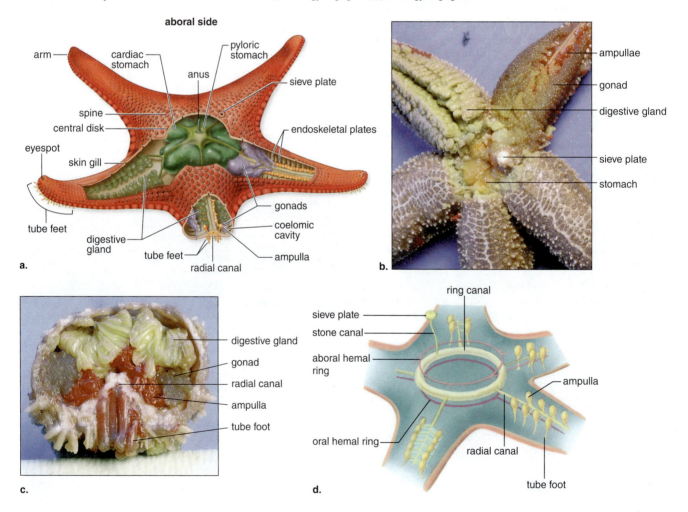

3. With the oral side uppermost, find the mouth, located in the center and protected by spines. Why is this side of the sea star called the oral side? _____

4. Locate the groove that runs along the middle of each arm and the **tube feet** (suctionlike disks) in rows on either side of the groove. Pluck away the tube feet from one area. How many rows of feet are there? _____

5. Turn the sea star over to its aboral side.

6. Locate the anal opening (Fig. 28.16). Why is this side of the sea star called the aboral side? _____
_____

7. Lightly run your fingers over the spines extending from calcium carbonate plates that lie buried in the body wall beneath the surface. The plates form an endoskeleton of the animal.

8. Identify the **sieve plate** (**madreporite**), a brownish, circular spot between two arms where water enters the water vascular system.

Internal Anatomy

1. Place the sea star so that the aboral side is uppermost. Refer to Figure 28.16a and b as you dissect the sea star following these instructions:

2. Cut the tip of one of the arms and, with scissors, carefully cut through the body wall along each side of this arm.

3. Carefully lift up the upper body wall. Separate any internal organs that may be adhering so that all internal organs are left intact.

4. Cut off the body wall near the central disk, but leave the sieve plate (madreporite) in place.

5. Remove the body wall of the central disk, being careful not to injure the internal organs.

6. Identify the digestive system. The mouth leads into a short **esophagus,** which is connected to the saclike **cardiac stomach.** When a sea star eats, the cardiac stomach sticks out through the sea star's mouth and starts digesting the contents of a clam or oyster. Above the cardiac stomach is the **pyloric stomach,** which leads to a short intestine. Each arm contains one pair of **digestive glands.**

   To which stomach do the digestive glands attach? _____

7. Cut off a portion of an arm, and examine the cut edge (Fig. 28.16c). Identify the digestive glands, ambulacral groove, radial canal, ampullae, and tube feet.

8. Remove the digestive glands of one arm.

9. Identify the **gonads** extending into the arm. What is the function of gonads? _____
_____

   It is not possible to distinguish male sea stars from females by this observation.

10. Remove both stomachs.

11. In the **water vascular system** (Fig. 28.16d), you have already located the sieve plate and tube feet. Now try to identify the following components:
    a. **Stone canal:** Takes water from the sieve plate to the ring canal.
    b. **Ring canal:** Surrounds the mouth and takes water to the radial canals.
    c. **Radial canals:** Send water into the ampulla. When the ampullae contract, water enters the tube feet. Each tube foot has an inner muscular sac called an ampulla. The ampulla contracts and forces water into the tube foot.

   What is the function of the water vascular system? _____
_____

_____ 1. What kind of body cavity is characteristic of the animals in the phyla covered in this lab exercise?

_____ 2. Is an animal a protostome or a deuterostome if the first opening becomes the mouth?

_____ 3. What do all molluscs have in addition to a muscular foot and a mantle?

_____ 4. What characteristic does an animal with a definite head region have?

_____ 5. What organ is found inside the pericardial sac (pericardium)?

_____ 6. What kind of skeleton do annelids have?

_____ 7. To what system do the crop and gizzard of an earthworm belong?

_____ 8. What is the location of the nerve cord in an earthworm?

_____ 9. What phylum of animals have chitinous exoskeletons?

_____ 10. What are the excretory structures in a crayfish called?

_____ 11. How is air taken directly to the muscles in a grasshopper?

_____ 12. What system of the echinoderms enables them to move about?

_____ 13. What kind of symmetry does an adult starfish have?

_____ 14. Where does water enter the water vascular system in a starfish?

Thought Questions

15. Explain the presence of specialized digestive organs in the animals studied in this laboratory given that they all have a coelom and a complete digestive system with a separate mouth and anus.

16. Explain why an earthworm will die if it dries out based on the type of skeleton annelids have and the absence of a respiratory organ or system in the earthworm.

17. Contrast an arthropod, the crayfish or the grasshopper, with the representative echinoderm, the starfish, with regard to signs of cephalization and how the type of symmetry each has is related.

29
The Vertebrates

Introduction

Vertebrates are **chordates.** All chordates have (1) a dorsal tubular nerve cord; (2) a dorsal supporting rod, called a notochord, at some time in their life history; (3) a postanal tail (e.g., tailbone or coccyx); and, (4) pharyngeal pouches. In chordates that breathe by means of gills, these pouches become gill slits. In terrestrial chordates, the pharyngeal pouches are modified for other purposes.

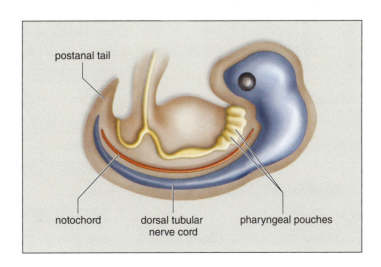

postanal tail

notochord

dorsal tubular nerve cord

pharyngeal pouches

29.1 Evolution of Chordates

The evolutionary tree of the chordates (Fig. 29.1) shows that some chordates, notably the **tunicates** and **lancelets,** are not vertebrates. These chordates retain the notochord and are called the **invertebrate chordates.**

Explain the term invertebrate chordates. _____

    The other animal groups in Figure 29.1 are **vertebrates** in which the notochord has been replaced by the vertebral column. Fishes include three groups: the **jawless fishes** were the first to evolve, followed by the **cartilaginous fishes** and then the **bony fishes.** The bony fishes include the **ray-finned** fishes (the largest group of vertebrates) and the **lobe-finned fishes.** The first lobe-finned fishes had a bony skeleton, fleshy appendages, and a lung. These lobe-finned fishes lived in shallow pools and gave rise to the amphibians. What three features called out in Figure 29.1 evolved among fishes? _____

_____

    The terrestrial vertebrates are all **tetrapods** because they have four limbs. The limbs of tetrapods are _____ appendages just like those of arthropods. Amphibians still return to the water to reproduce, but **reptiles** are fully adapted to life on land because among other features they produce an **amniotic egg.** The amniotic egg is so named because the embryo is surrounded by an amniotic membrane that encloses amniotic fluid. Therefore, amniotes develop in an aquatic environment of their own making. Do all animals develop in a water environment? _____ Explain. _____

_____

In placental mammals, such as humans, the fertilized egg develops inside the female, where the unborn mammal receives nutrients via the placenta. Reptiles (including birds) and mammals have many other adaptations that are suitable to living on land, as we will stress in later sections.

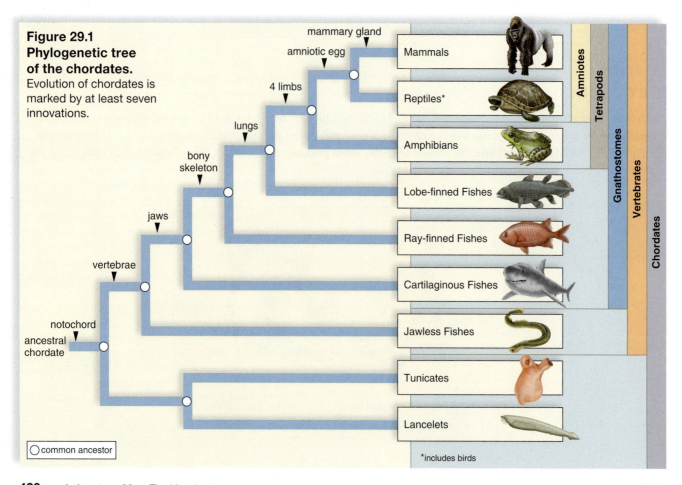

Figure 29.1 Phylogenetic tree of the chordates.
Evolution of chordates is marked by at least seven innovations.

29.2 Invertebrate Chordates

Among the chordates (phylum Chordata), two groups contain invertebrates, and the others contain the vertebrates.

Invertebrate Chordates

The two types of invertebrate chordates are urochordates and cephalochordates.

1. **Urochordates.** The tunicates, or sea squirts (Fig. 29.2), come in varying sizes and shapes, but all have incurrent and excurrent siphons. **Gill slits** are the only remaining chordate characteristic in adult tunicates. Examine any examples of tunicates on display.

2. **Cephalochordates.** Lancelets, also known as amphioxus (*Branchiostoma*), are small, fishlike animals that occur in shallow marine waters in most parts of the world. They spend most of their time buried in the sandy bottom, with only the anterior end projecting.

Figure 29.2 Urochordates.
The gill slits of a tunicate are the only chordate characteristic remaining in the adult.   © Andrew J. Martinez/Science Source

Observation: Lancelet Anatomy

Preserved Specimen

1. Examine a preserved lancelet (Fig. 29.3).
2. Identify the **caudal fin** (enlarged tail) used in locomotion, the **dorsal fin,** and the short **ventral fin.**
3. Examine the lancelet's ∨-shaped muscles.

Figure 29.3   Anatomy of the lancelet, *Branchiostoma*.
Lancelets feed on microscopic particles filtered out of the constant stream of water that enters the mouth and exits through the gill slits into a protective atrium formed by body folds. The water exits at the atriopore.   © Heather Angel/Natural Visions

4. Find the tentacled **oral hood,** located anterior to the mouth and covering a vestibule. Water entering the mouth is channeled into the **pharynx,** where food particles are trapped before the water exits at the **atriopore.** Lancelets are filter feeders. Has cephalization occurred? _____

Explain your answer. _____

_____

29.3 Vertebrates

In vertebrates (subphylum Vertebrata) the embryonic notochord is replaced by a vertebral column composed of individual vertebrae that protect the nerve cord. The internal jointed skeleton consists not only of a vertebral column, but also of a skull that encloses and protects the well-developed brain. There is an extreme degree of cephalization with complex sense organs. The eyes develop as outgrowths of the brain. The ears are primarily equilibrium devices in aquatic vertebrates; in land vertebrates, they also function as sound wave receivers.

The vertebrates are extremely motile and have well-developed muscles and usually paired appendages. They have bilateral symmetry and are segmented, as witnessed by the vertebral column. There is a large body cavity, a complete gut with both a mouth and an anus (or instead, a cloacal opening), and the circulatory system consists of a well-developed heart and many blood vessels. They have an efficient means of extracting oxygen from water (gills) or air (lungs) as appropriate. The kidneys are important excretory and water-regulating organs that conserve or rid the body of water as necessary. The sexes are generally separate, and reproduction is usually sexual.

Figure 29.4 Vertebrate groups.

Cartilaginous fishes
Lack operculum and swim bladder; tail fin usually asymmetrical (sharks, skates, and rays)

Tiger shark

Bony fishes
Operculum; swim bladder or lungs; tail fin usually symmetrical: lung-fishes, lobe-finned fishes, and ray-finned fishes (herring, salmon, sturgeon, eels, and sea horse)

Banded butterflyfish

Amphibians
Tetrapods with nonamniotic egg; nonscaly skin; some show metamorphosis; three-chambered heart (salamanders, frogs, and toads)

Northern leopard frog

Reptiles
Tetrapods with amniotic egg; scaly skin (snakes, lizards, turtles, and tortoises)

Pearl River redbelly turtle

Birds
Now grouped with reptiles; tetrapods with feathers; bipedal with wings; double circulation (sparrows, penguins, and ostriches)

Scissor-tailed flycatcher

Mammals
Tetrapods with hair, mammary glands; double circulation; teeth differentiated: monotremes (spiny anteater and duckbill platypus), marsupials (opossum and kangaroo), and placental mammals (whales, rodents, dogs, cats, elephants, horses, bats, and humans)

Gray fox

Anatomy of the Frog

In this laboratory the anatomy of the frog will be considered typical of vertebrates. Frogs are amphibians, a group of animals in which metamorphosis occurs. Metamorphosis includes a change in structure, as when an aquatic tadpole becomes a frog with lungs and limbs (Fig. 29.5). Amphibians were the first vertebrates to be adapted to living on land; however, they typically return to the water to reproduce. Underline every structure mentioned in the following Observation that represents an adaptation to a land environment.

Figure 29.5 External frog anatomy.

© Rod Planck/Science Source

lateral fold external naris eye shank hindlimb forelimb foot tympanum

Observation: External Anatomy of the Frog

1. Place a preserved frog (*Rana pipiens*) in a dissecting tray.
2. Identify the bulging eyes, which have a nonmovable upper and lower lid but can be covered by a **nictitating membrane** that serves to moisten the eye.
3. Locate the **tympanum** behind each eye (Fig. 29.5). What is the function of a tympanum? _____
   _____
4. Examine the external **nares** (sing., **naris,** or **nostril**). Insert a probe into an external naris, and observe that it protrudes from one of the paired small openings, the internal nares (Fig. 29.6), inside the mouth cavity.

   What is the function of the nares? _____
5. Identify the paired limbs. The bones of the fore- and hindlimbs are the same as in all tetrapods, in that the first bone articulates with a girdle and the limb ends in phalanges. The hind feet have five phalanges, and the forefeet have only four phalanges. Which pair of limbs is longest? _____

   How does a frog locomote on land? _____
   _____

   What is a frog's means of locomotion in the water? _____ _____
   _____

Observation: Internal Anatomy of the Frog

Mouth

1. Open your frog's mouth very wide (Fig. 29.6), cutting the angles of the jaws if necessary.
2. Identify the tongue attached to the lower jaw's anterior end.
3. Find the **auditory (eustachian) tube** opening in the angle of the jaws. These tubes lead to the ears. Auditory tubes equalize air pressure in the ears.
4. Examine the **maxillary teeth** located along the rim of the upper jaw. Another set of teeth— **vomerine teeth**—is present just behind the midportion of the upper jaw.
5. Locate the **glottis,** a slit through which air passes into and out of the **trachea,** the short tube from glottis to lungs. What is the

   function of a glottis? _____
   _____

6. Identify the **esophagus,** which lies dorsal and posterior to the glottis and leads to the stomach.

a.

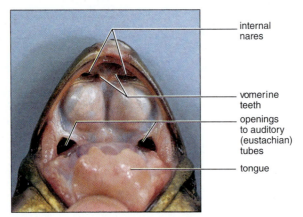

b.

Figure 29.6 Mouth cavity of a frog.
a. Drawing. b. Dissected specimen.
(b) © Carolina Biological Supply/Phototake

Opening the Frog

1. Place the frog ventral side up in the dissecting pan. Lift the skin with forceps, and use scissors to make a large, circular cut to remove the skin from the abdominal region as close to the limbs as possible. Cut only skin, not muscle.
2. Now, remove the muscles by cutting through them in the same circular fashion. At the same time, cut through any bones you encounter. A vein, called the abdominal vein, will be slightly attached to the internal side of the muscles.
3. Identify the **coelom,** or body cavity. Recall from Laboratory 27 that vertebrates are deuterostomes in which the first embryonic opening becomes the anus and the second opening becomes the mouth.
4. If your frog is female, the abdominal cavity is likely to be filled by a pair of large, transparent **ovaries,** each containing hundreds of black and white eggs. Gently lift the left ovary with forceps, and find its place of attachment. Cut through the attachment, and remove the ovary in one piece.

Respiratory System and Liver

1. Insert a probe into the glottis, and observe its passage into the trachea. Enlarge the glottis by making short cuts above and below it. When the glottis is spread open, you will see a fold on either side; these are the vocal cords used in croaking.
2. Identify the **lungs,** two small sacs on either side of the midline and partially hidden under the liver (Fig. 29.7). Sequence the organs in the respiratory tract to trace the path of air from the external nares to the lungs. _____

_____

3. Locate the **liver,** the large, prominent, dark-brown organ in the midventral portion of the trunk (Fig. 29.7). Between the right half and left half of the liver, find the **gallbladder.**

Circulatory System

1. Lift the liver gently. Identify the **heart,** covered by a membranous covering (the **pericardium**). With forceps, lift the covering, and gently slit it open. The heart consists of a single, thick-walled **ventricle** and two (right and left) anterior, thin-walled **atria.**
2. Locate the three large veins that join together beneath the heart to form the **sinus venosus.** (To lift the heart, you may have to snip the slender strand of tissue that connects the atria to the pericardium.) Blood from the sinus venosus enters the right atrium. The left atrium receives blood from the lungs.
3. Find the **conus arteriosus,** a single, wide arterial vessel leaving the ventricle and passing ventrally over the right atrium. Follow the conus arteriosus forward to where it divides into three branches on each side. The middle artery on each side is the **systemic artery,** which fuses behind the heart to become the **dorsal aorta.** The dorsal aorta transports blood through the body cavity and gives off many branches. The **posterior vena cava** begins between the two kidneys and returns blood to the sinus venosus. Which vessel lies above (ventral to) the other? _____

_____

Digestive Tract

1. Identify the **esophagus,** a very short connection between the mouth and the stomach. Lift the left liver lobe, and identify the stomach, whitish and J-shaped. The **stomach** connects with the esophagus anteriorly and with the small intestine posteriorly.
2. Find the **small intestine** and the **large intestine,** which enters the **cloaca.** The cloaca lies beneath the pubic bone and is a general receptacle for the intestine, the reproductive system, and the urinary system. It opens to the outside by way of the anus. Sequence the organs in the digestive tract to trace the path of food from the mouth to the cloaca. _____

_____

Accessory Glands

1. You identified the liver and gallbladder previously. Now try to find the **pancreas,** a yellowish tissue near the stomach and intestine.
2. Lift the stomach to see the **spleen,** a small, pea-shaped body.

Figure 29.7   Internal organs of a female frog, ventral view.   © Ken Taylor/Wildlife Images

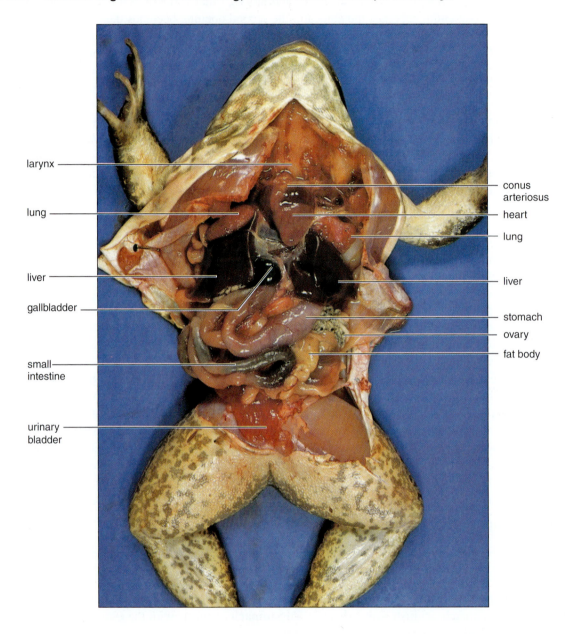

Urogenital System

1. Identify the **kidneys,** long, narrow organs lying against the dorsal body wall (Fig. 29.8).
2. Locate the **testes** in a male frog (Fig. 29.8). Testes are yellow, oval organs attached to the anterior portions of the kidneys. Several small ducts, the **vasa efferentia,** carry sperm into kidney ducts that also carry urine from the kidneys. **Fat bodies,** which store fat, are attached to the testes.

3. Locate the ovaries in a female frog. The ovaries are attached to the dorsal body wall (Fig. 29.9). Fat bodies are also attached to the ovaries. Highly coiled **oviducts** lead to the cloaca. The ostium (opening) of the oviduct is dorsal to the liver.

4. Find the **mesonephric ducts**—thin, white tubes that carry urine from the kidney to the cloaca. In female frogs, you will have to remove the left ovary to see the mesonephric ducts.

5. Locate the **cloaca.** You will need to split through the bones of the pelvic girdle in the midventral line and carefully separate the bones and muscles to find the cloaca.

6. Identify the urinary bladder attached to the ventral wall of the cloaca. In frogs, urine backs up into the bladder from the cloaca.

7. Explain the term *urogenital system.* _____

   _____

8. The cloaca receives material from (1) _____,

   (2) _____, and (3) _____.

9. Compare the frog's urogenital system to the human urinary system, which in females has no connection to the genital system. *Beside each organ listed on the right, tell how the comparable frog organ differs from that of a human.*

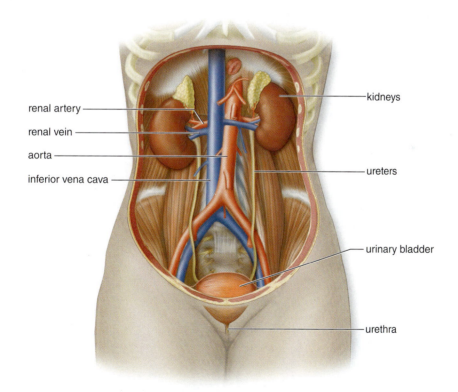

renal artery

renal vein

aorta

inferior vena cava

kidneys

ureters

urinary bladder

urethra

Figure 29.8 Urogenital system of a male frog.

a. Drawing. b. Dissected specimen. (b) © Ken Taylor/Wildlife Images

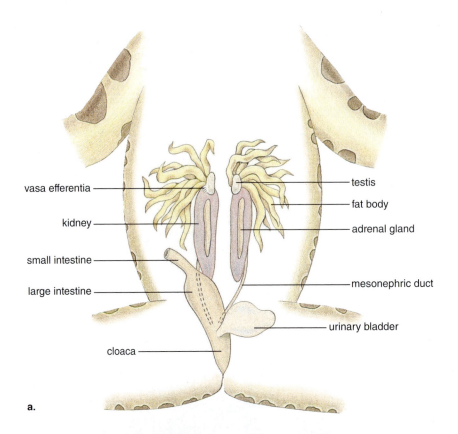

vasa efferentia

kidney

small intestine

large intestine

cloaca

testis

fat body

adrenal gland

mesonephric duct

urinary bladder

a.

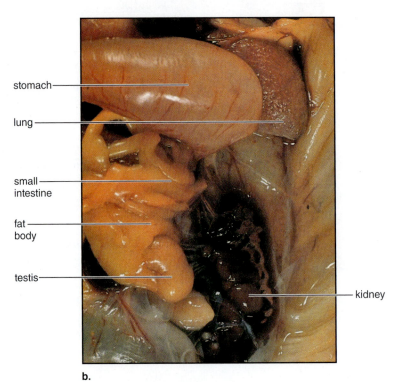

stomach

lung

small intestine

fat body

testis

kidney

b.

Figure 29.9 Urogenital system of a female frog.
a. Drawing. **b.** Dissected specimen.
(b) © Ken Taylor/Wildlife Images

ostium

oviduct

kidney

ovary

fat body

adrenal gland

large intestine

mesonephric duct

urinary bladder

uterus

cloaca

a.

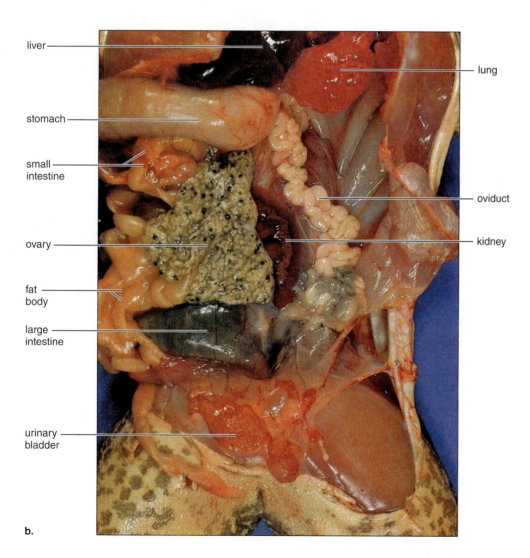

liver

stomach

small intestine

ovary

fat body

large intestine

urinary bladder

lung

oviduct

kidney

b.

In the frog demonstration dissection, identify the **brain,** lying exposed within the skull. With the help of Figure 29.10, find the major parts of the brain.

Figure 29.10   Frog brain, dorsal view.
a. Drawing. **b.** Dissected specimen.   (b) © Ken Taylor/Wildlife Images

olfactory
nerve

olfactory
bulb

cerebral
hemisphere

optic lobe

cerebellum

fourth ventricle

medulla

spinal cord

a.

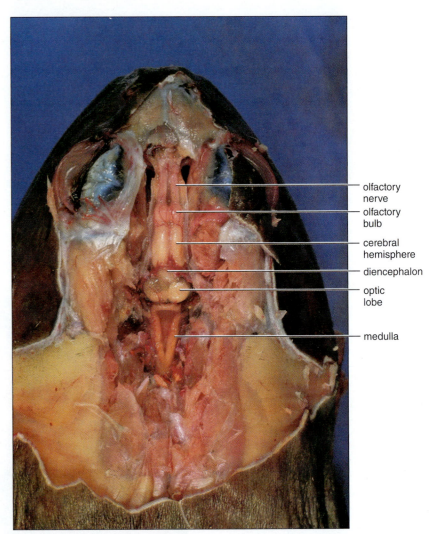

olfactory
nerve

olfactory
bulb

cerebral
hemisphere

diencephalon

optic
lobe

medulla

b.

29.4 Comparative Vertebrate Anatomy

In addition to the frog (amphibian), examine the perch (fish), pigeon (reptile), and rat (mammal) on display.

Observation: External Anatomy of Vertebrates

1. Compare the external features of the perch, frog, pigeon, and rat by examining specimens in the laboratory. Answer the following questions and record your observations in Table 29.1.
   a. Is the skin smooth, scaly, hairy, or feathery?
   b. Is there any external evidence of segmentation?
   c. Are all forms bilaterally symmetrical?
   d. Is the body differentiated into regions?
   e. Is there a well-defined neck?
   f. Is there a postanal tail?
   g. Are there nares (nostrils)?
   h. Is there a cloaca opening, or are there urogenital and anal openings?
   i. Are eyelids present? How many?
   j. How many appendages are there? (Fins are considered appendages.) (Fig. 29.11)

Table 29.1 Comparison of External Features	Perch	Frog	Pigeon	Rat
a. Skin				
b. Segmentation				
c. Symmetry				
d. Regions				
e. Neck				
f. Postanal tail				
g. Nares				
h. Cloaca				
i. Eyelids				
j. Appendages				

2. Evidence that birds are reptiles: Birds
   a. have feathers, which are modified scales.
   b. have scales on their feet.
   c. and reptiles both lay eggs.
   d. and reptiles have similar internal organs.
   e. and reptiles also show some skeletal (skull) similarities.

   Which of these can you substantiate by external examination? _____
   _____

3. The perch, which lives in fresh water, and the pigeon and rat, which live on land, have a nearly impenetrable covering. Why is this an advantage in each case? _____
   _____

4. A frog uses its skin for breathing. Is the skin of a frog thick and dry or thin and moist? Explain. _____
   _____

Figure 29.11   Perch anatomy.
All fins are shown except the pectoral fin.

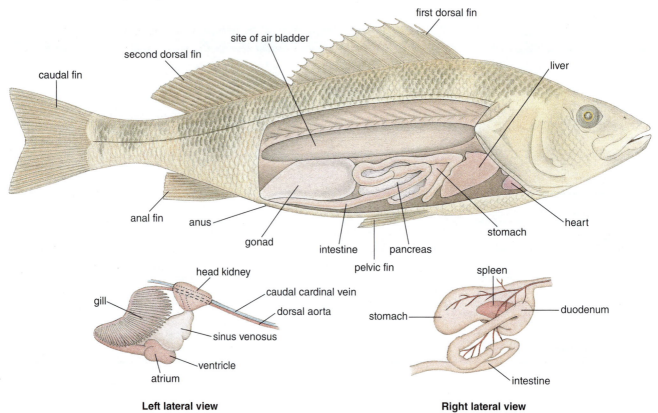

Left lateral view   Right lateral view

Observation: Internal Anatomy of Vertebrates

1. Examine the internal organs of the perch, frog, pigeon (Fig. 29.12) and rat (Fig. 29.13).
2. If necessary, make a median longitudinal incision in the ventral body wall, from the jaws to the anus. The body cavity is called a coelom because it is completely lined by mesoderm (see page 397).
3. Which of these animals has a **diaphragm** dividing the body cavity into **thorax** and **abdomen?** _____
   _____

Figure 29.12   Pigeon anatomy.

esophagus

trachea

crop

left lung

rib cage

heart

left lobe of liver

glandular stomach

left kidney

pancreas

gizzard

small intestine

duodenum

ureter

cloaca

rectum

Figure 29.13   Rat anatomy.

thyroid gland

trachea

esophagus

thymus gland

right atrium

left atrium

ventricles

right lung

left lung

diaphragm

liver

stomach

spleen

small intestine

Digestive Systems and Urogenital Systems

1. All the vertebrates have a stomach, small intestine, and large intestine where food is processed. They also have a liver and pancreas. Which is the larger, more prominent organ (liver or pancreas) in the pigeon and rat? _____

2. All vertebrates have an anus. As noted in Table 29.1, which vertebrates studied have a cloaca (receptacle for the urogenital and digestive systems? _____

    In the other vertebrates studied, the urogenital and digestive systems are separate.

3. Urogenital systems. All the animals have gonads and kidneys. Which of these organs is involved in urine production? _____

    As mentioned in Laboratory 16, the kidneys help maintain the proper balance of fluid and salts in the blood. Which of these organs is involved in reproduction? _____

    The sexes are separate in vertebrates: Females have ovaries and males have testes. In reptiles and mammals, males usually have a penis to pass sperm to the female. What is the chief biological benefit of the penis in terrestrial animals? _____

Circulatory Systems

Study heart models for a fish, amphibian, bird, and mammal. Trace the path of the vessel that leaves the ventricle(s), and determine whether the animals have a **pulmonary system** (Fig. 29.14). The word *pulmonary* comes from the Latin *pulmonarius,* meaning "of the lungs."

Figure 29.14 Cardiovascular systems in vertebrates.
a. In a fish, the blood moves in a single loop. The heart has a single atrium and ventricle, which pumps the blood into the gill region, where gas exchange takes place. **b.** Amphibians have a double-loop system in which the heart pumps blood to both the lungs and the body. **c.** In birds and mammals, the right side pumps blood to the lungs, and the left side pumps blood to the body.

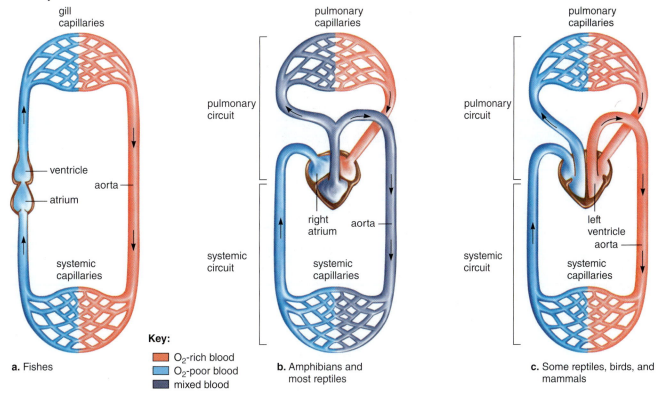

Complete Table 29.2.

Table 29.2 Comparative Circulatory Systems

Animal	Number of Heart Chambers	Pulmonary Circuit (Yes or No)
Perch		
Frog		
Pigeon		
Rat		

1. Do fish have a blood vessel that returns blood from the gills to the heart? _____ Would you expect blood pressure to be high or low after blood has moved through the gills? _____

2. What animals studied have pulmonary vessels that take blood from the heart to the respiratory organ and back to the heart? _____ What is the advantage of a pulmonary circuit? _____

3. Which of these animals has a four-chambered heart? _____
   What is the advantage of having separate ventricles? _____
   _____

4. The circulatory system distributes the heat of muscle contraction in birds and mammals. Is the anatomy of birds and mammals conducive to maintaining a warm internal temperature? _____
   Explain your answer. _____
   _____

Respiratory Systems

Compare the respiratory systems of the perch, frog, pigeon, and rat, and complete Table 29.3 by checking the anatomical features that appear in each animal.

Table 29.3 Respiratory Systems

	Gills	Trachea	Lungs	Rib Cage*	Diaphragm	Air Sacs
Perch						
Frog						
Pigeon						
Rat						

*A rib cage consists of ribs plus a sternum. Some ribs are connected to the sternum, which lies at the midline in the anterior portion of the rib cage.

1. Among the animals studied, only a perch breathes by _____. Can the particular respiratory organ be related to the environment of the animals? _____ Explain your answer. _____

   _____

2. Knowing that gills are attached to the pharynx (throat in humans), explain why fish have no trachea. _____

   _____

   _____

3. A rib cage is present in the rat and pigeon but missing in the frog. Can this difference be related to the fact that frogs breathe by positive pressure, while birds and mammals breathe by negative

   pressure? _____ (A frog swallows air and then pushes the air into its lungs; in birds and mammals, the thorax expands first, and then the air is drawn in.) Explain. _____

   _____

4. A diaphragm is present only in mammals (e.g., rat). Of what benefit is this feature to the expansion of lungs

   in mammals? _____

   _____

   _____

5. Air sacs (not shown in Fig. 29.12) are present only in birds. This feature allows air to pass one way through the lungs of a bird and greatly increases the bird's ability to extract oxygen from the air.

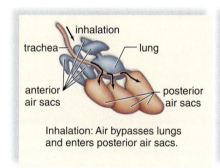

Inhalation: Air bypasses lungs and enters posterior air sacs.

Exhalation continues: Air passes through lungs and enters anterior air sacs.

Laboratory Review 29

_____ 1. What kind of development, protostome or deuterostome, is common to chordates and echinoderms?

_____ 2. Is the nerve cord dorsal or ventral in the invertebrate chordates, tunicates and lancelets?

_____ 3. What structure is replaced by the vertebral column in vertebrate chordates?

_____ 4. Do cartilaginous fishes or ray-finned fishes have a bony skeleton?

_____ 5. What term refers to terrestrial vertebrates that have four limbs?

_____ 6. What do reptiles produce that ensures their embryos won't dry out on land?

_____ 7. What term refers to the changes that occur when an aquatic tadpole becomes a frog with limbs and lungs?

_____ 8. What is the name of the membranous covering associated with the heart?

_____ 9. What structure in the frog receives materials from the large intestine, mesonephritic ducts, and the reproductive organs?

_____ 10. Which animals have modified scales covering their bodies, have scales on their feet, lay shelled eggs, and have internal organs and skulls that provide evidence to classify them as reptiles?

_____ 11. What organs are present if a vertebrate has a pulmonary system of circulation with pulmonary capillaries?

_____ 12. How many ventricles are present in the heart of an amphibian and most reptiles?

_____ 13. Which vertebrates have air sacs attached to their lungs that make the lungs more efficient?

_____ 14. What organs, present in all the vertebrates, produce urine and maintain the proper balance of fluids and salts in the blood?

Thought Questions

15. a. Does the circulatory system of fishes have a pulmonary circuit? Explain your answer.

   b. Given the attachment of gills to the pharynx, explain the absence of a trachea in fishes and the need for a trachea in mammals.

16. What is the role of the four-chambered hearts of birds and mammals in maintaining a warm body temperature?

17. a. Which of the animals studied in this laboratory exercise have a rib cage? Explain the importance of a rib cage to respiration.

   b. Which of the animals studied in this laboratory exercise have a diaphragm? Explain the importance of a diaphragm to respiration.

30

Sampling Ecosystems

Learning Outcomes

Introduction
- Define ecology and an ecosystem. Identify the abiotic and biotic components of an ecosystem.   439–440

30.1 Terrestrial Ecosystems
- Define and give examples of producers in terrestrial ecosystems.   440–443
- Define and give examples of consumers in terrestrial ecosystems.   440–443
- Define and give examples of decomposers in terrestrial ecosystems.   440–443

30.2 Aquatic Ecosystems
- Define aquatic ecosystem.   444
- Give examples of producers, consumers, and decomposers in aquatic ecosystems.   444–447

Introduction

Ecology is the study of interactions between organisms and their physical environment within an **ecosystem.** The **abiotic** (nonliving) components of an ecosystem include soil, water, light, inorganic nutrients, and weather variables. The **biotic** (living) components can be organized according to the **trophic** (feeding) level in which each organism belongs. This includes producers, consumers, and decomposers.

Producers are autotrophic organisms with the ability to carry on photosynthesis and to make food for themselves (and indirectly for the other populations as well). In terrestrial ecosystems, the predominant producers are green plants, while in freshwater and saltwater ecosystems, the dominant producers are various species of algae.

Consumers are heterotrophic organisms that eat available food. Three types of consumers can be identified, according to their food source:

1. **Herbivores** feed directly on green plants and are termed primary consumers. A caterpillar feeding on a leaf is a herbivore.
2. **Carnivores** feed on other animals and are therefore secondary or tertiary consumers. A blue heron feeding on a fish is a carnivore.
3. **Omnivores** feed on both plants and animals. A human who eats both leafy green vegetables and beef is an omnivore.

Decomposers and **detritivores** are organisms of decomposition, such as bacteria, fungi, and millipedes, that break down **detritus** (nonliving organic matter) to inorganic matter, which can be used again by producers. In this way, the same chemical elements are constantly recycled in an ecosystem.

The trophic structure (feeding relationships) of an ecosystem is represented in the form of a pyramid, such as the one shown in Figure 30.1. **Biomass** is the weight of all the organisms at each trophic level.

Figure 30.1 Ecological pyramid.

Organisms at lower trophic levels are higher in number and have greater biomass than organisms at higher trophic levels. The examples of organisms on the left form an aquatic food chain in which herons eat fish, which eat zooplankton, which eat phytoplankton. The examples of organisms on the right form a terrestrial food chain in which owls eat shrews, which eat beetles, which eat plants.

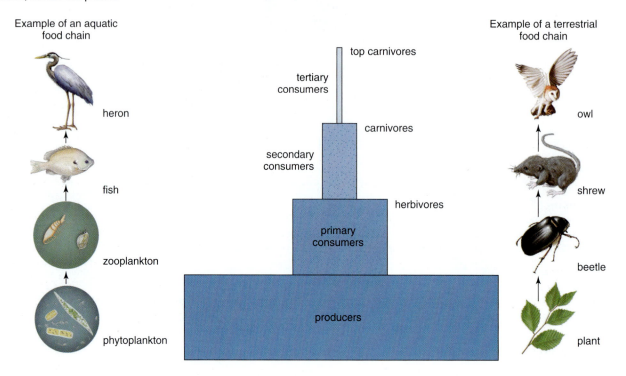

30.1 Terrestrial Ecosystems

Examining an ecosystem in a scientific manner requires concentrating on a representative portion of that ecosystem and recording as much information about it as possible. Representative areas or plots should be selected randomly. For example, random sampling of a terrestrial ecosystem often involves tossing a meterstick gently into the air in the general area to be sampled and then sampling the square meter where the stick lands. A **terrestrial ecosystem** sampling should include samples from the air above and from the various levels of plant materials growing on and in the ground beneath the selected plot.

Study of Terrestrial Sampling Site

The objective of the next Experimental Procedure is to characterize the abiotic and biotic components of a terrestrial ecosystem and to determine how those factors affect the trophic structure of an ecosystem. Although terrestrial ecosystems include deciduous forest, prairie, scrubland, and desert, a weedy field, if dominated by annual and perennial herbaceous plants up to a meter or so in height, is also a good site choice.

1. Gather all necessary equipment, such as metersticks, jars and bags (brown paper and plastic) with labels for collecting specimens, nets, pH paper, thermometers, and other testing equipment, to take with you to the site. Do not forget data-recording materials. Number your collection jars and bags so that you can use these numbers when recording data later (see Table 30.2).

2. When you arrive at the site, take several minutes to observe the general area. Describe what you observe, including weather conditions. _____

_____

_____

3. Choose two sampling locations at the site. The two locations should differ in significant features (e.g., northern or southern exposure, high- or low-slope position, time since last disturbance, native versus exotic species composition). Formulate hypotheses about differences between the two sampling locations in regard to the following variables:

a. Air temperature, humidity, and light intensity _____

_____

b. Soil temperature, moisture, and pH _____

_____

c. Producer biomass and diversity (variety of producers and how many of each type) _____

_____

d. Consumer biomass and diversity _____

_____

Experimental Procedure: Terrestrial Ecosystems

Your instructor will organize class members into teams, and each team will be assigned specific tasks at each of the two chosen sampling locations. At each sampling location (e.g., shaded versus not shaded), you will randomly choose three 1 m square plots. Within each of these plots, randomly choose one 0.1 m square area (subplot) for all of the following abiotic variables and record your measurements in Table 30.1. For each of the biotic variables, randomly choose separate 0.1 m square subplots to sample and complete Table 30.2. Keep outside the 1 m square plots when sampling to minimize disturbance.

Abiotic Components

1. Measure air temperature, relative humidity, and light intensity at 0.5 m above the surface of the ground. Calculate the average for all three subplots (replicates).
2. Measure soil temperature, soil moisture, and soil pH (and any standing water) at 0.2 m below the surface using a soil corer. Calculate the average for all three replicates.

Biotic Components—Plants

In each subplot:

1. Count the total number of live plants.
2. Harvest each entire living plant and wash any soil from the roots. Place plant material in labeled brown paper bags.

Biotic Components—Animals

In each subplot:

1. Sweep with a net as thoroughly as possible about three to five times to capture different organisms on the vegetation in each plot. Empty the contents of the net into a labeled jar of alcohol for later sorting and identification.
2. Collect leaf litter samples in labeled plastic bags for Berlese (or Tullgren) funnel analysis.

Table 30.1   Abiotic Components of a Terrestrial Ecosystem

		Location 1				Location 2			
	Abiotic Factor								
		a	b	c	avg	a	b	c	avg
Air	Temperature								
	Humidity								
	Light intensity								
Soil	Temperature								
	Moisture								
	pH								

Experimental Procedure: Terrestrial Ecosystems

Laboratory Work

1. Examine collected plants and animals using a stereomicroscope, or a compound microscope, if appropriate. Group organisms into different types based on morphological features (morphotypes) and classify them as producers, consumers, or decomposers. Further classify invertebrates into herbivores, detritivores, and carnivores. Complete Table 30.2.

2. Determine the dry biomass (weight) of the plant material, or wet biomass if a drying oven is not available. Dry biomass, although more time consuming to measure, is preferable when comparing biomasses among sites and between trophic levels.

3. Determine the dry biomass of the animal material (or wet weight if plant wet weight was used).

4. Construct graphs comparing the abiotic conditions of each terrestrial sampling location.

5. Construct a pyramid for biomass, morphotype, and the total number of producers, herbivores (including detritivores), and carnivores.

6. Select any three producers and any three consumers from the organisms collected, and explain how each has adapted to its terrestrial environment.

   Producer 1 _____

   Producer 2 _____

   Producer 3 _____

   Consumer 1 _____

   Consumer 2 _____

   Consumer 3 _____

7. Return all organisms and litter samples to their respective collection sites, as explained by your instructor. If any organisms were preserved, ask your instructor what to do with them.

Table 30.2 Biotic Components of a Terrestrial Ecosystem

	Biotic Factor		Location 1				Location 2			
			a	b	c	avg	a	b	c	avg
Plants	Producers	Total number								
		Number of morphotypes								
		Biomass								
Animals on vegetation	Herbivores	Total number								
		Number of morphotypes								
		Biomass								
	Carnivores	Total number								
		Number of morphotypes								
		Biomass								
Animals in litter	Herbivores	Total number								
		Number of morphotypes								
		Biomass								
	Carnivores	Total number								
		Number of morphotypes								
		Biomass								
	Detritivores	Total number								
		Number of morphotypes								
		Biomass								

Conclusions: Terrestrial Ecosystems

Compare the results of this terrestrial ecosystem analysis with the hypotheses you formulated (see page 441).

- Air temperature, humidity, and light intensity _____

_____

- Soil temperature, moisture, and pH _____

_____

- Producer biomass and diversity _____

_____

- Consumer biomass and diversity _____

_____

30.2 Aquatic Ecosystems

An **aquatic ecosystem** sampling should include samples from the air above the water column, the column of water itself, and the soil beneath the water column.

Study of Aquatic Sampling Site

The objective of the next Experimental Procedure is to characterize the abiotic and biotic components of an aquatic ecosystem and to determine how those factors affect the trophic structure of an ecosystem. Aquatic ecosystems consist of freshwater ecosystems (e.g., lakes, ponds, rivers, and streams) and marine ecosystems (e.g., oceans). A good site for this study is a large pond, small lake, or reservoir (with a shallow margin having rooted aquatic plants and a deeper zone with water 1–2 m deep).

1. Gather all necessary equipment, such as metersticks, collection jars and bags with labels, nets, pH paper, thermometers, and other testing equipment to take with you to the site. Do not forget data-recording materials. Number your collection jars and bags so that you can use these numbers when recording data in Table 30.3.

2. When you arrive at the site, take several minutes to observe the general area. Describe what you observe, including weather conditions. _____

   _____

   _____

3. Choose two sampling locations that differ in significant features (e.g., sheltered by trees versus unsheltered, near stream inflow versus far from stream inflow). Plan to sample conditions near shore (shallow-water zone) and in deeper water away from shore (or only one or the other, if logistics are limiting).

   Formulate hypotheses about differences between the two sampling locations in regard to the following variables:

   a. Temperature, humidity, and light intensity above the surface _____

   _____

   b. Temperature, dissolved oxygen, pH, and visibility below the surface _____

   _____

   c. Producer biomass and diversity (variety of producers and how many of each type) _____

   _____

   d. Consumer biomass and diversity _____

   _____

Table 30.3 Abiotic Components of an Aquatic Ecosystem

	Abiotic Factor	Location 1				Location 2			
		a	b	c	avg	a	b	c	avg
Air	Temperature								
	Humidity								
	Light intensity								
Water	Temperature								
	Dissolved oxygen								
	pH								
	Visibility								

Experimental Procedure: Aquatic Ecosystem

Your instructor will organize class members into teams, and each team will be assigned specific tasks at each of the two sampling locations. At each sampling location (e.g., shaded versus not shaded), you will randomly choose three 1 m square plots. Within each of these plots, randomly choose one 0.1 m square area (subplot) for all of the following abiotic variables and record your measurements in Table 30.3. For each of the biotic variables, randomly choose separate 0.1 m square subplots to sample and complete Table 30.4. Keep outside the 1 m square plots when sampling to minimize disturbance.

Abiotic Components

1. Measure air temperature, relative humidity, and light intensity at 0.5 m above the surface of the water.
2. Measure water temperature, dissolved oxygen, pH, and visibility at 0.5 m below the surface.

Biotic Components: Plants

In each subplot:

1. Count the total number of live plants in each subplot.
2. Harvest each entire plant and wash any sediment from the roots. Place plant material in labeled brown paper bags.

Biotic Component: Plankton

Lower a plankton net with attached collecting bottle into the water to a depth of 0.5 m. Slowly raise the net vertically two to three times to collect plankton in the water column. Pour the sample into a labeled collecting jar and preserve with Lugol's solution for later sorting and identification.

Biotic Component: Animals

Choose an area of the plot not disturbed by other sampling but near emergent vegetation. Sweep through the water inside the plot three to five times. Be sure to sample around the vegetation and bump the substrate several times to dislodge benthos from the sediment. Empty the contents of the net into a sieve placed over a bucket of water. Examine, wash, and collect any macroinvertebrates and place them in the bucket. Transfer collected organisms into a labeled jar of alcohol for later sorting and identification.

1. Examine collected plants, plankton, and animals using a stereomicroscope, or a compound microscope, if appropriate. Group organisms into different types based on morphological features (morphotypes) and classify them as producers, consumers, or decomposers. Further classify plankton and invertebrates into herbivores, detritivores, and carnivores. Complete Table 30.4.

Table 30.4 Biotic Components of an Aquatic Ecosystem

	Biotic Factor		Location 1				Location 2			
			a	b	c	avg	a	b	c	avg
Plants	Producers	Total number								
		Number of morphotypes								
		Biomass								
Plankton	Producers	Total number								
		Number of morphotypes								
		Biomass								
	Herbivores	Total number								
		Number of morphotypes								
		Biomass								
	Carnivores	Total number								
		Number of morphotypes								
		Biomass								
Animals	Herbivores	Total number								
		Number of morphotypes								
		Biomass								
	Carnivores	Total number								
		Number of morphotypes								
		Biomass								
	Detritivores	Total number								
		Number of morphotypes								
		Biomass								

2. Determine the dry biomass (weight) of the plant material, or wet biomass if a drying oven is not available.
3. Determine the dry biomass of the plankton and animal material (or wet weights if plant wet weight was used).
4. Construct graphs comparing the abiotic conditions of each aquatic sampling location.
5. Construct a pyramid for biomass, morphotype, and the total number of individuals of producers, herbivores (including detritivores), and carnivores.
6. Select any three producers and any three consumers from the organisms collected, and explain how each has adapted to its aquatic environment.

Producer 1 _____

Producer 2 _____

Producer 3 _____

Consumer 1 _____

Consumer 2 _____

Consumer 3 _____

7. Return all organisms and water samples to their respective collection sites, as explained by your instructor. If any organisms were preserved, ask your instructor what to do with them. Place any exposed petri dishes in a designated area for incubation until the next laboratory.

Conclusions: Aquatic Ecosystems

Compare the results of this aquatic ecosystem analysis with the hypotheses you formulated (see page 444).

- Temperature, humidity, and light intensity above the surface _____

    _____

- Temperature, dissolved oxygen, pH, and visibility below the surface _____

    _____

- Producer biomass and diversity _____

    _____

- Consumer biomass and diversity _____

    _____

_____ 1. What is being studied when organisms and their physical environment are observed?

_____ 2. Are soil, minerals, sunlight, water, and weather variables biotic or abiotic components of an ecosystem?

_____ 3. What is the general name for organisms at the first trophic level in an ecosystem?

_____ 4. What are the dominant producers in freshwater and saltwater ecosystems?

_____ 5. Is a black bear that eats berries, nuts, insects, and fish a herbivore, omnivore, or carnivore?

_____ 6. What organisms break down nonliving organic matter into organic matter that producers can use?

_____ 7. What term is used for the weight of all the organisms at each trophic level?

_____ 8. At what consumer level is a hawk that ate a snake that ate a mouse that ate some grain?

_____ 9. What is an example of a terrestrial ecosystem?

_____ 10. In a terrestrial ecosystem, what organisms are predominant producers?

_____ 11. What is an example of an abiotic component measured above the surface of the ground when surveying a terrestrial ecosytem?

_____ 12. What kinds of organisms are most likely to be found in samples of leaf litter?

_____ 13. In which type of ecosystem, aquatic or terrestrial, will phytoplankton and zooplankton be found?

_____ 14. What is an example of an abiotic component measured below the surface of the water in an aquatic ecosystem?

Thought Questions

15. What is the importance of many bacteria, fungi, and insects to the ability of plants to acquire the elements they need for making organic molecules?

16. Which organisms should have the greatest biomass in a health ecosystem? Justify your reply.

17. a. Will a herbivore ever be a secondary or higher consumer? Explain your reply.

    b. Will a carnivore ever be a primary consumer? Explain your reply.

31

Effects of Pollution on Ecosystems

Introduction

This laboratory will consider three causes of aquatic pollution: thermal pollution, acid pollution, and cultural eutrophication. **Thermal pollution** occurs when water temperature rises above normal. As water temperature rises, the amount of oxygen dissolved in water decreases, possibly depriving organisms and their cells of an adequate supply of oxygen. Deforestation, soil erosion, and the burning of fossil fuels contribute to thermal pollution, but the chief cause is use of water from a lake or the ocean as a coolant for the waste heat of a power plant.

When sulfur dioxide and nitrogen oxides enter the atmosphere, usually from the burning of fossil fuels, they are converted to acids, which return to Earth as **acid deposition** (acid rain or snow). Acid deposition kills plants, aquatic invertebrates, and also decomposers, threatening the entire ecosystem.

Cultural eutrophication, or overenrichment, is due to runoff from agricultural fields, wastewater from sewage treatment plants, and even excess detergents. These sources of excess nutrients cause an algal bloom seen as a green scum on a lake (Fig. 31.1). When algae overgrow and die, decomposition robs the lake of oxygen, causing a fish die-off.

Figure 31.1 Cultural eutrophication.
Eutrophic lakes tend to have large populations of algae and rooted plants.

© Pat Watson/ McGraw-Hill Education

31.1 Studying the Effects of Pollutants

We are going to study the effects of pollution by observing its effects on hay infusion organisms, on seed germination, and on an animal called *Gammarus*.

Study of Hay Infusion Culture

A hay infusion culture (hay soaked in water) contains various microscopic organisms, but we will be concentrating on how the pollutants in our study affect the protozoan populations in the culture. We will consider both of these aspects:

species composition: number of different types of microorganisms.
species diversity: composition and the abundance of each type of microorganism.

Experimental Procedure: Effect of Pollutants on a Hay Infusion Culture

During this Experimental Procedure you will examine, by preparing a wet mount, hay infusion cultures that have been treated in the following manner.

1. **Control culture:** This culture simulates the species composition and diversity of an untreated culture. Prepare a wet mount and answer the following questions:

   With the assistance of Figure 31.2 and any guides available in the laboratory, identify as many different types of microorganisms as possible in the hay infusion culture. State whether species composition is high, medium, or low. Record your estimation in the second column of Table 31.1. Do you judge species diversity to be high, medium, or low? Record your estimation in the third column of Table 31.1.

2. **Oxygen-deprived culture:** Thermal pollution causes water to be oxygen deprived; therefore, when we study the effects of low oxygen on a hay infusion culture, we are studying an effect of thermal pollution. Prepare a wet mount of this culture and determine if there is a change in species composition and diversity. Again record the species composition and species diversity as high, medium, or low in Table 31.1.

Figure 31.2   Microorganisms in hay infusion cultures.
Organisms are not to size.

a. Amoeboid (*Amoeba*)   **b.** Flagellate (*Euglena*)   **c.** Ciliate (*Paramecium*)   **d.** Ciliate (*Vorticella*)   **e.** Ciliate (*Euplotes*)

f. Rotifer (*Philodina*)   **g.** Ciliate (*Tetrahymena*)   **h.** Ciliate (*Stentor*)   **i.** Ciliate (*Colpoda*)   **j.** Ciliate (*Didinium*)

Table 31.1   Effect of Pollution on a Hay Infusion Culture

Type of Culture	Species Composition (High, Medium, or Low)	Species Diversity (High, Medium, or Low)	Explanation
Control			
Oxygen-deprived			
Acidic			
Enriched			

3. **Acidic culture:** In this culture, the pH has been adjusted to 4 with sulfuric acid (H_2SO_4). This simulates the effect of acid rain on a hay infusion culture. Prepare a wet mount of this culture and determine if there is a change in species composition and diversity. Again record the species composition and species diversity as high, medium, or low in Table 31.1.

4. **Enriched culture:** More organic nutrients have been added to this culture. These nutrients will cause the algae population, which is food for most protozoans, to increase. In the short term, their species composition should increase. Eventually, as the algae die off, decomposition will rob the water of oxygen and the protozoans may start to die off. Prepare a wet mount of this culture and determine if there is a change in species composition and diversity. Again record the species composition and species diversity as high, medium, or low in Table 31.1.

Conclusions

- What could be a physiological reason for the adverse effects of oxygen deprivation on a hay infusion culture? If consistent with your results, enter this explanation in the last column of Table 31.1.
- What could be a physiological reason for the adverse effects of a low pH on a hay infusion culture? If consistent with your results, enter this explanation in the last column of Table 31.1.
- What could be an environmental reason for the adverse effects of an enriched culture? If consistent with your results, enter this explanation in the last column of Table 31.1.

Effect of Acid Rain on Seed Germination

Seeds depend on favorable environmental conditions of temperature, light, and moisture to germinate, grow, and reproduce. Like any other biological process, germination requires enzymatic reactions that can be adversely affected by an unfavorable pH.

Experimental Procedure: Effect of Acid Rain on Seed Germination

In this Experimental Procedure we will test whether there is a negative correlation between acid concentration and germination. In other words, it is hypothesized that as acidity increases, it becomes more likely seeds

will _____.

Your instructor has placed 20 sunflower seeds in each of five containers with water of increasing acidity: 0% vinegar (tap water), 1% vinegar, 5% vinegar, 20% vinegar, and 100% vinegar.

1. Test and record the pH of solutions having the vinegar concentrations noted above. Record the pH of each solution in Table 31.2.
2. Count the number of germinated sunflower seeds in each container, and complete Table 31.2.

Table 31.2 Effect of Increasing Acidity on Germination of Sunflower Seeds

Concentration of Vinegar	pH	Number of Seeds that Germinated	Percent Germination
0%			
1%			
5%			
20%			
100%			

Conclusions: Effect of Increasing Acidity on Germination of Sunflower Seeds

- As you know, each enzyme has an optimum pH. Explain why acid rain is expected to inhibit metabolism, and therefore, seedling development. _____
- Do the data support or falsify your hypothesis? _____

Study of *Gammarus*

A small crustacean called *Gammarus* lives in ponds and streams (Fig. 31.3) where it feeds on debris, algae, or anything smaller than itself, such as some of the microorganisms in Figure 31.2. In turn, fish like to feed on *Gammarus*.

Experimental Procedure: Gammarus

- Add 25 ml of spring water to a beaker and record the pH of the water. _____ pH
- Add four *Gammarus* to the container. Do they all use their legs in swimming? _____
- Which legs are used in jumping and climbing? _____
- What do *Gammarus* do when they "bump" into each other? _____

Control Sample

After observing *Gammarus,* decide what behaviors are most often observed. During a 5-minute time span, total the amount of time spent doing each of these behaviors.

Behaviors	Amount of Time	Total Time
1. _____	_____	_____
2. _____	_____	_____
3. _____	_____	_____

Test Sample

If so directed by your instructor, put a *Gammarus* in a beaker of spring water adjusted to pH 4 by adding vinegar. During a 5-minute time span, total the amount of time spent doing each of these behaviors.

Behaviors	Amount of Time	Total Time
1. _____	_____	_____
2. _____	_____	_____
3. _____	_____	_____

Figure 31.3 *Gammarus.*
Gammarus is a type of crustacean classified in a subphylum that also includes shrimp.
© William Amos/Photoshot

Conclusion

- Draw a conclusion from this study: _____

_____

- Create a food chain that shows who eats whom when the food chain includes algae, protozoans, *Gammarus,*

fish, and humans. _____

  a. What would happen to this food chain if the water were oxygen deprived?_____

  _____

  Acidic? _____

  Enriched with inorganic nutrients (short term and long term)? _____

  _____

Conclusions: Studying the Effects of Pollutants

- Give an example to show that the hay infusion study pertains to real ecosystems. _____

_____

- What are the potential consequences of acid rain on crops that reproduce by seeds? _____

_____

  On the food chains of the ocean? _____

- How does the addition of nutrients affect species composition and species diversity of an ecosystem

  over time? _____

31.2 Studying the Effects of Cultural Eutrophication

Chlorella, the green alga used in this study, is considered to be representative of algae in bodies of fresh water. The crustacean *Daphnia* feeds on green algae such as *Chlorella* (Fig. 31.4). First, you will observe how *Daphnia* feeds, and then you will determine the extent to which *Daphnia* could keep the effects of cultural eutrophication from occurring in a hypothetical example. Keep in mind that this case study is an oversimplification of a generally complex problem.

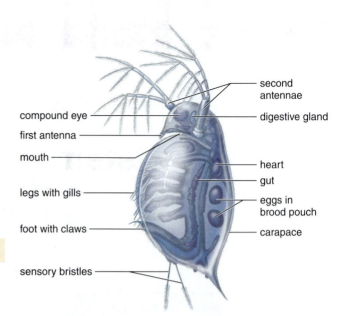

Figure 31.4   Anatomy of *Daphnia.*

Observation: Daphnia *Feeding*

1. Place a small pool of petroleum jelly in the center of a small petri dish.
2. Use a dropper to take a *Daphnia* from the stock culture, place it on its back (covered by water) in the petroleum jelly, and observe it under the stereomicroscope (Fig. 31.4).
3. Note the insectlike carapace and the legs waving rapidly as the *Daphnia* filters the water.
4. Add a drop of *carmine solution,* and observe how the *Daphnia* filters the "food" from the water and passes it through the gut. The gut is more visible if you push the animal onto its side. In this position, you may also observe the heart beating in the region above the gut and just behind the head.
5. Allow the *Daphnia* to filter-feed for up to 30 minutes, and observe the progress of the carmine particles

through the gut. Does the carmine travel completely through the gut in 30 minutes? _____

Experimental Procedure: Daphnia *Feeding on* Chlorella

This exercise requires the use of a spectrophotometer. Absorbance will be a measure of the algal population level; the greater the number of algal cells, the greater the absorbance. The higher the absorbance, the greater the amount of light absorbed and *not* passed through the solution.

1. Obtain two spectrophotometer tubes (cuvettes) and a Pasteur pipet.
2. Fill one of the cuvettes with distilled water, and use it to zero the spectrophotometer. Save this tube for step 6.
3. Use the Pasteur pipet to fill the second cuvette with *Chlorella.* Gently aspirate and expel the sample several times (without creating bubbles) to give a uniform dispersion of the algae.
4. Add ten hungry *Daphnia,* and following your instructor's directions, immediately measure the absorbance with the spectrophotometer. If a *Daphnia* swims through the beam of light, a strong deflection should occur; do not use any such higher readings—instead, use the lower figure for the absorbance. Record your reading in the first column of Table 31.3.
5. Remove the cuvette with the *Daphnia* to a safe place in a test tube rack. Allow the *Daphnia* to feed for 30 minutes.
6. Rezero the spectrophotometer with the distilled water cuvette.
7. Measure the absorbance of the experimental cuvette again. Record your data in the second column of Table 31.3, and explain your results in the third column.

Table 31.3 Spectrophotometer Data/*Daphnia* Feeding on *Chlorella*		
Absorbance Before Feeding	**Absorbance After Feeding**	**Explanation**

Experimental Procedure: Case Study in Cultural Eutrophication

The following problem will test your understanding of the ecological value of a single species—in this case, *Daphnia.* Please realize that this is an oversimplification of a generally complex problem.

1. Assume that developers want to build condominium units on the shores of Silver Lake. Homeowners in the area have asked the regional council to determine how many units can be built without altering the nature of the lake. As a member of the council, you have been given the following information:

    The current population of *Daphnia,* 10 animals/liter, presently filters 24% of the lake per day, meaning that it removes this percentage of the algal population per day. This is sufficient to keep the lake essentially clear. Predation—the eating of the algae—will allow the *Daphnia* population to increase to no more than 50 animals/liter. Therefore, 50 *Daphnia*/liter will be available for feeding on the increased number of algae that would result from building the condominiums.

    Using this information, complete Table 31.4.

Table 31.4   *Daphnia* Filtering

Number of *Daphnia*/Liter	Percent of Lake Filtered
10	24%
50	

2. The sewage system of the condominiums will add nutrients to the lake. Phosphorus output will be 1 kg per day for every 10 condominiums. This will cause a 30% increase in the algal population. Using this information, complete Table 31.5.

Table 31.5   Cultural Eutrophication

Number of Condominiums	Phosphorus Added	Increase in Algal Population
10	1 kg	30%
20		
30		
40		
50		

Conclusion: Cultural Eutrophication

- Assume that phosphorus is the only nutrient that will cause an increase in the algal population and that *Daphnia* is the only type of zooplankton available to feed on the algae. How many condominiums would you allow the developer to build? _____

_____

- What other possible impacts could condominium construction have on the condition of the lake? _____

_____

_____

_____ **1.** What results when sulfur dioxide and nitrogen oxides, formed by burning fossil fuels, enter the atmosphere?

_____ **2.** What kind of pollution results when water from a lake or the ocean is used to disperse waste heat?

_____ **3.** Does the amount of oxygen dissolved in water increase or decrease when the temperature of the water increases?

_____ **4.** What occurs to a body of water when there is runoff from agricultural fields?

_____ **5.** Which population experiences growth referred to as a "bloom" when excess nutrients enter a body of water?

_____ **6.** After excess nutrients cause algae to overgrow and then die, what causes the fish to die?

_____ **7.** What do we call the number of different microorganisms observed and their relative abundance in the hay infusion culture?

_____ **8.** What chemical was used to simulate the effect of acid rain on a hay infusion culture?

_____ **9.** What does an unfavorable pH adversely affect that inhibits seed germination?

_____ **10.** In the Experimental Procedure on the effect of acid rain on seed germination, if 18 of the sunflower seeds germinated in the 1% vinegar solution, what percent germinated?

_____ **11.** What kind of animal is _Gammarus?_

_____ **12.** What will _Daphnia_ do that might prevent cultural eutrophication from occurring?

_____ **13.** Are there more or fewer algal cells present in a water sample as the absorption of light decreases?

_____ **14.** What element from the Silver Lake condominiums' sewage output contributes to the increased algal population?

Thought Questions

15. A "dead zone" forms where the Mississippi River empties into the Gulf of Mexico due to nutrient runoff into the river that is carried to the Gulf. Explain how the dead zone forms and what impact it may have on humans.

16. Pollutants often affect the producers in an ecosystem first and then have far-ranging effects. Use acid rain and a food chain to illustrate your understanding of the impact pollutants have after they enter an ecosystem.

17. Describe how the cultural eutrophication case study illustrates the need to have a balance of population sizes in an ecosystem.

A

Preparing a Laboratory Report/ Laboratory Report Form

A laboratory report has the sections noted in the outline that follows. Use this outline and a copy of the Laboratory Report Form on page A–3 to help you write a report assigned by your instructor. In general, do not use the words *we, my, our, your, us,* or *I* in the report. Use scientific measurements and their proper abbreviations. (For example, cm is the proper notation for centimeter and sec is correct for seconds.)

1. **Introduction:** Tell the reader what the experiment was about.
   a. **Background information:** Begin by giving an overview of the topic. Look at the Introduction to the Laboratory (and/or at the introduction to the section for which you are writing the report). Do not copy the information, but use it to get an idea about what background information to include.

   > For example, suppose you are doing a laboratory report on "Solar Energy" in Laboratory 8 (Photosynthesis). You might give a definition of photosynthesis and explain the composition of white light.

   b. **Purpose:** Think about the steps of the experiment and state what the experiment was about. Tell the independent and dependent variable.

   > For example, you might state that the purpose of the photosynthesis experiment was to determine the effect of white light versus green light on the photosynthetic rate. The independent variable was the color of light and the dependent variable was the rate of photosynthesis.

   c. **Hypothesis:** Consider the expected results of the experiment in order to state the hypothesis. It's possible that the Introduction to the Laboratory might hint at the expected results. State this in the form of a hypothesis.

   > For example, you might state: It was hypothesized that white light would be more effective than green light for photosynthesis.

2. **Method:** Tell the reader how you did the experiment.
   a. **Equipment and sample used:** Use any illustrations in the laboratory manual that show the experimental setup to describe the equipment and the sample (subject) used.

   > For example, for the photosynthesis experiment look at Figure 8.4 and describe what you see. You might state that a 150-watt lamp was the source of white light directed at *Elodea*, an aquatic plant, placed in a test tube filled with a solution of sodium bicarbonate ($NaHCO_3$). A beaker of water placed between the lamp and the test tube was a heat absorber.

   b. **Collection of data:** Think about what you did during the experiment, such as what you observed or what you measured. Look at any tables you filled out in order to recall how the data were collected and what control(s) were used.

   > For example, for the photosynthesis experiment, you might state that the rate of photosynthesis was determined by the amount of oxygen released and was measured by how far water moved in a side arm placed in a stopper of the test that held *Elodea*. A control was the same experimental setup, except the test tube lacked *Elodea*.

3. **Results:**   Present the data in a clear manner.
   a. **Graph or table:**   If at all possible, show your data in table or graph form. You could reproduce a table you filled in or a graph you drew to show the results of the experiment. Be sure to include the title of the table; do not include any interpretation of the data column in the table.

For example, for the photosynthesis experiment you might reproduce Table 8.3.

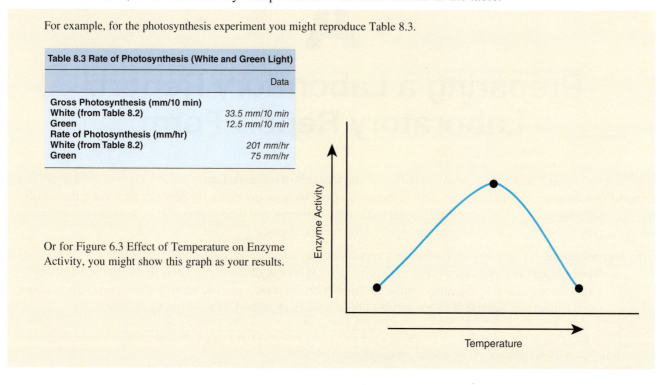

Table 8.3 Rate of Photosynthesis (White and Green Light)	
	Data
Gross Photosynthesis (mm/10 min)	
White (from Table 8.2)	33.5 mm/10 min
Green	12.5 mm/10 min
Rate of Photosynthesis (mm/hr)	
White (from Table 8.2)	201 mm/hr
Green	75 mm/hr

Or for Figure 6.3 Effect of Temperature on Enzyme Activity, you might show this graph as your results.

b. **Description of data:**   Examine your data, and decide what they tell you. Then, below any table or graph, add a description to help the reader understand what the table or graph is showing. Define any terms in the table that are not readily understandable.

For example, below Table 8.3 you might state that these data indicate that the rate of photosynthesis with white light is faster than with green light. Also, you should define gross photosynthesis. Or below the graph that shows the effect of temperature on enzyme activity, you might state that these data show that the rate of enzymatic activity speeds up until boiling occurs and then it drops off.

4. **Conclusion:**   Tell if the data support or do not support the hypothesis.
   a. **Compare the hypothesis with the data:**   Do your data agree or disagree with the hypothesis?

For example, for the photosynthesis experiment you might state: These results support the hypothesis that white light is more effective for photosynthesis than green light.

b. **Explanation:**   Explain why you think you obtained these results. Look at any questions you answered while in the laboratory, and use them to help you decide on an appropriate explanation.

For example, the answers to the questions in 8.2 Solar Energy might help you state that white light gives a higher rate of photosynthesis because it contains all the visible light rays. Green light gives a lower rate because green plants such as *Elodea* do not absorb green light.

If your results do not support the hypothesis, explain why you think this occurred.

In this instance you might state that while white light contains all visible light rays and green light is not absorbed by a green plant, the experiment did not support the hypothesis because of failure to use a heat absorber when doing the green light experiment.

Laboratory Report for _____

1. Introduction

   a. Background information

   b. Purpose

   c. Hypothesis

2. Method

   a. Equipment used

   b. Collection of data

3. Results

a. Graph or table
(Place these on attached sheets.)

b. Description of data

4. Conclusion

a. Compare the hypothesis with the data

b. Explanation

c. Conclusion

B

Metric System

Unit and Abbreviation	Metric Equivalent	Approximate English-to-Metric Equivalents	Units of Temperature

Length

nanometer (nm) = 10^{-9} m (10^{-3} μm)
micrometer (μm) = 10^{-6} m (10^{-3} mm)
millimeter (mm) = 0.001 (10^{-3}) m
centimeter (cm) = 0.01 (10^{-2}) m

1 inch = 2.54 cm
1 foot = 30.5 cm
1 foot = 0.30 m
1 yard = 0.91 m
1 mi = 1.6 km

meter (m) = 100 (10^2) cm
 = 1,000 mm
kilometer (km) = 1,000 (10^3) m

Weight (mass)

nanogram (ng) = 10^{-9} g
microgram (μg) = 10^{-6} g
milligram (mg) = 10^{-3} g
gram (g) = 1,000 mg

1 ounce = 28.3 g
1 pound = 454 g
 = 0.45 kg
1 ton = 0.91 t

kilogram (kg) = 1,000 (10^3) g
metric ton (t) = 1,000 kg

Volume

microliter (μl) = 10^{-6} l (10^{-3} ml)
milliliter (ml) = 10^{-3} l
 = 1 cm^3 (cc)
 = 1,000 mm^3

1 tsp = 5 ml
1 fl oz = 30 ml

liter (l) = 1,000 ml

1 pint = 0.47 l
1 quart = 0.95 l
1 gallon = 3.79 l

kiloliter (kl) = 1,000 l

Common Temperatures*

°C	°F	
100	212	Water boils at standard temperature and pressure.
71	160	Flash pasteurization of milk
57	134	Highest recorded temperature in the United States, Death Valley, July 10, 1913
41	105.8	Average body temperature of a marathon runner in hot weather
37	98.6	Human body temperature
13.7	56.66	Human survival is still possible at this temperature.
0	32.0	Water freezes at standard temperature and pressure.

* See temperature scale

Units of Temperature

°F / °C

212° — 210 / 100 — 100°
160° — 160 / 70 — 71°
134° — 130 / 60 — 57°
105.8° — 110 / 40 — 41°
98.6° — 100 / 40 — 37°
56.66° — 60 / 13.7°
32° — 30 / 0 — 0°

To convert temperature scales:

$$°C = \frac{5}{9}(°F - 32)$$

$$°F = \frac{9}{5}(°C + 32)$$

Practical Examination Answer Sheet

1.	_____	26.	_____
2.	_____	27.	_____
3.	_____	28.	_____
4.	_____	29.	_____
5.	_____	30.	_____
6.	_____	31.	_____
7.	_____	32.	_____
8.	_____	33.	_____
9.	_____	34.	_____
10.	_____	35.	_____
11.	_____	36.	_____
12.	_____	37.	_____
13.	_____	38.	_____
14.	_____	39.	_____
15.	_____	40.	_____
16.	_____	41.	_____
17.	_____	42.	_____
18.	_____	43.	_____
19.	_____	44.	_____
20.	_____	45.	_____
21.	_____	46.	_____
22.	_____	47.	_____
23.	_____	48.	_____
24.	_____	49.	_____
25.	_____	50.	_____

Index

Note: Page references followed by *f* and *t* refer to figures and tables, respectively.

Anticodon, 306–307
Antigen(s), 327–329, 327*f*
Anus
of chick embryo, 260
of clam, 400*f*, 401
of crayfish, 410*f*
of deuterostomes, 397
of earthworm, 405, 405*f*, 406
of fetal pig, 155, 155*f*, 159*f*, 164
of frog, 256, 425
of human, 165*f*, 168*f*
of lancelet, 421*f*
of perch, 432*f*
of rotifers, 395, 395*f*
of sea star, 416, 417
of squid, 402*f*
of vertebrates, 422, 435
Aorta, 178, 179*f*, 181*f*, 183*f*, 208*f*
of adult mammal, 184
of clams, 398
dorsal
of fetal pig, 185, 186*f*–188*f*
of frog, 425
of perch, 432*f*
of fetal pig, 185, 186*f*–188*f*, 194*f*, 197*f*
of grasshopper, 412*f*
of vertebrates, 435*f*
Aortic arch, 179*f*
in chick embryo, 261*f*
of earthworm, 406, 407*f*
of fetal pig, 186*f*–187*f*
Aortic semilunar valve, 180
Apex of heart, 179*f*, 181*f*
Apical meristem, 103, 103*f*
Apicomplexans, 342
Apicoplast, 342
Apis mellifera, 409*f*
Apostichopus japonicus, 415*f*
Appendicular skeleton, 242, 244–245
Appendix, of human, 165*f*
Aquatic ecosystems, 444–447
acid deposition in, 449
cultural eutrophication of, 449, 449*f*, 454–455
pollution of, 449–456
studying sampling site of, 444–447
Aqueous humor, 231*f*, 231*t*
Arcella, 345*f*
Archaea (domain), 331
Archaeplastids, 337*f*
Archegonia (sing., archegonium), 354
of ferns, 360, 360*f*, 363, 363*f*
of liverworts, 357, 357*f*
of moss, 354*f*, 355, 355*f*
of pines, 368*f*
Archenteron, 256, 256*f*, 258*f*
Argiope bruennichi, 409*f*
Arm(s)
bones of, 243, 244
of echinoderms, 415
in human embryo, 262*f*
muscles of, 246*f*, 247*t*
of sea star, 416, 416*f*
of squid, 402*f*
Armadillidium vulgare. See Pillbug
Arterial duct (ductus arteriosus), 184, 184*t*

Arterioles, 177, 208
afferent, 217, 217*t*, 218*f*
efferent, 217, 217*t*, 218*f*
Artery(ies), 177, 208, 208*f*
of adult mammal, 184
of fetal pig, 153, 184–189, 186*f*–188*f*, 205
structure of, 190–191, 190*f*
veins *vs.,* 190–191
Arthropods, 380*f*, 397, 409–415, 409*f*
Articular cartilage, 240
Ascaris, 393, 393*f*
Asexual reproduction
in cnidarians, 384, 384*f*, 385, 385*f*
in fungi, 346, 347, 347*f*, 348*f*
in green algae, 339, 339*f*
mitosis in, 57
in sponges, 381
Asplanchna, 345*f*
Aster, 58*t*, 59, 59*f*
Asymmetry, of animals, 381
Athlete's foot, 349, 349*f*
Atom(s), 27
ATP. *See* Adenosine triphosphate
Atria (sing., atrium), 178
contraction of, 209–210, 209*f*
of fetal pig, 161*f*
of frog, 425
left, 178, 179*f*, 181*f*, 183*f*
of perch, 432*f*
of rat, 434*f*
right, 178, 179*f*, 181*f*, 183*f*
of vertebrates, 435*f*
Atriopore, of lancelet, 421*f*, 422
Atrioventricular valve, 180, 181*f*
Auditory association area, 227*f*
Auditory canal, 234*f*, 234*t*, 242*f*
Auditory tube, 234*f*, 234*t*
of frog, 424, 424*f*
Aurelia, 387*f*
Auricle, of planarians, 388*f*, 390, 390*f*
Autosomal disorders, 288–290, 289*f*
Autosomal dominant traits, 286–287, 286*f*
Autosomal recessive traits, 286–287, 286*f*
Autosomes, 281, 285–287
Autotrophs, 439
AV valve. *See* Atrioventricular valve
Axial skeleton, 242–244
Axillary artery, of fetal pig, 186*f*
Axillary buds, of plants, 102, 102*f*, 113, 113*f*, 116
Axillary vein, of fetal pig, 186*f*
Axons, 149, 149*f*, 223, 229*f*
Axon terminal, 223

Bacillus (pl., bacilli), 335, 335*f*
Bacteria (bacterium)
cell, 42, 42*f*
classification (taxonomy) of, 331, 331*f*
colony of, 333–334, 334*f*
pathogenic, 332–333
shape of, 335, 335*f*
structure of, 332–333, 333*f*
Bacteria (domain), 331, 332–337
Bacterium (pl., bacteria). *See* Bacteria
Balance, 233, 234
Banded butterflyfish, 423*f*

Bark, 114, 114*f*
Barnacles, 409, 409*f*
Basal disk, of cnidarians, 385–386, 385*f*
Base(s)
chemical, pH of, 54, 54*f*
nucleotide, 299–303, 300*f*, 303*f*
Basement membrane, 140, 140*f*–142*f*
Base pairing, complementary, 300*f*, 301–302, 304, 306
Bases, nitrogenous, 300–304, 300*f*, 303*f*, 306
Basidia (basidium), 348–349, 348*f*
Basidiocarp, 348–349, 348*f*
Basidiospores, 348–349, 348*f*
Basophils, 145*f*
Bath sponge, 383*f*
Bean seed, 133–134, 133*f*
Beaumont, William, 167
Bee(s), pollination by, 119, 119*f*, 126–127
Benedict's reagent, 32, 49, 173
Benedict's test, for sugars, 32–34, 33*t*, 173
Berry, 131
Biceps brachii muscle, 239, 239*f*, 245, 246*f*, 247*t*
Biceps femoris muscle, 246*f*, 247*t*
Bicuspid valve, 180, 181*f*
Bilateral symmetry, 381
of echinoderms, 415
of flatworms, 388
of roundworms, 392
of vertebrates, 422
Bile, 171–172, 171*f*
Bile duct(s), 205*f*
of fetal pig, 164, 204
Bile salts, 35
Biomass, 439
Biomolecules, 27. *See also* Carbohydrates; Lipid(s); Protein(s)
Biotechnology, 299, 308–309
Biotic components
of aquatic ecosystem, 444–446
of ecosystem, 439
of terrestrial ecosystem, 441, 443
Birds, 423*f*
anatomy of, comparison with other vertebrates, 431–437, 433*f*
cardiovascular system of, 435–436, 435*f*
embryo of, 322, 322*f*
extraembryonic membranes of, 266
fossils of, 318, 318*f*
as reptiles, 432
respiratory system of, 436–437, 437*f*
Biuret reagent, 29
Bivalves, 398–403, 398*f*, 399*f*, 400*f*
Blackberry, fruit of, 130, 130*f*
Black bread mold, 347–348, 347*f*, 348*f*
Bladder worms, 391, 391*f*
Blade, of leaf, 102, 102*f*
Blastocoel, 253
of frog, 256, 256*f*
Blastocyst, 255, 255*f*
Blastopore, 256, 256*f*
Blastula, 253, 254
of frog, 256*f*
of human, 255, 255*f*
of sea star, 254, 254*f*
Blepharisma, 345*f*

Siphons
    of clam, 399, 400f
    of urochordates, 421, 421f
Sister chromatids, 57, 59, 59f, 60f–61f
Skeletal muscle, 138f, 147, 147f, 245–250
    actions of, 239, 239f, 247t, 248–249, 248f
    antagonistic pairs of, 239, 249
    contraction of, 147, 239, 249–250
        isometric, 249
        isotonic, 249
        mechanism of, 250, 251f
    major human, 246f, 247t
    naming of, 245
    superficial, 246f
Skeletal system, 239. See also Bone(s)
Skeleton, 242–245
    appendicular, 242, 244–245
    axial, 242–244
    chimpanzee and human, comparison of,
        322–325, 323f, 324f
Skin, 151, 151f
    sensory receptors in, 229, 229f,
        235–236, 235f
Skin pigmentation, genetics and, 286f, 287
Skull, 242f, 243–244, 243f
    chimpanzee and human, comparison of,
        324, 324f
    hominid, comparison of, 325–326, 325f
    vertebrate, 422
Small intestine
    anatomy of, 175, 175f
    of fetal pig, 161f, 162, 163f, 204
    of frog, 425, 426f, 428f, 429f
    of human, 165f, 168f
    nutrient absorption in, 175
    of pigeon, 433f
    of rat, 434f
    of vertebrates, 435
Smell, sense of, 237
Smell receptors, 237
Smooth endoplasmic reticulum, 43t
Smooth muscle, 138f, 148, 148f, 239
Snail, 398f
Sodium hydroxide, 29
Soft palate
    of fetal pig, 156–157, 156f
    of human, 165f
Soil erosion, 449
Solar energy, in photosynthesis, 91, 91f, 94–97
Somatosensory association area, 227f
Somites, in chick embryo, 259, 260f, 261f
Sorus (pl., sori), of ferns, 360f, 361–362,
        361f, 362f
Sound, locating, 235
Soybeans, cellular respiration in, 83–86,
        84f–86f
Sperm, 57
    of cnidarians, 384f, 385
    of crayfish, 411
    of earthworm, 406
    of ferns, 360f, 363, 363f
    of flowering plants, 122–124, 122f–124f,
        373, 374f, 375
    of moss, 354f, 355, 355f
    of nonvascular plants, 354
    of pines, 368f

of seed plants, 366, 366f
of sponges, 381
transport of, 195, 198
of vascular seedless plants, 357
of vertebrates, 435
Spermatic artery, of fetal pig, 188f, 197f
Spermatic cord, of fetal pig, 197f
Spermatic vein, of fetal pig, 186f, 189f
Sperm duct
    of crayfish, 410f
    of earthworm, 405, 406, 407f
    of planarians, 388f
Sphenoid bone, 242f, 244
Spicules, of sponges, 381–383, 382f, 383f
Spiders, 409, 409f
Spinal cord, 223, 223f, 228, 228f, 229f, 244
    development, in chick embryo, 260
    of frog, 430f
    of sheep, 225f
Spinal nerves, 223, 223f, 228, 229f
Spinal reflexes, 228–230
Spindle, mitotic, 58t, 59, 59f, 60f–61f
Spine, chimpanzee and human, comparison of,
        322–323
Spines
    of echinoderms, 415
    of sea star, 416f
Spiracles, of grasshopper, 412, 412f
Spiral colon, of fetal pig, 163f, 164
Spiral organ. See Cochlea
Spirillum (pl., spirilla), 335, 335f
Spirillum volutans, 335f
Spirogyra, 338, 338f
Spirometer, 214–215, 214f
Spleen
    of fetal pig, 161f, 162
    of frog, 426
    of human, 165f
    of perch, 432f
    of rat, 434f
Sponges, 381–383
    anatomy of, 381–382, 382f
    diversity of, 383, 383f
    evolution of, 380f, 381
Spongy bone, 240f, 241
Spongy mesophyll, 115
Sporangium (pl., sporangia), 122–124, 123f
    of ferns, 360f, 361–362, 362f
    of horsetails, 359, 359f
    of land plants, 353, 353f
    of lycophytes, 358
    of moss, 354f, 356
    of pines, 368f
    of seed plants, 366, 366f
    of whisk ferns, 358, 359f
Spores
    of ferns, 360f, 361–362, 362f
    of fungi, 346, 346f
    of lycophytes, 358, 358f
    of moss, 354f
    of pines, 368f, 369
    of seed plants, 366, 366f
    of sporozoans, 342
    of vascular seedless plants, 357
Sporophyte(s)
    dispersal by seed plants, 365

of ferns, 360f, 361, 361f
of flowering plants, 122–125, 122f–124f,
        373, 374f
of land plants, 353, 353f
of moss, 354f, 355–356, 356f
of nonvascular plants, 354
of pines, 368f
of seed plants, 365, 366, 366f
of vascular seedless plants, 357
Sporozoans, 342
Spruces, 367
Squid, 401–403, 402f
S stage, 58, 58f
Staining, of tissue, 137
Stalk
    of flower, 372, 372f
    of mushroom, 349
Stamens, 120–122, 120f, 121f, 124f, 372–373,
        372f, 374f
Stapes, 234, 234f
Starch, 30, 30f
    digestion, by pancreatic amylase, 173–174
    test for, 31–32, 173
Starfish. See Sea star
Stem(s), plant, 101–105, 101f–102f, 110–114
    arrangement of leaves on, 116, 116f
    diversity of, 111–112, 112f
    herbaceous, 110–111, 110f–111f
    internodes of, 102, 102f
    lycophyte, 358, 358f
    monocot vs. eudicot, 104–105, 105f
    nodes of, 102, 102f, 113, 113f
    woody, 113–114, 113f–114f
Stem cells, epidermal, 151f
Stentor, 345f, 450f
Stereocilia, in ear, 234t
Stereomicroscope, 9, 14, 15–17
    focusing, 17
    parts of, 15–17, 16f
Sternocleidomastoid muscle, 246f, 247t
Sternum, 243f
Steroids, 34
Stethoscope, 210
Stigma, of flower, 120f, 124f, 372–373,
        372f, 374f
Stipe, of algae, 339–340, 340f
Stolon, 112, 112f
Stoma (pl., stomata), plant, 104, 104f,
        115, 115f
    of pine trees, 369, 369f
Stomach
    of clam, 400f, 401
    of crayfish, 410f, 411
    of fetal pig, 161f, 162, 204
    of frog, 425, 426f, 428f, 429f
    of grasshopper, 412f
    of human, 165f, 168f
    of perch, 432f
    of pigeon, 433f
    of rat, 434f
    of rotifers, 395f
    of sea star, 416, 416f, 417
    of squid, 402f
    of vertebrates, 435
Stone canal, of sea star, 416f, 417
Stonewort, 352–353, 353f

Stony coral, 387
Strawberry plant, stems of, 112, 112*f*
Streptococcus thermophilus, 335*f*
Strobili, 358, 358*f,* 359, 359*f*
Stroma, in chloroplasts, 91, 91*f*
Strongylocentrotus pranciscanus, 415*f*
Style, of flower, 120, 120*f,* 124*f,* 372,
        372*f,* 374*f*
Stylonychia, 345*f*
Subclavian artery(ies), 179*f,* 181*f*
    of fetal pig, 185, 186*f*–187*f*
Subclavian vein(s), of fetal pig, 185, 186*f*–187*f*
Subcutaneous layer, 151
Submucosa, of intestine, 150, 150*f*
Subscapular artery, of fetal pig, 187*f*
Substrates, 73–74, 73*f*
Suckers
    of leeches, 404*f*
    of squid, 402*f*
    of tapeworm, 391–392, 391*f,* 392*f*
Sucrose, 87
Sudoriferous (sweat) glands, 151, 235*f*
Sugar(s), 30, 30*f*
    in fermentation, 87–88
    intestinal absorption of, 175
    test for, 32–34, 32*t,* 173
Sunflower, 331*f*
    acid rain and germination of, 451–452
Sunlight. *See* Solar energy; White light
Superior vena cava, 178, 179*f,* 181*f,* 183*f,* 208*f*
Supraorbital ridge, chimpanzee and human,
        comparison of, 324, 324*f*
Suspensor, of eudicot embryo, 128, 128*f*–129*f*
Suspensory ligament, of eye, 231*f*
Sutures, 243
Sweat glands, 151, 235*f*
Swimmerets, of crayfish, 410*f,* 411
Symmetry, 381
    in cnidarians, 384
    in echinoderms, 415
    in flatworms, 388
    in roundworms, 392
    in vertebrates, 422
Synapsis, 66, 68*f,* 70*f*
Synthesis reaction, 73*f,* 74
Systemic artery, of frog, 425
Systemic capillaries, 208, 208*f*
Systemic capillary exchange, 211–212, 212*f*
Systemic circuit, 177, 182, 183*f*
    of adult mammals, 184
    of fetal pig, 184–189
    in vertebrates, 435*f*
Systole, 209–210, 209*f*

Table, in laboratory report, 458
Tadpoles, 423
Taenia, 391*f,* 392, 392*f*
Taenia pisiformis, 392, 392*f*
Tail
    of chordates, 419, 419*f*
    of fetal pig, 154, 155*f,* 159*f*
    of human embryo, 262*f,* 263*f,* 268*f*
    of vertebrate embryos, 321
Tamarack (larch), 367
Tapeworms, 391–392, 391*f,* 392*f*
Taproot, 109, 109*f*

Tarsals, 243*f,* 245
Taste, sense of, 237
Taste buds, 237
Taste receptors, 237
Tay–Sachs disease, 290
Teeth
    chimpanzee and human, comparison of,
        324, 324*f*
    of fetal pig, 156, 156*f*
    of frog, 424, 424*f*
    of squid, 401
Telophase, in mitosis, 58*f,* 61, 61*f,* 71*f*
Telophase I, in meiosis, 66, 69*f,* 71*f*
Telophase II, in meiosis, 67, 69*f*
Telson, of crayfish, 410*f*
TEM. *See* Transmission electron micrograph
Temperature, and enzyme activity, 75–76,
        76*f,* 167
Temperature receptors, 235–236, 235*f*
Temperature scales, 13, 13*f,* 461
Template, DNA, 305, 305*f*
Temporal bones, 242*f,* 243*f,* 244
Temporalis muscle, 246*f*
Temporal lobe
    of human, 227, 227*f*
    of sheep, 224, 225*f*
Tendons, 143, 143*f*
Tentacles
    of cephalopods, 398, 398*f*
    of cnidarians, 384*f*
    of *Hydra,* 385–386, 385*f,* 386*f*
    of lancelet, 421*f,* 422
    of squid, 401, 402*f*
Terminal bud (shoot tip), 102, 102*f,* 103, 103*f,*
        113, 113*f*
Terminal bud scar, 113, 113*f*
Terrestrial ecosystems, 440–443
    abiotic components of, 441, 442
    biotic components of, 441, 443
    studying sampling site of, 440–444
Testes, 193, 196, 196*t*
    of cnidarians, 385
    of crayfish, 411
    of fetal pig, 196, 197*f*
    of frog, 426, 428*f*
    of human, 198*f*
    of planarians, 388*f*
    of roundworms, 393*f*
    of tapeworm, 392*f*
    of vertebrates, 435
Testosterone, 34
Tetrahymena, 345*f,* 450*f*
Tetrapods, 154, 420
Thalamus, of sheep, 224, 225*f*
Thallus, of liverworts, 357, 357*f*
Theory
    definition of, 3
    scientific, development of, 3, 3*f*
Thermal pollution, 449, 450
Thigh
    bones of, 243, 245
    muscles of, 246*f,* 247*t*
Thoracic cage, 243*f*
Thoracic cavity, of fetal pig, 158, 160, 163*f*
Thoracic vein(s), of fetal pig, 186*f*–187*f*
Thoracic vertebrae, 244

Thorax
    of crayfish, 410–411, 410*f*
    of grasshopper, 412, 412*f*
    of insects, 412
    of vertebrates, 432
Thrush, 349, 349*f*
Thylakoids, 42, 91, 91*f*
Thymine, 301
Thymus gland
    of fetal pig, 160, 161*f*
    of human, 165*f*
    of rat, 434*f*
Thyroid gland
    of fetal pig, 157*f,* 160, 161*f,* 163*f*
    of human, 165*f*
    of rat, 434*f*
Tibia, 243, 243*f,* 245
Tibialis anterior muscle, 246*f,* 247*t*
Tidal volume (TV), 214, 215*t*
Tiger shark, 423*f*
Tinea, 349, 349*f*
Tissue(s). *See also* Connective tissue;
            Epithelial tissue; Muscular tissue;
            Nervous tissue
    definition of, 137
    embedding of, 137
    fixation of, 137
    of intestinal wall, 150–151, 150*f*
    major, in human body, 138*f*–139*f*
    sectioning of, 137
    staining of, 137
Tissue stages, of development, 256
    in frog, 256, 256*f*
    in human, 256, 257*f*
Tobacco seedlings, color of, 273–274, 273*f*
Toe(s)
    bones of, 243*f,* 245
    of rotifers, 395*f*
Toe rays, in human embryo, 268*f*
Tomato, fruit of, 130, 130*f*
Tongue
    of fetal pig, 155*f,* 156, 156*f*
    of frog, 424, 424*f*
    of human, 165*f*
Tonicella, 398*f*
Tonicity
    in cells, 50–53, 51*f*
    definition of, 50
Torso, chimpanzee and human, comparison of,
        322–323
Touch, sense of, 236
Touch receptors, 235–236, 235*f*
Trabeculae, 240*f,* 241
Trachea
    of fetal pig, 156–157, 157*f,* 160, 203, 203*f*
    of frog, 424, 425
    of human, 165*f*
    of pigeon, 433*f*
    of rat, 434*f*
Tracheae, of grasshoppers, 412, 412*f*
Tracheids, 117, 117*f*
Transcription, 304–305, 305*f*
Transfer RNA (tRNA), 299, 304, 306–307,
        306*f,* 307*f*
Translation, 304, 306–307, 306*f,* 307*f*
Transmission electron micrograph (TEM), 14*f*

forelimbs of, comparison of, 320–321, 320*f*
  fossils, 318, 318*f*
  groups of, 423*f*
  respiratory system of, 436–437, 437*f*
  urogenital system of, 435
Vesicles, 43, 43*t*. *See also specific types*
Vessel elements, of xylem, 117, 117*f*
Vestibular nerve, 234, 234*f*
Vestibule of ear, 234, 234*t*
Villi, of small intestine, 175
Vinegar eels, 394
Visceral mass
  of clam, 398, 401
  of molluscs, 398
Visceral muscle, 148
Vision, 230–233
Visual association area, 227*f*
Vital capacity (VC), 214–215, 215*t*
Vitelline arteries, of chick embryo, 259, 260*f*, 261*f*
Vitelline veins, of chick embryo, 259, 260*f*, 261*f*
Vitreous humor, 231*f*, 231*t*
Vocal cords, of frog, 425
Voice box. *See* Larynx
Volume, metric units of, 12–13, 461
Volumeter, 84–85, 85*f*, 94–95, 95*f*
Voluntary muscle, 147
Volvox, 337*f*, 338–339, 339*f*
Vomer bone, 242*f*
Vomerine teeth, of frog, 424, 424*f*
von Recklinghausen disease, 289
Vorticella, 345*f*, 450*f*

Walking legs, of crayfish, 410*f*, 411
Water
  diffusion across plasma membrane, 49–53, 50*f*

transpiration from leaves, 117
  transport in plant, 117
Water vascular system, 416*f*
  of echinoderms, 415
  of sea star, 417
Waxy cuticle, of plants, 351
Webbed toes, in human embryo, 268*f*
Weight, metric units of, 11, 461
Wet mount, 23, 23*f*
Whisk ferns, 357, 358–359
White blood cell(s), 139*f*, 145, 145*f*
  in urine, testing for, 220–221, 221*f*
Whitefish blastula, mitosis in, 59, 63
White light, and photosynthesis, 94–95, 94*f*
White matter
  of sheep brain, 224
  of spinal cord, 228, 228*f*, 229*f*
Widow's peak, 286–287, 286*f*, 295, 295*f*
Windpipe. *See* Trachea
Wing bud, in chick embryo, 261*f*
Wing length, in *Drosophila*, 279–280
Winter twig, anatomy of, 113–114, 113*f*
Womb. *See* Uterus
Wood, 113–114, 114*f*
Woody stems, 113–114, 113*f*–114*f*
Worms. *See also specific types*
  segmented, 403
Wrist
  bones of, 243, 244*f*
  of fetal pig, 154, 155*f*, 159*f*
  in human embryo, 268*f*

Xanthophylls
  action spectrum for, 96, 96*f*
  on paper chromatography, 92*f*
X chromosome, 281, 290, 294
Xestospongia testudinaria, 383*f*
X-linked crosses, 281–283

X-linked disorders, 290–291, 291*f*
X-linked recessive disorders, 290–291, 291*f*
Xylem, 103*t*, 104
  of plant root, 107–108, 107*f*–108*f*
  of plant stem, 110–111, 110*f*–111*f*
  secondary, 113–114, 114, 114*f*
  structure of, 117, 117*f*
  transport in, 117

Y chromosome, 281, 290, 294
Yeast fermentation, 87–88
Yellow bone marrow, 240*f*, 241
Yolk, 253, 258*f*, 260
Yolk gland, of planarians, 388*f*
Yolk sac, 257*f*, 260, 264–266, 265*f*, 266*f*

Z line, 251*f*
Zone of cell division, of plant root, 106, 106*f*
Zone of elongation, of plant root, 106, 106*f*
Zone of maturation, of plant root, 106, 106*f*
Zygomatic bones, 242*f*, 243*f*, 244
Zygomaticus muscle, 246*f*, 247*t*
Zygospore, 347, 347*f*
Zygotes
  of animals, 253, 253*f*, 254
  of charophytes, 351
  of cnidarians, 384*f*
  of eudicots, 128*f*
  of ferns, 360*f*
  of flowering plant, 122–125, 122*f*–124*f*
  formation of, 254
  of green algae, 338
  of humans, 255, 255*f*
  of land plants, 353, 353*f*
  of moss, 354*f*, 355
  of pines, 368*f*
  of sea star, 254, 254*f*
  of seed plants, 366, 366*f*